ELECTRODYNAMICS

Electrodynamics

Lectures on Theoretical Physics, Vol. III

BY ARNOLD SOMMERFELD

University of Munich

Translated by
EDWARD G. RAMBERG

NEW YORK, N. Y.
ACADEMIC PRESS INC., PUBLISHERS

1952

PREFACE

Heinrich Hertz's great paper on the "Fundamental Equations of Electrodynamics for Bodies at Rest" has served as model for my lectures on electrodynamics ever since my student days (see §1). Following this example I proceed in Part I from Maxwell's equations as an axiomatic basis, expressed not as with Hertz, in the coordinates and in differential form, but in vectorial integral form. In Part II the several classes of phenomena, in static, stationary, quasistatic, and rapidly variable fields, are derived from these equations, as in Hertz's paper. After I had heard Hermann Minkowski's lecture on "Space and Time" in 1909 in Cologne, I carefully developed the four-dimensional form of electrodynamics as an apotheosis of Maxwell's theory and at the same time as the simplest introduction to the theory of relativity; in return, this has always met with an enthusiastic reception on the part of my audience. This four-dimensional electrodynamics is presented in Part III. Its title "Theory of Relativity and Electron Theory" requires the comment that it is limited to the *special* theory of relativity on the one hand and to the theory of the *individual* electron on the other. The statistics of electrons in metals and electrons in insulators belong in Vols. IV* and V* of the Lectures. Schwarzschild's principle of action, which establishes the fundamental relationship between Maxwell's theory and the dynamics of the individual electron (or individual electrons), is presented at the end of Part III with certain modifications which appear necessary from our point of view. In Part IV is developed the electrodynamics of moving media, again following Minkowski rather closely. As the most important application the fields of unipolar induction are discussed and are calculated as an exercise for a particularly simple example.

Part II constitutes the main portion of the Lectures. I distinguish between summation and boundary-value problems in electrostatics and magnetostatics. The computation of the electric potential for given charge distribution and the calculation of the magnetic potential for given magnetization are examples of the first; the theory of the permanent magnet, insofar as it falls within the competence of Maxwell's theory rather than atomic theory, becomes simple and clear from this standpoint. On the other hand, the solution of the electric and magnetic boundary-value problems properly belongs in Vol. VI; only the most important cases are treated in the present volume. The calculation of stationary fields for a given distribution of the current density, either by the method of the vector potential in §15 or that of the magnetic shell in §16, is also a simple summation prob-

---

* See p. xii for list of *Lectures on Theoretical Physics.*

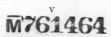

lem. Among the rapidly variable fields those of the wire-wave type are treated with some completeness. The principal wave on a single wire in §22 (symmetric electrical type) serves as primary example; however, because of their recent practical applications (theory of wave guides in §24) and their utilization in the theory of the Lecher system, the magnetic type and the asymmetric secondary waves as well as wire waves on nonconductors are also dealt with in §23. As conclusion of Part II the Lecher system is treated fully for arbitrary separation and dimensions of the two parallel wires, employing bipolar coordinates for the exterior of the wires and ordinary polar coordinates for their interior. It is only assumed that the two wires are rather good conductors.

The dimensional character of the field entities is taken seriously throughout. We do not accept Planck's position, according to which the question of the real dimension of a physical entity is meaningless; Planck states in §7 of his Lectures on Electrodynamics that this question has no more meaning that that of the "real" name of an object. Instead, we derive from the basic Maxwell equations the fundamental distinction between entities of *intensity* and entities of *quantity*, which has heretofore been applied consistently in the excellent textbooks of G. Mie. The Faraday-Maxwell induction equation shows that the magnetic induction B is an entity of intensity along with the electric field strength E; B, rather than H, deserves the name *magnetic field strength*. H, like D, is best designated as "excitation." div H represents the magnetic density, just as div D represents the electric charge density. Hertz's distinction between "true" and "free" electricity becomes pointless, since div E is, dimensionally, not a charge, but a divergence of lines of force. The same applies to the distinction between "true" and "free" magnetism, particularly since div B is everywhere zero. The current density J, the electric polarization P, and the magnetization M, are entities of quantity like H and D. Energy quantities always take the form of products of an entity of quantity and an entity of intensity, e.g. $\frac{1}{2}D \cdot E$, $\frac{1}{2}H \cdot B$, $J \cdot E$, $E \times H$. The fact that B and E, and H and D, belong together follows unambiguously from the theory of relativity, in which the quantities cB and $-iE$, and H and $-icD$, respectively, are coupled together in a six-vector (antisymmetric tensor). We call the first the field tensor F, the second, the excitation tensor f.

The introduction of a *fourth electric unit*, independent of the mechanical units, is decisive for the fruitfulness of these dimensional considerations. We choose for this the unit of charge Q, which, as a matter of convenience we may identify with the coulomb if we wish. In this manner we avoid the "bed of Procrustes" of the cgs-units, in which the electromagnetic quantities are forced to take on the well known unnatural dimensions. Since we must definitely give up the hope of a mechanical interpretation of electrical quantities, we must regard the charge as a basic, irreducible entity which can claim a dimension of its own. We shall refer to the "electrostatically" or "electromagnetically" measured charge only in passing, and exclusively for historical reasons. With the particular unit of charge Q = 1 coulomb the electric current has the customary unit amperes = Q/sec.

As mechanical units, following the suggestion of G. Giorgi, we shall employ the units meter M, kilogram (mass) K, and second S. The unit of energy then becomes 1 joule (without a power of ten as factor!) and that of power 1 joule S^{-1} = 1 watt. Furthermore, the powers of ten disappear also for the electric units volt, ohm, farad, and henry; we have 1 volt = 1 joule/Q, 1 ohm = 1 joule S/Q^2, 1 farad = 1 Q^2/joule and 1 henry = 1 joule S^2/Q^2.

On the other hand powers of ten must appear as factors when the units of the magnetic field strength $\mathbf{B}$ and the magnetic excitation $\mathbf{H}$ are expressed in terms of the gauss and the oersted, respectively, which as we shall see in §8, have been adapted to the cgs-system. As expected, the unit Q automatically drops out of the energy densities $\frac{1}{2}\mathbf{D}\cdot\mathbf{E}$, $\frac{1}{2}\mathbf{H}\cdot\mathbf{B}$, and $\mathbf{J}\cdot\mathbf{E}$ referred to above; their dimension becomes joule/M^3 directly, whereas that of the energy flux $\mathbf{E}\times\mathbf{H}$ is joule/(M^{2}S).

With our dimensional differentiation between entities of intensity and entities of quantity the dielectric constant and the permeability evidently become *dimensional* quantities and therefore cannot be set equal to 1 in vacuum. Their choice, in which we accept electrical engineering practice, happily permits us to meet the demand for "rational units" without difficulty. It is only necessary to set,

$$\mu_0 = 4\pi\cdot 10^{-7}\,\frac{\Omega S}{M}$$

in accord with international conventions, and to derive ε_0 from the relation $\varepsilon_0\mu_0 = 1/c^2$, verified by Hertz's experiments. With this choice the 4π's disappear wherever they do not belong, as in Poisson's equation and the energy expressions in Poynting's theorem, and appear where they belong, as in Coulomb's law and for the spherical condenser. We thus avoid the desperate expedient by which Lorentz achieves rationalization in his articles in the Enzyklopaedie, namely the introduction of the factor $\sqrt{4\pi}$ in the definition of the charge and of magnetism.

At the same time, with this choice of ε_0 and μ_0, the square root of the ratio of μ_0 and ε_0 evidently becomes a resistance, namely the so-called "wave resistance of vacuum." This quantity occurs in Part II as a factor wherever the wave fields $\mathbf{E}$ and $\mathbf{H}$ enter into formulas of the same dimensions. It occurs again in the theory of relativity in the relation between the excitation tensor f and the field tensor F, which in vacuum assumes the simple form for all six components

$$f = \sqrt{\frac{\mu_0}{\varepsilon_0}}\, F.$$

These questions of units, dimensions, and rationalization, often discussed to excess in recent years, are disposed of as briefly as possible in the lectures; however, the reader is repeatedly urged in them to convince himself of the dimensional logic of formulas. In numerical computations our MKSQ system of units is found convenient throughout since it is adapted to the practical and legal units, volt, ampere, etc. We leave the question open as to whether it is also appropriate for atomic physics. So as to permit an effortless transition to the Gaussian system ($\varepsilon_0 = \mu_0 = 1$), which is customary in this case, we explain Cohn's system in §9, which in our interpretation is based on the five units MKSQP (P = magnetic unit pole).

The wonderful simplicity and beauty of the Maxwell equations, which is most striking in their relativistic formulation for vacuum, lead to the conviction that these equations, along with the equations of gravitation (§38), are the demonstration of an all-inclusive world geometry. Approaches to this by no means resolved problem are summarily discussed in §37. The amazingly simple representation of the general theory of relativity in §38 is based on a derivation of Schwarzschild's line element kindly made available to me by W. Lenz. In this manner the three tests of the theory open to astronomical observation may be treated without tensor calculus.

This volume is based on lecture notes prepared by H. Welker in the winter semester of 1933/34, at which time I first abandoned the cgs system and passed over to the more general system of the four units. In the final formulation of Parts I and II I have had the benefit of the constant advice of Professor J. Jaumann, whose electrotechnical experience and point of view have been of great advantage to this volume. I am grateful to Messrs. P. Mann and E. Gora and to my colleague F. Bopp for critical remarks and suggestions for improvements. Dr. W. Becker has kindly assisted me in reading the proof of this, as of preceding volumes.

Munich, April 1948

Arnold Sommerfeld

Translator's Note

A minimum number of changes has been made in this translation of Sommerfeld's "Elektrodynamik" (the third volume of the Lectures on Theoretical Physics) to adapt it for use in English-speaking countries. As far as possible, the same conventions regarding notation are employed as in G. Kuerti's translation of Volume II, "Mechanics of Deformable Bodies." Thus vectors are represented by bold-face letters, vector components and scalars (as well as tensors and their components) by italics; this in spite of the fact that the Gothic letters employed in the original text for both vectors and vector components were used even in Maxwell's Treatise. To avoid confusion a few additional changes of symbols were required in consequence of this major change.

E. G. R.

CONTENTS

Lectures on Theoretical Physics

VOLUME I: Mechanics. 1952. Translated by Martin O. Stern

VOLUME II: Mechanics of Deformable Bodies. 1950. Translated by G. Kuerti

VOLUME IV: Optics. 1953. Translation in preparation

VOLUME V: Thermodynamics and Statistical Mechanics

VOLUME: VI: Partial Differential Equations in Physics. Translated by Ernst G. Straus

PART I

FUNDAMENTALS AND BASIC PRINCIPLES OF MAXWELL'S ELECTRODYNAMICS

§1. Historical Review. Action at a Distance and Action by a Field

I can best give you an idea of the sweeping changes in viewpoint brought about by the theory of Faraday and Maxwell by telling you of the time I spent as a student, 1887–1891.

My native city, Königsberg, was the earliest fountainhead of mathematical physics in Germany, thanks to the activity of the revered Franz Neumann, 1798–1894. At the University of Königsberg he taught, in addition to crystallography, theoretical physics which was not at the time given elsewhere in Germany. His students, of whom Gustav Kirchhoff of Königsberg was the most prominent, spread the teachings of the master to the other German universities. Through the seminar in mathematical physics, founded by him and C. G. J. Jacobi, he also saw to it that the East Prussian secondary-school teachers received a particularly thorough preparation. This may bear some relation to the fact that the Gymnasium in the Altstadt graduated the mathematician Hermann Minkowski and the physicists Max and Willy Wien shortly before my final examination, while at the same time the only slightly older David Hilbert and Emil Wiechert were attending other Königsberg schools. Neumann's greatest successes in research were achieved in the elastic theory of light and in the physics of crystals; his mathematical formulation of the induction currents discovered by Faraday will be discussed in §15. Simultaneously with Neumann and Jacobi, and almost outshining them, F. W. Bessel taught in Königsberg.

My time of study coincided with the period of Hertz's experiments. At first, however, electrodynamics was still presented to us in the old manner—in addition to Coulomb and Biot-Savart, Ampère's law of the mutual action of two elements of current and its competitors, the laws of Grassmann, Gauss, Riemann, and Clausius, and as a culmination the law of Wilhelm Weber, all of which were based on the Newtonian concept of action at a distance. The total picture of electrodynamics thus presented to us was awkward, incoherent, and by no means self-contained. Teachers and students made a great effort to familiarize themselves with Hertz's experiments step by step as they became known and to explain them with the aid of the difficult original presentation[1] in Maxwell's Treatise.

It was as though scales fell from my eyes when I read Hertz's great paper:[2] "Über die Grundgleichungen der Elektrodynamik für ruhende Körper." Here Maxwell's equations, purified by Heaviside and Hertz, were made the axioms and the starting point of the theory. The totality of electromagnetic phenomena is derived from them systematically by deduction. Coulomb's law, which formerly provided the basis, now appears as a necessary consequence of the all-inclusive theory. Electric currents are always closed. Current elements arise only as mathematical increments of line integrals. All effects are transmitted by the electromagnetic field, which may be represented by force-line models. *Action at a distance* gives way to field action,[3] the "constructable representation" of a space-time propagation postulated already by Gauss.[4]

I have held to the order of Hertz's paper in all my lectures on Maxwell's theory. In this presentation, too, we shall not begin with electrostatics, as is done so commonly and also in Maxwell's Treatise, but treat it merely as an extreme simplification of the general field theory. We shall deviate from Hertz only insofar as we shall start not from Maxwell's equations in differential form, but in integral form. It goes without saying that we shall replace the rather extensive coordinate calculations of Hertz by vector algebra, which is perfectly suited to the electromagnetic field. We shall see that this algebra, extended to four dimensions, leads directly to the special theory of relativity. The latter will provide an approach to the electrodynamics of moving bodies, which Hertz unsuccessfully sought to master in the second paper cited. In agreement with Hertz we see in Maxwell's *equations* the essence of his theory. We need not discuss the *mechanical pictures*, which guided Maxwell in the setting up of his equations. We have discussed one such picture in Vol. II, §15 of these lectures.

---

[1] The great student of electrolysis, Wilhelm Hittorf, who had heard much of the new theory of electricity, in advanced years attempted to study the Treatise, but was unable to find his way through the unfamiliar mass of equations and concepts. He was thus led into a state of deep depression. His colleagues in Münster persuaded him to take a vacation trip to the Harz Mountains. However when just before his departure they checked his luggage they found in it—the two volumes of the Treatise on Electricity and Magnetism by James Clerk Maxwell. (As told by A. Heidweiller.)

[2] Göttinger Nachr. March 1890 and Ann. Physik, Vol. 40; continued in Ann. Physik, Vol. 41: "Über die Grundgleichungen der Elektrodynamik für bewegte Körper."

[3] We avoid the alternative term "near action" which signifies merely action at a small distance, and by our notation direct attention to the medium transmitting the effect, namely the field.

[4] In a letter to Wilhelm Weber, of 1845. See Collected Works, Vol. V, p. 627.

Biographical Notes

Michael Faraday, 1791–1867

He was born as son of a blacksmith in impecunious circumstances. The family belonged to the pious sect of the Sandemanians, to which Faraday remained faithful to his death. His high ethical concept of life and human kindness derived from the religious spirit of his family. He was first newspaper carrier, then bookbinder. In science and letters he was entirely self-taught. The lectures of Sir Humphry Davy at the Royal Institution were decisive for his career; he wrote them up carefully and found an opportunity to present them to the great chemist. He became his laboratory assistant in the Royal Institution. His first important work was "the rotation of a current about a magnet and the rotation of a magnet about a current," and also the liquefaction of chlorine. This work brought about his election as Fellow of the Royal Society and later the indirect succession to Davy at the Royal Institution. In 1832 he began the publication of the "Experimental Researches." His discoveries recorded in these extend to the most diverse fields of physics, electrochemistry, and the study of materials. We mention as most significant for us: The discoveries of the law of electromagnetic induction in 1831, the dielectric constant, para- and diamagnetic behavior, and the picture of electric and magnetic lines of force. His magneto-optical discoveries are discussed in Vol. IV. The failing of his memory forced many pauses in his work, as well as the repetition of experiments made at an earlier date. It is uncertain whether this is to be attributed to mental overexertion or, as is commonly assumed today, to mercury poisoning in the poorly ventilated basement rooms of the Royal Institution. Certainly his purely intuitive method of working, devoid of any mathematical aid, required tremendous mental concentration. In his last years a restful summer retreat in the royal palace, Hampton Court, was made available to him at the suggestion of the Prince Consort, Albert. At his death there were found ninety-five honorary diplomas of learned societies, bound with his own hand.

James Clerk Maxwell, 1831–1879

He came from a prominent Scottish family (the father's name was Clerk, the added name Maxwell being derived from his mother) and was given the best in education that his time offered, both in the field of letters and that of science and mathematics. Thus, at an early date, he could translate Faraday's pictures of lines of force into a mathematical form which could be generally understood. See his paper of 1855 "On Faraday's Lines of Force" (translated into German by Boltzmann in Ostwald's Klassiker Nr. 69). In the preface to his Treatise he states: "Faraday, in his mind's eye, saw lines of force traversing all space where the mathematicians (from the preceding discussion it is apparent that he refers particularly to Gauss, Wilhelm Weber, Riemann, Franz and Carl Neumann) saw centres of force attracting at a distance: Faraday saw a medium where they saw nothing but distance: Faraday sought the seat of the phenomena in real actions going on in the medium, they were satisfied that they had found it in a power of action at a distance impressed on the electric fluids. When I had translated what I considered to be Faraday's ideas into a mathematical form, I found that in general the results of the two methods coincided, . . . but that . . . several of the most fertile methods of research discovered by the mathematicians could be expressed much better in terms of ideas derived from Faraday than in their original form.''

The Treatise appeared in 1873. Its greatest achievement is the unification of optics and electrodynamics. The simplified form of the Maxwell equations, later rediscovered by Heaviside and Hertz, is to be found already in Part III of his paper for the Royal Society of 1864. Almost as important as his electromagnetic papers are those on the kinetic theory of gases (Maxwellian velocity distribution) and on general statistics, to which belongs also his theory of the rings of Saturn. He is also the author of purely mathematical papers (on cycloidal surfaces, the theory of the top, and the determination of magnitudes in Helmholtz's color triangle) and of an important paper on lattice structures (see Vol. II of these Lectures, p. 310).

After a brief teaching engagement in Aberdeen he became the first director of the newly founded Cavendish Laboratory in Cambridge; he died there at an early age.

ANDRÉ MARIE AMPÈRE, 1775–1836

We shall add a biographical note on Ampère not on account of the fundamental law already mentioned, nor because of the classical experiments, which enabled him to derive it with the simplest possible means, but for his discovery of the general relationship between the magnetic field and electric currents.

Born in Lyon, as a precocious boy he occupied himself with philological and mathematical studies. His father was a victim of the Revolution. Because of his mathematical papers he was named professor at the École Polytechnique in Paris in 1804. Here he soon directed his attention to chemistry, where he was able to compete with Avogadro in the field of atomism. There follow five years in which he is concerned primarily with psychology and metaphysics, though accepted as a mathematician into the Academy of Sciences of Paris. His interest in physics is not awakened until 1820, when he hears of Oersted's discovery. In a few weeks he verifies his belief that electricity in motion, and not electricity at rest, has a magnetic effect. The years 1820–1826 he spent elaborating his concept of the connection between the magnetic field and the electric current, which is equivalent to half of Maxwell's equations provided that the concept of the electric current is extended by the addition of Maxwell's displacement current. We shall hence denote this portion of the Maxwell equations (in integral form) in §3 directly as Ampère's law. From this point of departure Ampère recognized the equivalence of a solenoid traversed by current to a permanent magnet. The strengthening of the magnetic field by a soft-iron core placed in the solenoid is also to be attributed to him. Ampère may thus be regarded as the father of the "electromagnet." We may mention in addition Ampère's molecular currents and the elegant method of the magnetic sheet.

When, in 1826, however, Ampère obtained a professorship in physics at the Collège de France his interests changed once more: he returned to philosophy and logic and devoted himself finally to biology and comparative anatomy. Altogether a scientific career of extraordinary breadth and depth, of intensity and versatility! (This material has been taken from an essay by Louis de Broglie in his book Continu et Discontinu, Paris, 1941.)

Heinrich Hertz, 1857–1894

He was born in Hamburg the son of a respected merchant family; his father was in later years Senator of the Free City. Initially his great modesty prevented Heinrich Hertz from entering upon the career of a scholar; instead, he turned to engineering at the Technische Hochschule in Munich. Soon, however, he begged his father to permit him to transfer to pure physics. He studied first in Munich, then in Berlin, and became the favorite student and assistant of Helmholtz. The relationship between teacher and student was the closest imaginable and finds touching expression in the memorial addressed to him by Helmholtz (reprinted in Vol. I of Hertz's Collected Works). A prize problem set up by Helmholtz directed him to the testing of Maxwell's theory. After a short term as Privatdozent in Kiel he was called to the Technische Hochschule in Karlsruhe.

Even the earliest papers of Hertz show his mastery in relating theory and experiment. Several of them received the warm recognition of his colleagues, as his quantitative determination of hardness among engineers, and his description of the condensation processes in rising air currents among meteorologists. His years in Karlsruhe, from 1885 to 1889, represent the high point in his creative activity. We mention in particular his paper of 1888: "Forces of electrical oscillations treated by Maxwell's theory." It provides the characteristic solution now generally designated as the Hertzian vector and shows the familiar force-line pictures of the

Hertzian dipole. It is amazing how much of the later development of radio telegraphy has been anticipated in this paper. We should also point out the great paper on "Rays of electric force." The theoretical papers (basic equations of electrodynamics) have already been discussed.

The discovery of the photoelectric effect also falls into this period. With his last experimental paper of 1891 "On the passage of cathode rays through thin metal films" he reached beyond the problems set by Maxwell's theory and without knowing it, blazed the path to the electron theory. The very thin metal films later designated as "Lenard windows" are described already in this paper.

In 1889 he was called to Bonn. Here he prepared his last work, "Principles of Mechanics," which we have discussed in Vol. I, §39. The introduction of non-holonomic auxiliary conditions, the polydimensional treatment of mechanical systems of many degrees of freedom, the principle of the straightest path attest the keen logic and the geometric intuition of their author. Increasing illness prevented experimental work. He died on January 1, 1894, 37 years of age.

§2. Introduction to the Basic Concepts of the Electromagnetic Field

We regard the existence of electric *charges* as an established fact, whether we produce them by rubbing a piece of amber, the godparent of electricity, or recognize them from the spark when connecting the poles of a battery. We interpret the observed attraction, repulsion, and heat generation as the result of charges which have been produced. We take care not to define the charge verbally or to ascribe a derived dimension to it by some arbitrary procedure. Instead we regard it as having its own dimension, as an entity beyond the range of mechanics. We call this quantity Q. We could choose as unit of charge, whether negative or positive, the familiar universal charge of the electron. We prefer however to let Q stand for the *coulomb*, the accepted unit in the practical system, in terms of which the electron charge is expressed by $e = 1.60 \cdot 10^{-19}$ coulomb. We assume that electrometer apparatus is available with which we can compare different charges with each other and with the coulomb as unit of charge. The atomistic nature of charge is disregarded in the Maxwell theory proper. The charge of the atoms and elementary particles is to a much higher degree an absolute constant than the mass (see §27J).

In addition to the electric unit Q we normally employ as mechanical units of length, mass, and time the Giorgi units M (meter), K (kilogram mass), and S (second), which have been established internationally by the decision of the appropriate commissions. As already pointed out in Vol. I, p. 8, there is the advantage that in this system the units of energy and power correspond exactly (without multiplying powers of ten) to the joule and watt introduced previously in the cgs system. We designate them as

$$1 \text{ joule } = 1 \text{ M}^2\text{KS}^{-2} = 10^7 \text{ cm}^2 \cdot \text{g} \cdot \text{sec}^{-2} = 10^7 \text{ erg}$$

$$1 \text{ joule/S} = 1 \text{ M}^2\text{KS}^{-3} = 10^7 \text{ cm}^2 \cdot \text{g} \cdot \text{sec}^{-3} = 10^7 \text{ erg/sec} = 1 \text{ watt}$$

and define correspondingly

$$1 \text{ newton } = 1 \text{ MKS}^{-2} = 10^5 \text{ cm} \cdot \text{g} \cdot \text{sec}^{-2} = 10^5 \text{ dynes}$$

This unit of force "newton" is seen to be conveniently comparable in size with the practical unit of force, the "kilogram" $= 9.81 \cdot 10^5$ dynes.

We will show presently that the annoying powers of ten vanish also for the practical units of the volt and ohm when the MKSQ system is employed.

We now proceed to examine in sequence the basic electromagnetic concepts. In most cases we shall be concerned with a dimensional description rather than with a complete definition; the latter will be derived from their interrelation through the basic equations of the theory, which can be tested by experiment. In the succeeding section we will follow directly the present day enumeration of the basic concepts.

We begin with the *electric fieldstrength*, for which a true definition is possible and is generally conventional. Let this quantity be denoted by $\mathbf{E}$.[1] We define it as the mechanical force exerted in an electric field on an (infinitesimally small) test body, divided by the charge of the test body. $\mathbf{E}$ is therefore a vector with the dimension[2]

$$\mathbf{E} = \frac{\text{Force}}{\text{Charge}} = \frac{\text{newton}}{\text{Q}}; \tag{1}$$

within the field it varies from point to point in direction and magnitude. In following everywhere the direction of $\mathbf{E}$ we describe an electric line of force.

We now consider the *line integral*

$$\int_A^B E_s \, ds = \int_A^B \mathbf{E} \cdot d\mathbf{s} \tag{2}$$

between two points A and B. E_s is the perpendicular projection of $\mathbf{E}$ on the direction of the line element $d\mathbf{s}$ and $d\mathbf{s}$ is the line element regarded as vector; $\mathbf{E} \cdot d\mathbf{s}$ denotes, as usual, the scalar product. We call this line in-

---

[1] Maxwell employed gothic letters (rather than bold-face letters) for the vectors of the electromagnetic field (see Vol. II of the Treatise, art. 618). Except for this distinction we use the symbols here given. $\mathbf{J}$ will denote the electric current density, I the total current in a wire.

[2] Here, and at many other points, we use the equality sign to indicate equality of dimension. Where, as in Eq. 2a, we wish to distinguish between actual numerical equality and mere dimensional equality, we write $= \ldots$, i.e., "equal except for a numerical factor."

tegral the "voltage" V:

$$V = \int_A^B \mathbf{E} \cdot d\mathbf{s} = \cdots \frac{\text{newton} \cdot \text{M}}{Q} = \cdots 10^7 \frac{\text{dyne} \cdot \text{cm}}{Q}$$

$$= \cdots 10^7 \frac{\text{erg}}{Q} .$$

(2a)

The conversion of the dimension from the MKSQ- to the cgs-system shows that our unit of voltage is identical with

$$1 \text{ volt} = 10^8 \text{ cgs units} \tag{2b}$$

if, as decided above, we fix Q at

$$1 \text{ coulomb} = \tfrac{1}{10} \text{ cgs unit.} \tag{2c}$$

For the definition of the voltage it is necessary that in addition to the *terminal points A, B*, the path between them be prescribed. Only in lamellar fields (see Vol. I, Eq. 6.16, and Vol. II, p. 137) is the independence of the line integral with respect to the path guaranteed by Stokes' law (see Vol. II, Eq. 3.6). In place of voltage we may then speak of difference of potential between the two points A and B, designated by V_{AB}.

We introduce as companion to the fieldstrength $\mathbf{E}$ a second electric vector $\mathbf{D}$. We shall call this preferably electric *"excitation,"* but shall also frequently employ, particularly in the first part of these Lectures, the customary term *"dielectric displacement"* (Maxwell's designation).

We make the introduction of $\mathbf{D}$ comprehensible by the following consideration: Charge, in its historical origin, is a concept based on the notion of action at a distance. To adapt it to the viewpoint of action by a field it is necessary to imagine an excitation of the surrounding medium proceeding from the charge centers, which excitation will be described by the vector $\mathbf{D}$. For a single point charge e we imagine "$\mathbf{D}$ lines" leaving e uniformly in all directions, with such density that the "$\mathbf{D}$-flux" becomes

$$\oint D_n \, d\sigma = e. \tag{3}$$

$d\sigma$ is an element of an arbitrary surface surrounding e. If, in particular, we choose a spherical surface of radius r, we find

$$4\pi r^2 D = e. \tag{3a}$$

For arbitrarily, including continuously, distributed charges, Eq. (3) is replaced by

$$\oint D_n \, d\sigma = \bar{e}, \qquad \bar{e} = \sum e \tag{3b}$$

where $\bar{e}$ indicates the total charge within σ, the algebraic sum of positive and negative charges. We will see in §4 that this description of **D**, for an arbitrary choice of σ, is selfconsistent, but does not suffice for a unique definition of **D**. We will also see there that in the simplest case (isotropic medium, linear relation between **D** and **E**) the "D-lines" are identical with the lines of force defined by the **E** vector.

From the preceding equations the dimension of **D** is seen to be

$$D = \frac{\text{charge}}{\text{area}} = \frac{Q}{M^2}. \tag{4}$$

This dimension is entirely different from the dimension of the fieldstrength **E**, given by Eq. (1). With regard to Maxwell's designation "dielectric displacement," we note that it fits strictly not the vector **D** itself, but only that fraction of **D** which arises from the presence of ponderable matter and which will later (see §11C) be designated as the *polarization* **P**. Thus this portion **P** vanishes for vacuum, the medium which is of greatest importance to us. Nevertheless the "displacement" **D** retains its individual meaning, distinct from **E**, in this case also.

We compare Eq. (4) with the dimension of the *electric current density* **J**. One knows that this is to be defined as the quantity of electricity traversing unit area in unit time in a conductor. Its dimension is therefore

$$J \equiv \frac{\text{charge}}{\text{area} \cdot \text{time}} = \frac{Q}{M^2 S}. \tag{4a}$$

Depending on whether the unit area is placed perpendicular to the direction of the current or at an angle thereto, the absolute magnitude of **J** or a component of it is obtained. **J** is thus a vector similar in character to **D**. Dimensionally, however, not **D**, but the time rate of change of **D**, the socalled *displacement current* $\dot{\mathbf{D}}$, corresponds to **J**.

We have here assumed a sharp distinction between conductor and nonconductor (dielectric medium). Actually, no perfect insulator exists since even the best nonconductor conducts to some extent, e.g. under the influence of cosmic radiation. Maxwell therefore supplements the displacement current to form the total current

$$\mathbf{C} = \dot{\mathbf{D}} + \mathbf{J}; \tag{5}$$

the designation **C** (current) was introduced by Maxwell. This notion of the equivalence of $\dot{\mathbf{D}}$ and **J** is a basically new idea of Maxwell, which is a prerequisite for the unified representation of electromagnetic phenomena. Similarly, he supplements in the metallic conductor the conduction current **J** by the addition of a hypothetical displacement current $\dot{\mathbf{D}}$, although here the first term completely outweighs the second.

We now pass to the *magnetic field*. This quantity exerts a mechanical force on a magnetic pole P, which, to begin with, may be thought of as isolated. With the same letter P we designate also the strength of the magnetic pole and with P the as yet undetermined dimension "pole strength." The mechanical force divided by P we should most properly call the *magnetic field strength*. We will however, at least in the beginning, adhere to custom and call this quantity the *magnetic induction* **B**:

$$B \equiv \frac{\text{force}}{\text{pole strength}} = \frac{\text{newton}}{\text{P}}. \tag{6}$$

We shall even go a step further in our adherence to customary notions and utilize the relation between current and magnetism elaborated by Ampère, whose systematic description must however be postponed until §17. Thus, for example, the magnetic field of a plane circulating current I about the area F is, at a great distance from I, equal to the field of a bar magnet placed normal to F at I, with the moment

$$m = IF. \tag{6a}$$

This relation, which in the conventional cgs system serves to measure the current I "magnetically," we shall here employ to define the pole strength P in terms of our electric unit of charge Q. We set

$$m = \text{pole strength} \cdot \text{pole separation} = Pl \tag{6b}$$

and obtain from Eq. (6a)

$$P = I \cdot \frac{F}{l} = \frac{Q}{S} \cdot \frac{M^2}{M} = Q \frac{M}{S}. \tag{7}$$

Our dimensional equation (6) thus becomes

$$B = \frac{\text{newton}}{Q} \frac{S}{M}. \tag{8}$$

A complete description of the magnetic field also requires in addition to **B** a second vector which we shall designate with **H**. We cannot however adhere to the customary notation "magnetic field strength," which, as we have seen, rightfully belongs to the vector **B**, but will call **H** the *magnetic excitation*. We follow here the carefully thought-out representation of electrodynamics of Mie.[1] With the name magnetic excitation we place **H** in parallel with the "electric excitation" **D**. Corresponding to Eq. (4) we therefore define **H** dimensionally by

$$H = \frac{\text{pole strength}}{\text{area}} = \frac{P}{M^2}, \tag{9}$$

[1] Gustav Mie, Lehrbuch der Elektrizität und des Magnetismus, 2nd Ed., Enke, Stuttgart, 1941, and Handbuch der Experimentalphysik Vol. XI, Part 1, Elektrodynamik.

which, in view of Eq. (7), we may write

$$H = \frac{Q}{MS}. \tag{9a}$$

This representation also justifies a designation which is commonly employed in engineering and which, though rather awkward, is more appropriate than the unfortunate name "magnetic field strength," namely, the designation "ampere turns per unit length." For further details see the end of §4.

The direction of the field vector **B**, varying from point to point, is represented by the form of the *magnetic lines of force*. As is well known, these are made evident by the automatic alignment of iron filings which are brought into the neighborhood of the magnet and were known long before the corresponding *electric* lines of force. Their expressive appearance still contributes greatly to the understanding of the field concept. Because of the equality of direction of **H** and **B** in air or any other isotropic medium the line patterns corresponding to the **H** vector are identical with the lines of force of the **B** vector.

We may indicate finally a subdivision of physical entities into entities of intensity and entities of quantity. **E** and **B** belong to the first class, **D** and **H**, to the second. The entities of the first class are answers to the question "how strong," those of the second class, to the question "how much." In the theory of elasticity, for example, the stress is an entity of intensity, the corresponding strain, one of quantity; in the theory of gases pressure and volume form a corresponding pair of entities. In **D** the quantity character is clearly evident as the quantity of electricity that has passed through; in **H** the situation is slightly obscured by the fact that there are no isolated magnetic poles (see §3). We are in general inclined to regard the entities of intensity as cause, the corresponding entities of quantity as their effect.

§3. Maxwell's Equations in Integral Form

After this very incomplete preparation we pass to the *axiomatic foundation of Maxwell's theory*. The axioms of electrodynamics, just as the Newtonian axioms of mechanics, rest on experience—more exactly on the ordering of the totality of experience into a simplified and idealized form. Thus the law of inertia of mechanics appears very different from what is observed in a particular case for terrestrial bodies. Similarly, our electromagnetic axioms are much more abstract and mathematically generalized than what is measured with coils, wires, and pointer instruments. Nevertheless like the mechanical axioms, they are simply a summary of diverse observations.

To begin with we set up two principal axioms which we shall then supplement by secondary axioms. One of these we shall call *Faraday's law of induction*. We shall state it, as far as practicable, in Faraday's own line-of-force language. The other axiom we shall name after Ampère, since he was the first to formulate the relationship between current and magnetic fields. The fact that *Ampère's law* also rests on experience has been emphasized by its author.[1] We shall, however, state both axioms in the universal form whose possibility was first realized by Maxwell.

For this purpose we consider an arbitrary surface σ with the boundary curve s. We provide the latter with a pointer indicating sense of travel and shall define that direction of the normal to the surface as positive which forms a *right-handed screw* with the s-pointer. We compute the surface integrals

$$\int B_n\, d\sigma \quad \text{and} \quad \int C_n\, d\sigma \tag{1}$$

extended over σ and shall call them *magnetic flux* and *electric current flux*. *Number of lines of force* and *number of lines of current* is another common designation. This notation is of course audacious since these bundles of lines are not countable. It is first necessary to group them in "tubes," just as in Vol. II (p. 136) the lines of turbulence were grouped in tubes of turbulence. The tubes must be constructed so that their cross section becomes inversely proportional to the magnitude of **B** and **C**, respectively, at the point in question. The counting of the tubes of force or current traversing our surface then amounts to the same as the evaluation of the integrals (1).

Next we compute the following line integrals extended over the boundary curve s:

$$\oint \mathbf{E}\cdot d\mathbf{s} \quad \text{and} \quad \oint \mathbf{H}\cdot d\mathbf{s}. \tag{2}$$

We call these the *electric* and *magnetic loop tension*. The first has also for a long time been called E.M.F. or electromotive force; included in this designation, it is true, are also other "electromotive" causes, such as differences in temperature and chemical effects. The word "force" is here used in its antiquated meaning of energy.

The remarkable thing in Maxwell's point of view is that the E.M.F., which to the experimenter had had meaning only for closed metallic circuits, is here defined for arbitrary loops, whether they pass through conductors, nonconductors, or through parts of both. The same geometric freedom then exists also for the magnetic loop tension or magnetomotive force. We now write down the two principal axioms which relate the

---

[1] In the title of his comprehensive paper: La théorie analytique des phénomènes électrodynamiques, *uniquement déduite de l'expérience*.

quantities defined in Eqs. (1) and (2) in this completely general sense. They are:

$$\frac{d}{dt} \int B_n \, d\sigma = -\oint \mathbf{E} \cdot d\mathbf{s}, \tag{3}$$

$$\int C_n \, d\sigma = \oint \mathbf{H} \cdot d\mathbf{s} \tag{4}$$

In words: *Every change in the number of magnetic lines of force which traverse a given surface σ produces in its boundary s an electric loop tension which is numerically equal to the rate of change, but opposite in sign* (Faraday's law of induction) and

The number of electric current lines, which traverse an arbitrary surface σ is accompanied by a magnetic loop tension in the bounding curve s of σ which is equal to it in both magnitude and direction (Ampère's law relating magnetic field and electric current).

Let us convince ourselves first that this equating of electric and magnetic quantities is dimensionally proper. The two surface integrals defined in (1) (in spite of their dimensionally incorrect designation as *numbers* of force lines or current lines) have, according to (2.8) and 2.4a), the dimensions

$$\frac{\text{newton MS}}{Q} = \frac{\text{joule S}}{Q} \quad \text{and} \quad \frac{Q}{S}, \text{ respectively.}$$

According to (2.9a) the latter dimension agrees with the dimension of the line integral in (4). The time rate of change of the first expression yields joule/Q, i.e. the dimension of an electric tension (expressible in volts), in agreement with the right side of Eq. (3). From this dimensional check our fundamentally different conception of **B** and **H** becomes apparent, and it is clear that our special introduction of the symbol Q for the dimension of charge is unavoidable.

Next we concern ourselves with the signs in Eqs. (3) and (4). They correspond to the rules of Lenz and Ampère. Ampère's rule is simply the right-handed screw rule, by which we correlated the positive normal of the surface σ with the sense of travel along the boundary s. The various rules of thumb commonly given in textbooks are merely specializations of our righthanded screw convention. To check Lenz's rule we imagine in Eq. (3) the boundary curve s to be realized by a wire loop, and the magnetic flux traversing the surface σ in the direction n as lines of force proceeding from the positive pole P of a bar magnet, the negative pole being assumed to be sufficiently far away. We bring (see Fig. 1) the bar magnet near to the wire loop and thus increase the magnetic flux, so that the left side of (3) becomes *positive*. Then, as shown by the equation, the line integral on the right side must become *negative*. The E.M.F. and the corresponding current induced in the wire loop then form a left-handed screw with the direction of motion of the bar magnet. The magnetic field corresponding to the induced current is, on the other hand, represented, according to our right-hand screw rule, by the arrow P' in Fig. 1. The positive pole of this magnetic field thus points in the direction from which the positive pole P of the magnet approaches the loop: The two poles repel each other or the induced current *inhibits* the motion of the inducing magnet. This is the meaning of Lenz's rule: The appearance of the induced current opposes the disturbance of equilibrium produced by the motion of the bar magnet.

We emphasized above that the bounding curve s may be fixed quite arbitrarily; the same remark applies also for fixed boundary to the surface σ. If two different surfaces σ_1 and σ_2 are passed through the same curve s, the left sides of Eqs. (3) and (4) computed for σ_1 and σ_2, must turn out to

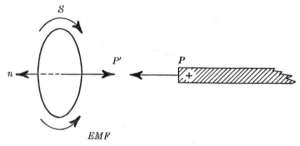

FIG. 1. Lenz's rule.

be equal. This is equivalent to stating that they must vanish for the closed surface formed by σ_1 and σ_2 if the positive normal (n always pointing outward) is defined in a uniform manner. We realize this fact also from the following: We consider a closed surface σ with a boundary curve which has contracted to a point. This does not contribute to the line integrals in Eqs. (3) and (4). If we indicate the integration over the now unbounded surface with $\oint$, we thus obtain

$$\frac{d}{dt} \oint B_n \, d\sigma = 0 \qquad \text{and} \qquad \oint C_n \, d\sigma = 0; \qquad (5)$$

by Eq. (2.5) the second equation may also be written

$$\oint J_n \, d\sigma + \frac{d}{dt} \oint D_n \, d\sigma = 0. \qquad (5a)$$

More particularly, if the surface σ lies entirely in nonconducting material and is hence traversed by no conduction currents,

$$\frac{d}{dt} \oint D_n \, d\sigma = 0. \qquad (5b)$$

The first equation (5) and Eq. (5b) state, in integrated form,

$$\oint B_n \, d\sigma = \text{const.}, \qquad \oint D_n \, d\sigma = \text{const.}, \qquad (6)$$

while the second Eq. (5) and Eq. (5a) show that the total electric current is always *closed* in Maxwell's theory: the quantities entering and leaving just compensate each other; the current lines traversing our surface σ form closed loops somewhere outside of it. Furthermore, the *magnetic* lines of force also are always closed. If a magnet (or electromagnet) is subdivided, north poles and south poles, which compensate each other as far as the total magnetic flux is concerned, are formed anew on every part. It follows that the constant in the first Eq. (6) *must be zero*, while in the second equation this constant is the algebraic sum $\bar{e}$ of the charges e enveloped by the surface σ. According to the above this must be a *constant in time for a nonconductor:*

$$\oint B_n \, d\sigma = 0, \qquad \oint D_n \, d\sigma = \bar{e}, \qquad \bar{e} = \sum e = \text{const.} \qquad (6a)$$

The first Eq. (6a) is a *supplementary axiom*, an addition to our principal axioms required by experience. The second Eq. (6a) agrees with our earlier Eq. (2.3b) and states the constancy in time of the charge in nonconductors. The **D**-lines and the **E**-lines coinciding with them geometrically originate at points of positive charge and end at points of negative charge. Eq. (5a) generalizing the second Eq. (6a), may be designated in hydrodynamic terminology as the *continuity equation of electricity*. If the definition of $\bar{e}$ in Eq. (6a) is employed it takes on the form

$$\frac{d\bar{e}}{dt} + \oint J_n \, d\sigma = 0. \qquad (6b)$$

This expresses the fact that the electricity within a surface σ may decrease as the result of flowing off through metallically conducting portions of σ.

The first Eq. (6a) may be expressed, with Hertz, in the form: *There is no true magnetism.* In this statement one proceeds from the assumption, formerly regarded as obvious, that **B** is the magnetic analogue of **D**. From our standpoint, however, this analogue is **H**, and not **B**. We shall hence have to relate the definition of "magnetism," in particular of the pole strength P (see §7), not to **B** but to **H**.

We now apply the first Eq. (6a) to the neighborhood of the *boundary surface between two bodies of different magnetic properties* such as iron and air. Let the closed surface σ be the surface of a very flat prism (Fig. 2), whose height Δh is very small compared to the base Δf, and let this base lie for example in iron, while the parallel top side is in air. Eq. (6a) then demands, with arbitrary accuracy in view of the arbitrary smallness of Δh,

$$(B'_{n'} + B_n)\, \Delta f = 0. \tag{7}$$

Let $\mathbf{B}'$ refer, for example, to iron, $\mathbf{B}$, to air. The normal (n' in iron, n in air) points outward on both surfaces Δf. Then, in view of Eq. (7),

$$B'_{n'} = -B_n \qquad \text{and hence also} \qquad B'_n = B_n,$$

provided that now n denotes the same direction in both media. We have thus obtained a first *boundary condition* for the magnetic field: At the transition between two magnetically different media the *normal component of the induction is continuous.*

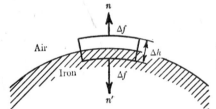

FIG. 2. Derivation of the continuity of B_n at the boundary between two media from the relation $\oint B_n\, d\sigma = 0$.

We will show that the same applies to the *tangential component of the excitation* $\mathbf{H}$. For this purpose we consider a very small rectangular loop s (Fig. 3), with the height Δh normal to the boundary surface and the side length Δs parallel to it. Here we assume that $\Delta h \ll \Delta s$ so that in the limit $\Delta h \to 0$ the area $\Delta \sigma = \Delta h \Delta s$ vanishes. With the assumption that the current density parallel to the boundary surface, referred to in Eq. (4), does not become infinitely large[1] we obtain from Eq. (4):

$$0 = (H'_{s'} + H_s)\, \Delta s \tag{8}$$

so that

$$H'_{s'} = -H_s \qquad \text{and hence also} \qquad H_s = H'_s \tag{8a}$$

where again s denotes the same direction in the two media.

From exactly the same figure and the same consideration for two electrically different media we are led from Faraday's law of induction to the conclusion that the *tangential components of the electric field strength* $\mathbf{E}$ are continuous along the boundary of the two media:

$$E'_s = E_s. \tag{9}$$

[1] This limiting case is the normal one for good conductors at high frequencies. Then H_s becomes discontinuous and B_n vanishingly small.

Nothing has been said regarding the *normal component of* **E**. Furthermore, the continuity of the normal component of **D** (unlike that of **B**) is not required by Eq. (6a). For, if D_n has a discontinuity at the boundary of two electrically different media (e.g. glass and air) or at any other surface, we say that a surface charge is present on the surface. If we call this surface charge ω (dimension Q/M^2), the charge present in the prism in Fig. 2 for the transition to the limit $\Delta h \rightarrow 0$ is,

$$\bar{e} = \omega \, \Delta f. \tag{10}$$

thus by the consideration leading to Eqs. (7) and (7a), the second Eq. (6a) demands

$$(D'_{n'} + D_n) \, \Delta f = \omega \, \Delta f, \tag{10a}$$

or, employing the same direction of the normal n:

$$D_n - D'_n = \omega. \tag{11}$$

Discontinuous behavior of the normal component of **D** *signifies that the boundary surface considered carries a surface charge; the magnitude of the discontinuity indicates the surface charge directly.*

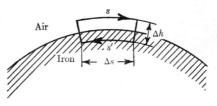

Fig. 3. Derivation of the continuity of H_s at the boundary between two media from the relation $\oint \mathbf{H} \cdot ds = 0$.

Finally, we obtain from Eq. (6b) for the boundary surface between a *conductor and a nonconductor* by utilizing Fig. 2 and Eq. (10),

$$\frac{d\omega}{dt} + J_n = 0, \tag{12}$$

that is, a *loss* of surface charge if electric current is possible in the conductor. In electrostatics, where the interior of conductors is fieldfree ($\mathbf{D} = 0$, $\mathbf{J} = 0$), Eq. (12) is fulfilled identically and Eq. (11) takes on the special form

$$\omega = D_n. \tag{12a}$$

In the static field conductors bear a surface charge varying from point to point and given by the normal component of **D**.

§4. *The Maxwell Equations in Differential Form and the Material Constants of the Theory*

We pass from the integral to the differential form by allowing the loops s in the integral form, and hence also the surfaces σ passed through them, to become arbitrarily small. If we call the latter $\Delta\sigma$ we can write in the limit:

$$\int B_n \, d\sigma = \Delta\sigma B_n, \qquad \int C_n \, d\sigma = \Delta\sigma C_n. \tag{1}$$

Furthermore we recall the definition of the vector operation "curl" by the transition to the limit of a loop integral (Vol. II, Eq. 2.21). For our infinitesimal loops this leads to

$$\oint E_s \, ds = \Delta\sigma \, \mathrm{curl}_n \, \mathbf{E}, \qquad \oint H_s \, ds = \Delta\sigma \, \mathrm{curl}_n \, \mathbf{H}. \tag{2}$$

We must form the time derivative of the first Eq. (1). We will here imagine the surface $\Delta\sigma$ to remain fixed, which obviously applies to media *at rest*, to which we shall confine ourselves initially. We then obtain

$$\frac{d}{dt} \int B_n \, d\sigma = \Delta\sigma \, \dot{B}_n, \qquad \dot{\mathbf{B}} = \frac{\partial \mathbf{B}}{\partial t}. \tag{3a}$$

At the same time, using Eq. (2.5), we write Eq. (1) in the form

$$\int C_n \, d\sigma = \Delta\sigma(J_n + \dot{D}_n), \qquad \dot{\mathbf{D}} = \frac{\partial \mathbf{D}}{\partial t}. \tag{3b}$$

With Eqs. (2) and (3a, b), cancelling the factor $\Delta\sigma$ which is common to all terms, as well as omitting the common index n, the principal axioms (3.3) and (3.4) lead to the two *vectorial differential equations*:[1]

$$\dot{\mathbf{B}} = - \, \mathrm{curl} \, \mathbf{E}$$

$$\dot{\mathbf{D}} + \mathbf{J} = \mathrm{curl} \, \mathbf{H}. \tag{4}$$

The universal importance and impressive beauty of these equations led Boltzmann[2] to quote: "Was it a god who wrote these lines ..."

---

[1] Our second equation (4) is usually called the first set of Maxwell's equations, our first Eq. (4), the second set. We prefer the sequence of the text since in our presentation the intensity entities **E** and **B** were introduced first as being more readily interpreted. We can also point to §7, where electrostatics will result from the specialization of the first, magnetostatics, from the specialization of the second Eq. (4), in support of our order. Since it would be improper to treat magnetostatics ahead of the simpler electrostatics the numbering of the Maxwell equations which differs from ours appears unsuitable.

[2] Motto of the second volume of his "Vorlesungen über Maxwells Theorie der Elektrizität und des Lichtes," München 1893. Our formulation, which deviates slightly from Boltzmann's (vector in place of coordinate notation), clearly only serves to enhance the beauty and simplicity of the equations.

We complete them by the supplementary axiom (3.6a) for **B**, and the relation between **D** and the charge, contained in the same equation. We shall now regard the latter as continuously distributed in accord with our differential point of view. Thus, we shall not speak of point charges e, but of finite densities in space ρ, so that the infinitesimal charge

$$\Delta e = \rho \,\Delta\tau$$

is contained in an element of volume $\Delta\tau$ which approaches zero in magnitude. At the same time we recall the vector operation "divergence" and its representation (in Vol. II, Eq. 2.20) by the limit of a volume integral.[1] For our present purposes we write this representation

$$\lim \frac{1}{\Delta\tau} \oint B_n \, d\sigma = \operatorname{div} \mathbf{B}, \qquad \lim \frac{1}{\Delta\tau} \oint D_n \, d\sigma = \operatorname{div} \mathbf{D}$$

and obtain for Eqs. (3.6a, b), omitting the factor $\Delta\tau$, *their differential form:*

$$\operatorname{div} \mathbf{B} = 0, \tag{4a}$$

$$\operatorname{div} \mathbf{D} = \rho, \tag{4b}$$

$$\frac{\partial\rho}{\partial t} + \operatorname{div} \mathbf{J} = 0. \tag{4c}$$

Our Eqs. (4) and (4a, b, c) set up the framework into which the phenomena of electrodynamics must be fitted. But this framework is still *too wide.* Five vectors **E**, **D**, **J**, **B**, and **H** occur in our equations, or altogether 15 unknown functions of time and space. (The scalar ρ is referred back to the vectors **D** and **J** by the Eqs. (4b) and (4c) respectively.) For their determination we have two vector equations (4), i.e., altogether only six differential equations. We must narrow down the framework to be able to fill it out with a unified electrodynamic model. The electromagnetic

---

[1] We contrast the volume divergence here introduced with the term *surface divergence.* Referring to Fig. 2 and the integration there carried out over a prism with a base Δf and vanishing height, we understand by this the result of the integration divided by Δf. According to Eq. 3.7 and with the meaning of the normals n and n' there given, the surface divergence of an arbitrary vector **A** is:

$$A_{n'} + A_n; \tag{4d}$$

Eqs. 3.7a and 3.10a then state simply: *The surface divergence of* **B** *vanishes, that of* **D** *equals the surface charge.*

Similarly, we can contrast the volume curl with the *surface curl.* Referring to Fig. 3 and the integration over a rectangle of base Δs and vanishing height carried out in Eq. 3.8, we understand by the surface curl the result of the integration divided by Δs. The surface curl of an arbitrary vector **A** is hence, according to Eq. 3.8,

$$A_{s'} + A_s; \tag{4e}$$

it represents the discontinuity of **A** at the surface in question.

material constants serve this purpose. We shall discuss them in the sequence *conductivity, dielectric constant, permeability.*

1. Conductivity and Ohm's Law

The electric current density $\mathbf{J}$ depends on the electric field strength $\mathbf{E}$ within the conductor. We assume a linear dependence

$$\mathbf{J} = \sigma\mathbf{E} \tag{5}$$

and call the real positive constant σ the *electric conductivity*. Eq. (5) expresses *Ohm's law* for unit length of a wire carrying a stationary current. To recognize this, we replace $\mathbf{J}$ by the total current $I = qJ$ (q = cross section of the wire) and multiply Eq. (5) with the length of the wire. We obtain

$$RI = V \cdot \begin{cases} R = \dfrac{1}{q\sigma} \\[2ex] V = lE = \displaystyle\int_0^l E\,ds = \text{voltage} \end{cases} \tag{5a}$$

The concept of voltage had been created already by Volta, while the concept of resistance was first introduced by Georg Simon Ohm in 1827. For us Ohm's law signifies the introduction of the material constant σ. According to Eq. (5) its dimension is:

$$\sigma = \frac{Q^2}{M^2S\ \text{newton}} = \frac{Q^2}{MS\ \text{joule}}. \tag{5b}$$

According to Eq. (5a) σ may also be designated as the reciprocal of the *specific electric resistance*, i.e. the resistance of a prism of the length $l = 1$ M and of the cross section $q = 1$ M^2. The dimension of the resistance is by Eqs. (5a, b):

$$R = \frac{\text{joule S}}{Q^2}. \tag{5c}$$

The unit of resistance in the practical system of units is the Ω (pronounced "ohm") $= 10^9$ cgs units. It is identical with the unit in our MKSQ system provided that we choose, according to our convention, Q equal to 1 coulomb $= \frac{1}{10}$ cgs unit We then obtain

$$1\,\frac{\text{joule S}}{Q^2} = 10^7\,\frac{\text{erg sec}}{Q^2} = 10^9\ \text{cgs units} = 1\Omega. \tag{5d}$$

Ohm's law applies only to macrophysical events, not to Ampère's molecular currents, electron paths in atoms, Larmor precessions; cathode rays in vacuum tubes are also resistance-free electric currents.

2. Dielectric Constant

The displacement **D** depends on the electric field strength **E** at the point in question. We assume the dependence to be linear:

$$\mathbf{D} = \varepsilon \mathbf{E} \tag{6}$$

and call the real positive constant ε the dielectric constant. Its dimension is, by Eqs. (2.4) and (2.1),

$$\varepsilon = \frac{Q^2}{M \text{ joule}}. \tag{6a}$$

We denote the dielectric constant of vacuum by ε_0. It also is a definite quantity of the dimension (6a). The relation

$$\mathbf{D} = \varepsilon_0 \, \mathbf{E}, \tag{6b}$$

valid for vacuum, was pointed out already in §2. Invariably $\varepsilon > \varepsilon_0$.

3. Permeability

A relation also exists between the two magnetic vectors **H** and **B**, which, as a first approximation, we shall also assume to be linear. We would like to write it in the form

$$\mathbf{H} = \mu' \mathbf{B},$$

since we regard **H** as analogue of **D** and **B** as analogue of **E**. However, we are unfortunately obliged to follow general usage and choose the form

$$\mathbf{B} = \mu \mathbf{H}. \tag{7}$$

The material constant μ is called *permeability* and has, according to Eqs. (2.8) and (2.9a), the dimension

$$\mu = \frac{\text{joule}}{Q^2} \frac{S^2}{M}. \tag{7a}$$

This introduction of μ, which is illogical in view of Eq. (6), leads to the obvious consequence that in later formulas, such as Coulomb's law, not μ, but its reciprocal μ' will take the place of ε. For vacuum we write

$$\mathbf{B} = \mu_0 \, \mathbf{H}; \tag{7b}$$

μ_0 also obviously has the dimension given in Eq. (7a). For *paramagnetic* bodies $\mu > \mu_0$, for *diamagnetic* bodies, $\mu < \mu_0$. Our formulas (5), (6), and (7) do not have the same degree of certainty and general validity as Maxwell's equations (4). This has long been known for the *ferromagnetic materials*, where a general functional relationship

$$\mathbf{B} = \mathbf{B}(H, T), \qquad T = \text{absolute temperature}$$

takes the place of the linear relation (7). Rochelle salts[1] show a dielectric behavior similar to that of the ferromagnetic materials, exhibiting, like the latter, both saturation and hysteresis phenomena.

For paramagnetic materials deviations from linearity occur only at extremely high field strengths or extremely low temperatures. Deviations from the linearity of Ohm's law have been expected at very high field strengths; the failure of this law for superconductors is obvious. Furthermore the simple proportionality between corresponding vectors expressed by Eqs. (5), (6), and (7) is true only for isotropic bodies. In *crystals* the dependence is expressed instead quite generally by a *linear vector function* (see Vol. II, Eq. 1.10). The varied and interesting phenomena of crystal optics, which we shall treat in Vol. IV, rest on this fact.

On the other hand, the general field equations (4) apply also for anisotropic bodies. Beyond this, they appear to hold true even in the face of all new proposals of a *generalized* electrodynamics, proposals which are concerned with extremely strong fields (such as must occur, for example, close to an electron), but intrinsically amount merely to a replacement of the linear relation (6) by a generalized variation of $\mathbf{D}$ with $\mathbf{E}$ (see the final section of this volume). The deeper reason for the remarkable vitality of the form of equation discovered by Maxwell will be found to rest in its invariance properties, which will not, however, be taken up until Part III.

We can now undertake the required contraction of our electrodynamic framework. If, in particular, we employ for this purpose our simple linear relations and treat σ, ε, and μ as quantities independent of t (restriction to media *at rest*), we obtain by substituting Eqs. (5), (6), and (7) in Eq. (4):

$$\mu \frac{\partial \mathbf{H}}{\partial t} = - \text{ curl } \mathbf{E},$$

$$\left(\varepsilon \frac{\partial}{\partial t} + \sigma \right) \mathbf{E} = \text{ curl } \mathbf{H},$$

(8)

i.e., six simultaneous differential equations of the first order for six unknowns, the $2 \cdot 3$ components of $\mathbf{E}$ and $\mathbf{H}$. Thus we find ourselves presented with a well-defined mathematical problem.[2]

---

[1] Also known as Seignette salts. Seignette was the name of a pharmacist in the French fortress La Rochelle. We are here concerned with hydrated sodium potassium tartrate:

$$NaOOC \cdot CHOH \cdot CHOH \cdot COOK + 24 \ H_2O.$$

[2] We could of course also have written Eqs. (8) as relations between $\mathbf{E}$ and $\mathbf{B}$, or also between $\mathbf{D}$ and $\mathbf{H}$. However, the form in the text is the customary one and, in general, also the most convenient one.

At the same time the conditions (4a, b, c) take on the form

$$\text{div}(\mu \mathbf{H}) = 0, \qquad \text{div}(\varepsilon \mathbf{E}) = \rho, \tag{8a, b}$$

$$\text{div}\left\{ \left(\varepsilon \frac{\partial}{\partial t} + \sigma \right) \mathbf{E} \right\} = 0. \tag{8c}$$

Eq. (8a) is to be regarded as a restrictive supplementary condition on Maxwell's equations, Eq. (8b), as defining equation for ρ. Eq. (8c) is obtained by forming the divergence of the second equation (8). Its simplest solution results from setting the parenthesis $\{\}$ equal to zero; it is represented by the exponential function

$$\mathbf{E} = \mathbf{E}_0 \exp\left(-\frac{\sigma}{\varepsilon} t \right), \quad \mathbf{E}_0 = \text{arbitrary function of space.} \tag{9}$$

We set

$$\frac{\varepsilon}{\sigma} = T, \tag{9a}$$

and call T the *relaxation time of the conductor*. Its dimension is the second by Eq. (5a) and (6a), as must be the case, its magnitude for good conductors a very small fraction of a second. The field decays within the conductor everywhere in accord with this relaxation time and is known, provided that $\mathbf{E}_0$ is given.

We might continue with the already discussed conditions at the boundary between two electromagnetically different media. To use the differential form of the Maxwell equations, however, it would be necessary to regard the transition between the two media as continuous, i.e., to speak of a "boundary layer" rather than a "boundary surface." We will carry out this procedure in problem I.1, where we shall find that the derivation becomes less straightforward than in the Eqs. 3.7 to 3.12, which followed from the integral form of Maxwell's equations.

The same conclusion is reached in other problems distinguished by a particular symmetry: *The general development of Maxwell's theory must proceed from its differential form; for special problems the integral form may, however, be more advantageous.*

The following two fundamental problems, which will be treated also by the differential method in problem I.2 and I.3, are examples of this:

1. An infinitely long wire in the form of a circular cylinder is traversed by current distributed uniformly over its cross section. The return of the current may take place through a similarly traversed hollow cylinder which is coaxial with the wire. The magnetic excitation is to be determined within the wire, within the hollow cylinder, and in the region between them.

2. An infinitely long, tightly wound coil is traversed similarly by stationary current. The magnetic excitation is to be determined at any point within the coil.

Regarding 1: Let a be the radius of the wire, b and c, the inner and outer radius of the cylindrical return conductor. We introduce a right-handed coordinate system about the center line of the wire as z-axis. Let the current density have the direction of the positive z-axis in the wire, that of the negative z-axis in the return conductor. Let the total current be I and $-I$ respectively:

$$I = \pi a^2 J_z, \qquad -I = \pi(c^2 - b^2) J_{-z}.$$

The symmetry of the problem indicates that $\mathbf{H}$ is independent of φ and has the direction of increasing φ. We write $H_\varphi = H$ and carry out the line

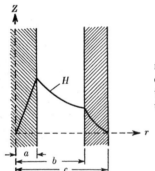

FIG. 4. A straight wire carrying a stationary current and a hollow cylinder surrounding it as return conductor. The magnetic excitation $H_\varphi = H$ within the wire, in the air space between the two conductors, and in the return conductor.

integral of $\mathbf{H}$ about any circle $r = $ const in any cross-section plane of the wire. Since the displacement current vanishes everywhere in view of the assumed stationary condition, we obtain

$$0 < r < a: \quad 2\pi r H = \pi r^2 J_z = \frac{r^2}{a^2} I, \qquad H = \frac{r}{a}\frac{I}{2\pi a}, \tag{10}$$

$$a < r < b: \quad 2\pi r H = I, \qquad H = \frac{I}{2\pi r} \tag{11}$$

$$b < r < c: \quad 2\pi r H = I + \pi(r^2 - b^2) J_{-z} = I\left(1 - \frac{r^2 - b^2}{c^2 - b^2}\right),$$
$$\tag{12}$$
$$H = \frac{I}{2\pi r}\frac{c^2 - r^2}{c^2 - b^2}$$

$$c < r \quad : \quad 2\pi r H = I - I, \qquad H = 0. \tag{13}$$

The boundary conditions for $\mathbf{H}$ at the surface of the wire $r = a$ and at the cylinder surfaces $r = b, c$ are satisfied automatically by Eqs. 10 to 13. The variation of H is plotted in Fig. 4.

Regarding 2: We use a right-handed system r, φ, z which has the center line of the coil as z-axis. For sufficient length of the coil and sufficiently close winding no magnetic lines of force penetrate to the exterior of the coil; the current I has the direction of increasing φ, the excitation $\mathbf{H}$ that of increasing z. We shall show that $H_z = H$ is constant within the coil.

For this purpose we consider the rectangular loop, of length l in the z-direction, shown in Fig. 5. Its plane intersects the coil in $N_1 l$ points, where N_1 is the number of turns per unit length of the coil. Since $H_r = 0$ both

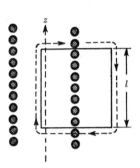

FIG. 5. The magnetic excitation $\mathbf{H}$ within an infinitely long coil.

within and outside of the coil and $H_z = 0$ outside of the coil, only one side of the loop contributes to the line integral. We find

$$Hl = N_1 l I, \qquad H = N_1 I. \tag{14}$$

The magnetic excitation within the coil is given by the "number of ampere turns per unit length" $N_1 I$. This explains the designation of H customary in engineering practice which was introduced on p. 12. The value of H given by Eq. 14 is independent of r, i.e. the same throughout the interior of the coil.

§5. Law of Conservation of Energy and Poynting Vector

Starting from Eqs. (4.4) we carry out a scalar multiplication of the first with $\mathbf{H}$, a scalar multiplication of the second with $\mathbf{E}$. We obtain as the sum of the two:

$$\mathbf{H} \cdot \dot{\mathbf{B}} + \mathbf{E} \cdot \dot{\mathbf{D}} + \mathbf{E} \cdot \mathbf{J} = \mathbf{E} \cdot \text{curl } \mathbf{H} - \mathbf{H} \cdot \text{curl } \mathbf{E}. \tag{1}$$

On the right-hand side we apply the transformation, valid for arbitrary vectors $\mathbf{U}$, $\mathbf{V}$:

$$\mathbf{V} \cdot \text{curl } \mathbf{U} - \mathbf{U} \cdot \text{curl } \mathbf{V} = \text{div}(\mathbf{U} \times \mathbf{V}). \tag{2}$$

We prove this relation most readily by utilizing the symbolic "nabla operator"

$$\nabla = \frac{\partial}{\partial x}, \frac{\partial}{\partial y}, \frac{\partial}{\partial z}$$

(see Vol. II, footnote 1 on p. 23) and interpreting the divergence as scalar multiplication, the curl as vector multiplication with this vector:

$$\text{div}(\mathbf{U} \times \mathbf{V}) = \nabla \cdot (\mathbf{U} \times \mathbf{V}) = \nabla_U \cdot (\mathbf{U} \times \mathbf{V}) + \nabla_V \cdot (\mathbf{U} \times \mathbf{V}), \quad (2a)$$

$$\text{curl } \mathbf{U} = \nabla \times \mathbf{U}, \qquad \text{curl } \mathbf{V} = \nabla \times \mathbf{V}. \quad (2b)$$

In Eq. (2a) the subscripts U, V indicate that the ∇-differentiations are to be carried out only on the vectors $\mathbf{U}$ and $\mathbf{V}$, respectively. Since the sequence of the vectors may be cyclically interchanged in the double products, Eq. (2a) may also be written

$$\text{div}(\mathbf{U} \times \mathbf{V}) = \mathbf{V} \cdot (\nabla \times \mathbf{U}) + \mathbf{U} \cdot (\mathbf{V} \times \nabla)$$
$$= \mathbf{V} \cdot (\nabla \times \mathbf{U}) - \mathbf{U} \cdot (\nabla \times \mathbf{V}). \quad (2c)$$

Here the right side, in view of Eq. (2b), is the same as the left side of Eq. (2), so that Eqs. (2c) and (2) become identical. This proof of Eq. (2) is only an abbreviated form for the direct, but much more involved, calculation with rectangular coordinates x, y, z.

Let us now set $\mathbf{V} = \mathbf{E}$ and $\mathbf{U} = \mathbf{H}$ in Eq. (2) and introduce the abbreviation

$$\mathbf{S} = \mathbf{E} \times \mathbf{H}, \quad (3)$$

Then Eq. (1) becomes

$$\mathbf{H} \cdot \dot{\mathbf{B}} + \mathbf{E} \cdot \dot{\mathbf{D}} + \mathbf{E} \cdot \mathbf{J} + \text{div } \mathbf{S} = 0. \quad (4)$$

Eq. (4) is Poynting's theorem, **S**, *the Poynting vector*. We shall show that **S** is the *energy flux* vector.

We consider first the dimension of the individual terms of Eq. (4). The first two terms have, according to Eqs. (2.9a) and (2.8), and (2.1) and (2.4), respectively, the dimension

$$\frac{\text{newton}}{\text{M}^2\text{S}} = \frac{\text{joule}}{\text{M}^3\text{S}} = \text{energy per unit volume and unit time.} \quad (4a)$$

The third term has, as must be the case, the same dimension (see Eqs. (2.1) and (2.4a)). The dimension of Eq. (3) is, by Eqs. (2.1) and (2.9a),

$$\frac{\text{joule}}{\text{M}^2\text{S}} = \text{energy per unit area and unit time.} \quad (4b)$$

The operation div, which indicates a differentiation with respect to the space coordinates, yields for the dimension of the fourth term in Eq. (4) the same result.

We see that our *electrical unit* Q does not occur in (4a, b). *It has discreetly withdrawn from the company of the mechanical units MKS.* The same will be noted in many later dimensional considerations in which we are dealing with purely mechanical quantities, which are independent of the choice of the electrical unit.

We pass to the physical interpretation of the individual terms in Eq. (4). It is simplest for the third term: this signifies the work done by the electric field on moving electric charge per unit volume and per unit time. It is generally converted into heat and is known as *Joule heat*. We designate it W_J, transferring the symbol W (work), which Maxwell generally employs for total energy, to *energy density*. Thus we obtain

$$W_J = \mathbf{E} \cdot \mathbf{J}. \tag{5}$$

We shall see right away that the two first terms of (4) are the time rate of change of the *magnetic* and *electric energy densities;* the latter are defined, in accord with Maxwell, by

$$W_m = \tfrac{1}{2}\mathbf{H} \cdot \mathbf{B}, \qquad W_e = \tfrac{1}{2}\mathbf{E} \cdot \mathbf{D}. \tag{6}$$

By this definition the energy is *localized* in the field; a definite electric and magnetic energy content $W_e\,d\tau$ and $W_m\,d\tau$ is ascribed to every element of volume $d\tau$. This constitutes a first step in the adaptation of the energy concept to the ideas of field theory.

The factor $1/2$ in the two defining equation (6) evidently points to a continuous generation of energy, comparable with the stretching of a spring. In accord with the pattern force $\times$ increase in path length $=$ intensity entity $\times$ change in quantity entity, we obtain

$$W_e = \int \mathbf{E} \cdot d\mathbf{D},$$

which, for a linear relationship between $\mathbf{E}$ and $\mathbf{D}$, reverts, in fact, to (6). The situation is slightly different for the magnetic energy. Here Poynting's theorem (4) directs us to start from

$$W_m = \int \mathbf{H} \cdot \dot{\mathbf{B}}\, dt = \int \mathbf{H} \cdot d\mathbf{B}. \tag{6a}$$

From the point of view of our general system ($\mathbf{B} = $ intensity entity, $\mathbf{H} = $ quantity entity) it would have seemed more reasonable to represent the energy density not by (6a), but by

$$W_m = \int \mathbf{B} \cdot d\mathbf{H}. \tag{6b}$$

For a linear relationship between $\mathbf{H}$ and $\mathbf{B}$ this of course leads again to Eq. (6); for a nonlinear variation, on the other hand, it leads to a result which differs from $\int \mathbf{H} \cdot d\mathbf{B}$, and is therefore *incorrect* by Poynting's theorem. From this we learn that work need not be expressible in the form intensity entity $\times$ change in quantity entity.

G. Mie, who takes the same standpoint throughout in respect to the meaning of **B** and **H** as we do, on p. 467 of his excellent textbook cited in §2, points to the following mechanical analogue: a moving body carries with it, in unit volume, the momentum (intensity entity) **p**. For its acceleration the force per unit volume $d\mathbf{p}/dt$ is required, and hence the work

$$\frac{d\mathbf{p}}{dt} \cdot d\mathbf{s} = d\mathbf{p}\cdot\mathbf{v} = \mathbf{v}\cdot d\mathbf{p};$$

This is a product of the type $\mathbf{H}\cdot d\mathbf{B}$, i.e. quantity entity $\times$ change in an intensity entity.[1]

In this representation the magnetic energy parallels the *kinetic* energy of mechanics. We shall meet the same correspondence in electron theory. Also in Helmholtz's analogy between vortices in fluids and electric currents the magnetic energy corresponds to the kinetic energy of the fluid. The same applies for our semielastic ether model in Vol. II, §15.

If we should refer to Maxwell in connection with Eqs. (6), we would find that in Maxwell's work the factor $1/2$ is replaced by $1/(8\pi)$, which from Maxwell has passed over into the major portion of the literature. It evidently lacks the simple logical basis of the factor $1/2$ and has only historical justification.

We must now belatedly give the proof that the quantities $\mathbf{H}\cdot\dot{\mathbf{B}}$ and $\mathbf{E}\cdot\dot{\mathbf{D}}$ occurring in (4) are identical with the time rates of change of the energy densities given by (6). For this purpose we deduce from (6)

$$\dot{W}_e = \tfrac{1}{2}\,\mathbf{E}\cdot\dot{\mathbf{D}} + \tfrac{1}{2}\,\dot{\mathbf{E}}\cdot\mathbf{D}. \tag{6c}$$

The two terms on the right are equal, to begin with, in an *isotropic* medium, where $\mathbf{D} = \varepsilon\mathbf{E}$. They are also equal in an *anisotropic crystal*, where a "linear vector function" replaces the simple proportionality (see p. 22):

$$D_i = \sum_k \varepsilon_{ik} E_k.$$

From this we calculate for the two expressions on the right side of (6)

$$\sum_i E_i \dot{D}_i = \sum_i E_i \sum_k \varepsilon_{ik}\dot{E}_k\,, \qquad \sum_k \dot{E}_k D_k = \sum_k \dot{E}_k \sum_i \varepsilon_{ki} E_i$$

$$= \sum_i E_i \sum_k \varepsilon_{ki}\dot{E}_k\,.$$

---

[1] For the elementary relationship between **p** and **v**, i.e. $\mathbf{p} = M\mathbf{v}$, and for constant mass we have again $d\mathbf{p}\cdot\mathbf{v} = \mathbf{p}\cdot d\mathbf{v}$. However, for a mass varying with time, in particular, the velocity-dependent mass of relativity theory, this is not the case. Then the form $d\mathbf{p}\cdot\mathbf{v}$ of the text expresses the energy change uniquely.

The two expressions are equal to each other since, irrespective of the crystal symmetry[1]

$$\varepsilon_{ik} = \varepsilon_{ki}. \tag{6d}$$

It follows from (6c) that for the anisotropic case, as for the isotropic case,

$$\dot{W}_e = \mathbf{E} \cdot \dot{\mathbf{D}}. \tag{6e}$$

The same applies for the magnetic energy density both for the isotropic medium (proportionality between $\mathbf{H}$ and $\mathbf{B}$) and for the magnetic crystal (linear vector function with $\mu_{ik} = \mu_{ki}$). Here also

$$\dot{W}_m = \tfrac{1}{2}\mathbf{H} \cdot \dot{\mathbf{B}} + \tfrac{1}{2}\dot{\mathbf{H}} \cdot \mathbf{B} = \mathbf{H} \cdot \dot{\mathbf{B}}. \tag{6f}$$

In view of (5) and (6e, f), (4) yields

$$\dot{W} + \operatorname{div} \mathbf{S} = - W_J, \qquad W = W_e + W_m. \tag{7}$$

In this form Poynting's theorem expresses the *energy balance* in the electromagnetic field. The Joule heat is recorded as a loss on the right side of the equation. The left side corresponds to the energy exchange between the volume element $d\tau$ in question and neighboring elements. This becomes even clearer if Eq. (7) is integrated over a given volume; then the application of Gauss's theorem leads to

$$\frac{\partial}{\partial t} \int W \, d\tau + \int S_n \, d\sigma = - \int W_J \, d\tau. \tag{7a}$$

The significance of $\mathbf{S}$ as energy flux through the surface of the volume considered is now evident.

With the introduction of this concept Poynting passes beyond Maxwell's localization of the energy. We now learn not merely how much energy exists at any place, but also where it will go or (for the opposite sign of $\mathbf{S}$) from where it has come.

---

[1] This restriction on the otherwise arbitrary coefficients ε_{ik} is necessary in order that the work done on an element of volume, $\mathbf{E} \cdot d\mathbf{D}$, may be a complete differential. Otherwise the electric energy density would not be a characteristic function of the state, as we postulate for ideal solid bodies. (It is true that for certain known crystals hysteresis phenomena occur which make the notion of a quantity characteristic of this state illusory). Compare the quite analogous situation in the case of the elastic body, Vol. II, p. 72 and p. 288.

In the crystal W_e is a general positive form of the second order in the E_i, not a simple sum of squares as for the isotropic case. The notation in the text as scalar product is in any case conceptually preferable, particularly since it becomes necessary for $\dot{W}_e$, where $\mathbf{E}$ and $\dot{\mathbf{D}}$ need not have the same direction even in the isotropic case.

In contrast with W_e and W_m W_J is not a state function. The condition $\sigma_{ki} = \sigma_{ik}$ should hence apply, in the crystalline conductor, only for a particular crystal symmetry.

In the ideal nonconductor the right side of (7) vanishes, so that (7) takes on the form of the *hydrodynamic equation of continuity* (see Vol. II, Eq. (5.4)): W replaces the hydrodynamic density ρ, $\mathbf{S}$ replaces $\rho\mathbf{v}$. Continuing with this hydrodynamic analogy, we may say that even in the insulator the energy flows not like an incompressible but like a compressible fluid. In a conductor it is absorbed furthermore, in the measure in which heat is generated in any element of volume.

In optics $\mathbf{S}$ plays a dominant role as *ray vector;* the emission and irradiation of a given surface element $d\sigma$ is distinguished by the positive and the negative sign of $\mathbf{S}$.

We know from mechanics that the law of conservation of energy is not only of fundamental importance physically, but is also highly useful mathematically as a first integral of the equations of motion. Something similar applies for our electrodynamic law of conservation of energy: From it may be derived the *uniqueness* of the integration of the Maxwell equations for given *initial condition* and suitably prescribed *boundary conditions* on the boundaries of the region considered.

As usual, the proof is indirect: We assume the existence of two solutions, form their difference, and deduce therefrom a contradiction.

Let the two solutions be $\mathbf{E}_1$, $\mathbf{H}_1$ and $\mathbf{E}_2$, $\mathbf{H}_2$ (by §4 the corresponding vectors $\mathbf{D}$, $\mathbf{B}$ are then also known).

We put

$$\mathbf{E} = \mathbf{E}_1 - \mathbf{E}_2, \qquad \mathbf{H} = \mathbf{H}_1 - \mathbf{H}_2. \tag{8}$$

In view of the *linearity* of Maxwell's Eqs. (4.8), $\mathbf{E}$ and $\mathbf{H}$ are solutions as well as $\mathbf{E}_1$, $\mathbf{H}_1$ and $\mathbf{E}_2$, $\mathbf{H}_2$. Hence Poynting's theorem, e.g. in the form (7a), applies formally also for them. However, the quantities W, $\mathbf{S}$, W_J, because of their *quadratic* character, are composed not merely of the corresponding quantities of the individual fields 1 and 2, but also of *mixed* terms involving 1 and 2. We show this for the quantity W_e as example, assuming isotropy for the sake of brevity.

$$W_e = \tfrac{1}{2}\mathbf{E}\cdot\mathbf{D} = \frac{\varepsilon}{2} E^2 = \frac{\varepsilon}{2} (\mathbf{E}_1 - \mathbf{E}_2)^2, \tag{9}$$

or expanded

$$W_e = \frac{\varepsilon}{2} E_1^2 + \frac{\varepsilon}{2} E_2^2 - \varepsilon\mathbf{E}_1\cdot\mathbf{E}_2. \tag{9a}$$

The last expression on the right is the mixed term mentioned above, while the first two terms denote the electric energy of the individual fields 1 and 2. However, we shall not need this expanded form and shall refer below to the representation in (9). Now, including the quantity W_m and the case of anisotropic media in our consideration, we can say: The quantity W in (7a) represents a *definitely positive* quadratic form, formed with the components of the difference field $\mathbf{E}$, $\mathbf{H}$. The same applies for the quantity W_J. Finally the quantity $\mathbf{S}$ is (irrespective of the difference terms arising in its calculation) the vector product $\mathbf{E} \times \mathbf{H}$ formed by the difference fields.

Let the domain over which (7a) is integrated be composed of partial domains a, b, ... j, ... with, in general, different material constants ε, μ, σ. We indicate this by replacing W and W_J by $\sum_j W$ and $\sum_j W_J$, which, according to the preceding, can, just like the individual W, *never become negative*. Consider now the term

$$\sum_j \int S_n \, d\sigma_j \tag{10}$$

which arises from (7a) in the same manner. Pairs of terms which refer to the same *inner* boundary surface cancel here because for them the S_n are equal and opposite—*opposite* because of the opposite direction of the normal n, *equal* because of the boundary conditions for the tangential components of the fields $\mathbf{E}_1$, $\mathbf{E}_2$ and $\mathbf{H}_1$, $\mathbf{H}_2$, from which follows the equality of the tangential components of the difference fields $\mathbf{E}$, $\mathbf{H}$ and of the component of $\mathbf{S}$ normal to the boundary surface. The sum (10) becomes, therefore, simply equal to the surface integral over the outer boundary of the region of integration

$$\int S_n \, d\sigma. \tag{10a}$$

Let the boundary condition to be prescribed for this outer boundary simply consist in the tangential component of either the electric or the magnetic field being given everywhere on it. For the difference field (8) this signifies that the tangential components of either $\mathbf{E}$ or of $\mathbf{H}$ vanish. In either case the vector product $\mathbf{S}$ formed with them and, hence, the integral (10a) vanish also.

Now (7a) applied to our case takes on the form

$$\frac{\partial}{\partial t} \sum_j \int W \, d\tau_j = - \sum_j \int W_J \, d\tau_j, \tag{11}$$

or, integrated with respect to t:

$$\sum_j \int W \, d\tau_j \Big|_0^t = - \int_0^t dt \sum_j \int W_J \, d\tau_j. \tag{11a}$$

Here the right side is less than or at most equal to zero. The left side vanishes at the lower limit $t = 0$, since for prescribed initial values of the fields 1 and 2 $\mathbf{E} = 0$ and $\mathbf{H} = 0$ in every one of the domains j, so that $W = 0$ also. At the upper limit t, on the other hand, the left side of (11a) is, in view of the meaning of W, certainly not negative; its least value is zero. Only then the inconsistency with the right side is resolved. For this value we must have for all $t > 0$

$$\mathbf{E} = 0, \quad \mathbf{H} = 0,$$

so that, by (8),

$$\mathbf{E}_1 = \mathbf{E}_2, \quad \mathbf{H}_1 = \mathbf{H}_2.$$

This proof of uniqueness satisfies any demand for rigor. An unrigorous proof may be deduced directly from the form of Eqs. (4.8). For these equations permit the determination of the change with *time* of $\mathbf{E}$ and $\mathbf{H}$ if their distribution in *space* is known at any one moment. This means in a sense: the values of $\mathbf{E}$ and $\mathbf{H}$ at the time $t + dt$ can be calculated from their values at the time t. This calculation is *unique* since the Maxwell equations are linear in $\mathbf{E}$ and $\mathbf{H}$.

In the preceding we have confined ourselves to a finite closed domain. Physically the unlimited domain is of course of greater interest. The uniqueness of the integration problem can be proved here for the static case as in §10D. We will consider the significance of the Poynting vector for the unique formulation of the problem of waves along wires in §22.

§6. The Role of the Velocity of Light in Electrodynamics

It appears reasonable to eliminate $\mathbf{H}$ from Eqs. (4.8) and to obtain in this manner a single vector equation for $\mathbf{E}$. For this purpose the operation curl is applied to the first Eq. (4.8), the operation $\mu\partial/\partial t$, to the second. Adding the two equations yields

$$\varepsilon\mu \frac{\partial^2 \mathbf{E}}{\partial t^2} + \sigma\mu \frac{\partial \mathbf{E}}{\partial t} = - \text{ curl curl } \mathbf{E}, \tag{1}$$

i.e., a linear differential equation of the second order in four coordinates of space and time.

We will convert this expression to a form which is more familiar to the mathematician. For this we utilize the general transformation (3.10) of Vol. II:

$$\text{curl curl } \mathbf{E} = \text{grad div } \mathbf{E} - \Delta\mathbf{E}. \tag{2}$$

As indicated there, this equation is to be applied with caution, since the Laplace operator Δ can, by its definition as div grad, only be applied to scalar quantities. Incidentally, (2) may also be derived from the well-known vector formula

$$\mathbf{A} \times (\mathbf{B} \times \mathbf{C}) = \mathbf{B}(\mathbf{A} \cdot \mathbf{C}) - \mathbf{C}(\mathbf{A} \cdot \mathbf{B}) \tag{2a}$$

by symbolic calculation with the nabla operator (see the beginning of §5), where it takes the form

$$\mathbf{\nabla} \times (\mathbf{\nabla} \times \mathbf{E}) = \mathbf{\nabla}(\mathbf{\nabla} \cdot \mathbf{E}) - (\mathbf{\nabla} \cdot \mathbf{\nabla})\mathbf{E}. \tag{2b}$$

This is identical with Eq. (2), term for term. We consider similar vector formulas in Problem I.4.

Equation (1) is valid in any coordinates, curvilinear as well as Cartesian. On the other hand, Eq. (2), according to the above, is restricted to the Cartesian coordinates x, y, z and the components E_x, E_y, E_z, since only these may be treated as scalar quantities. With this restriction we find from (1) and (2)

$$\varepsilon\mu \frac{\partial^2 \mathbf{E}}{\partial t^2} + \sigma\mu \frac{\partial \mathbf{E}}{\partial t} = \Delta\mathbf{E} - \text{grad div } \mathbf{E}. \tag{3}$$

This can be further simplified if we specify that $\mathbf{E}$ represents a solution for a medium of uniform dielectric constant and *free of charge*. Then Eq. (4.4b), with $\varepsilon = $ const and $\rho = 0$, becomes div $\mathbf{D} = \varepsilon$ div $\mathbf{E} = 0$. Thus the last term on the righthand side of Eq. (3) vanishes and Eq. (3) assumes the form of the wave equation:

$$\varepsilon\mu \frac{\partial^2 \mathbf{E}}{\partial t^2} + \sigma\mu \frac{\partial \mathbf{E}}{\partial t} = \Delta\mathbf{E}. \tag{4}$$

The same equation evidently applies, under similar restricting conditions, also for $\mathbf{H}$ (as well as for $\mathbf{D}$ and $\mathbf{B}$).

The first coefficient in (4) is, as can be read directly out of Eq. (4), the reciprocal square of a velocity. Correspondingly, we find from Eqs. (4.6a) and (4.7a):

$$\varepsilon\mu = \frac{Q^2}{M \text{ joule}} \cdot \frac{\text{joule S}^2}{Q^2 M} = \left(\frac{M}{S}\right)^{-2}. \tag{5}$$

What is the meaning of this velocity? Maxwell's answer is: *It is the velocity of propagation of electromagnetic waves, which in vacuum is identical with that of light:*

$$(\varepsilon_0 \mu_0)^{-\frac{1}{2}} = c = (2.9978 \pm 0.0002) \, 10^8 \, \frac{M}{S} \sim 3 \cdot 10^8 \, \frac{M}{S}. \tag{6}$$

Even at an early date the velocity of light c, then denoted as "critical velocity," maintained an elusive existence in electrodynamics, as in the theorem of Wilhelm Weber and the numerous measurements of the ratio of an "electromagnetically" and "electrostatically" determined charge on a condenser (§16D). However the role of c in electrodynamics was first clarified by Maxwell's theory of light and Hertz's experiments.

If we pass from vacuum to an arbitrary electromagnetic medium, the velocity $(\varepsilon\mu)^{-\frac{1}{2}}$ appearing in (5) signifies, according to Maxwell, the velocity of light (more precisely, the "phase velocity of the light") in a ponderable body characterized by ε and μ

$$(\varepsilon\mu)^{-\frac{1}{2}} = v, \qquad \frac{c}{v} = n = \text{refractive index.} \qquad (7)$$

It is true that the statement (7) has by no means the same certainty as statement (6). For it does not account for dispersion phenomena and hence cannot even explain the prismatic colors. We will learn in Vol. IV how these are to be fitted into electromagnetic optics.

Eq. (6) is evidently a supplementation of Maxwell's theory derived from experiment, which establishes a relationship between the two material constants ε_0, μ_0 of vacuum. In the following section we will discuss how the constants are to be determined individually.

We now turn to the integration of Eq. (4) specialized for vacuum

$$\frac{1}{c^2} \frac{\partial^2 \mathbf{E}}{\partial t^2} = \Delta \mathbf{E} \qquad (8)$$

with the auxiliary condition already made use of

$$\text{div } \mathbf{E} = 0. \qquad (8a)$$

We seek, in particular, solutions of (8) which are independent of y and z. For purely periodic time dependence these represent monochromatic plane waves which advance along the x-axis. We shall show that they are necessarily transverse. In view of the assumed independence of y and z of the function Eq. (8a) reduces to

$$\frac{\partial E_x}{\partial x} = 0.$$

Equation (8) yields accordingly:

$$\frac{\partial^2 E_x}{\partial t^2} = 0. \qquad (8b)$$

E_x would thus be a linear function of t, which is inconsistent with the periodic dependence on t. Hence $E_x = 0$. Thus we already note a decided advantage of electromagnetic optics over the old elastic optics. As we saw in Vol. II, §45, the latter could never get rid of the longitudinal component of the plane wave: Even if it was originally absent, a reflection or refraction would cause its appearance along with the transverse component. In contrast to this we have proved that the plane wave of the electromagnetic theory of light is necessarily transverse. We can designate Eq. (8a) as the condition of transversality.

If the wave has a single electrical component or, in the usual terminology, is plane polarized, we can take its direction of vibration[1] as the y-axis, so that, in addition to $E_x = 0$, also $E_z = 0$. Eq. (8) then becomes

$$\frac{1}{c^2} \frac{\partial^2 E_y}{\partial t^2} = \frac{\partial^2 E_y}{\partial x^2}. \tag{9}$$

The solution which is purely periodic in time is

$$E_y = a \cos (kx - \omega t + \alpha). \tag{10}$$

According to (9) the wave number k introduced here and the angular frequency ω are related by

$$\frac{\omega}{k} = c; \tag{10a}$$

in terms of the wave length λ and the period τ we have

$$k = \frac{2\pi}{\lambda}, \qquad \omega = \frac{2\pi}{\tau}. \tag{10b}$$

Omitting the sign Re, denoting "real part of," we shall write (10) in a form which will prove more convenient for what follows:

$$E_y = A e^{ikx - i\omega t}, \qquad A = ae^{i\alpha}. \tag{11}$$

This is permissible as long as we are dealing with linear relations, such as the Maxwell differential equations; in dealing with energetic quantities which are quadratic in the field components we must obviously return to real expressions such as (10).

We next investigate the magnetic component of the plane wave. It may be derived from the first vector equation (4.8), specialized for vacuum:

$$\mu_0 \frac{\partial \mathbf{H}}{\partial t} = - \text{curl } \mathbf{E}.$$

Since $E_x = E_z = 0$ and $\dfrac{\partial}{\partial y} = \dfrac{\partial}{\partial z} = 0$, this leads to

$$H_x = H_y = 0$$

and furnishes the following equation for H_z:

$$\mu_0 \frac{\partial H_z}{\partial t} = - \frac{\partial E_y}{\partial x} = - ikA e^{ikx - i\omega t}. \tag{12}$$

---

[1] It should be noted that we are here dealing with the direction of vibration of the electric field, not with the direction of any material displacement.

For purely periodic time dependence its integration with respect to t is carried out simply by dividing the right side by $-i\omega$. Accordingly,

$$\mu_0 H_z = \frac{k}{\omega} A e^{ikx - i\omega t} = \frac{1}{c} A e^{ikx - i\omega t},$$

and, in view of (6),

$$H_z = \sqrt{\frac{\varepsilon_0}{\mu_0}} \, A e^{ikx - i\omega t}. \tag{13}$$

The dimension of the coefficient $(\varepsilon_0/\mu_0)^{\frac{1}{2}}$ is that of a reciprocal resistance, i.e. Ω^{-1}. For, by (4.6a), (4.7a), and (4.5c),

$$\frac{\varepsilon}{\mu} = \frac{Q^2}{\text{joule M}} \Big/ \frac{\text{joule S}^2}{Q^2 M} = \left(\frac{Q^2}{\text{joule S}}\right)^2 = \frac{1}{\Omega^2}. \tag{14}$$

$(\mu_0/\varepsilon_0)^{\frac{1}{2}}$ is designated as "wave resistance of vacuum." We shall see in §18D that this quantity actually assumes the role of a resistance (voltage/current) in the telegraph equation.

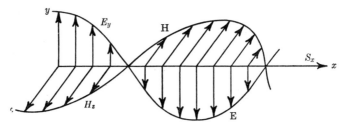

FIG. 6. The relative orientation of **E**, **H**, and **S** for a plane wave progressing in the x-direction.

Fig. 6 shows the orientation of **E** and **H** relative to each other and relative to the Poynting vector **S** at a given instant. In this sequence they form a right-handed system. With increasing t the figure is displaced with the velocity of light in the direction of the positive x-axis. It may not be superfluous to point out that **E** and **H** become zero at the same point and attain their maxima at the same point. The situation differs from that of pendulum vibrations in mechanics, where the energy appears in turn in its kinetic and in its potential form.

For the experiments of Hertz and many optical experiments air and our vacuum are equivalent. A distinction between air and vacuum need only be made in high-precision wave-length determinations.

We have continually employed the term vacuum in preference to the term "ether" ("light ether"), which is frequently used elsewhere. This negative term appears to have more significance than the latter scholastic word, which gives rise to false notions that cannot be reconciled with the theory of relativity.

We can indicate the material constants of ponderable bodies by their relative values referred to vacuum instead of by ε, μ, setting

$$\varepsilon = \varepsilon_{rel}\varepsilon_0, \qquad \mu = \mu_{rel}\mu_0. \tag{15}$$

ε_{rel} and μ_{rel} then are pure numbers, which in general do not differ greatly from 1. In a ponderable nonconductor Eq. (10a) must of course be replaced by

$$\frac{\omega}{k} = v = \frac{c}{n} \tag{16}$$

and Eq. (13) by

$$H_z = \sqrt{\frac{\varepsilon}{\mu}}\, A e^{ikx - i\omega t} \tag{17}$$

Plane transverse waves are possible also in an absorbing medium ($\sigma \neq 0$). The general wave equation (4) is satisfied by the form (11), for given ω, by subjecting k to the condition generalizing Eq. (16):

$$k^2 = \varepsilon\mu\omega^2 + i\sigma\mu\omega, \qquad k = \sqrt{\varepsilon'\mu}\,\omega, \qquad \varepsilon' = \varepsilon + \frac{i\sigma}{\omega}. \tag{18}$$

ε' is the "complex dielectric constant" frequently employed in the optics of absorbing media. If the relaxation time introduced in (4.9a) is employed, we obtain

$$\frac{\varepsilon'}{\varepsilon} = 1 + \frac{i\sigma}{\varepsilon\omega} = 1 + \frac{i}{2\pi}\frac{\tau}{T_r}. \tag{18a}$$

If $T_r \gg \tau$ the added imaginary term of k is merely a correction term; if $T_r \ll \tau$ the real and imaginary parts of k become equal (because $\sqrt{i} = (1 + i)/\sqrt{2}$). In both cases the wave is damped exponentially as it progresses along the positive x-axis.

§7. *The Coulomb Field and the Fundamental Constants of Vacuum. Rational and Conventional Units*

On the basis of their time dependence we classify fields as *static, stationary, quasistationary,* and *rapidly varying fields.*

In *static fields* not only field and density variations, but also currents of electricity and energy are to be zero. Hence we demand

$$\dot{B} = 0, \qquad \dot{D} = 0, \qquad \dot{\rho} = 0, \qquad J = 0, \qquad S = 0.$$

According to Eqs. (4.4) and the succeeding equations these conditions are fulfilled if we set:

A. Electrostatics

$$\text{curl } \mathbf{E} = 0, \qquad \text{div } \mathbf{D} = \rho \text{ in nonconductors,}$$

$$\mathbf{D} = \mathbf{E} = 0 \text{ in conductors,} \qquad (1)$$

$$\mathbf{H} = 0 \text{ in all cases.}$$

B. Magnetostatics

$$\text{curl } \mathbf{H} = 0, \qquad \text{div } \mathbf{B} = 0 \text{ always,} \qquad \text{but eventually div } \mathbf{H} = \rho_m, \qquad (2)$$

$$\mathbf{E} = 0 \text{ in all cases.}$$

An explanation of the "magnetic density" ρ_m here introduced will be given in connection with Eq. (9a) below.

In *stationary fields* we retain the conditions $\dot{\mathbf{B}} = 0$, $\dot{\mathbf{D}} = 0$, $\dot{\rho} = 0$, but prescribe current fields $\mathbf{J}$ in the conductors, which according to Eq. (4.4c) must be free of sources. The electric field must still satisfy, both within and outside the currents, curl $\mathbf{E} = 0$; on the other hand, curl $\mathbf{H} = 0$ only outside the currents.

In *quasistationary fields* we shall determine the fields as in the stationary case, but take account of their time dependence in the first approximation. The system of the Maxwell equations is fully utilized only for *rapidly varying fields*.

A. Electrostatics

We defer all problems requiring the use of the theory of functions. These are the *boundary-value problems* with conductors or nonconductors of different dielectric constant present in the field. We shall therefore deal first of all only with a *uniform* dielectric, so that we may set $\varepsilon = $ const. In this case we are faced with a simple *summation problem* instead of a boundary-value problem.

Eqs. (1) then take the simpler form

$$\text{curl } \mathbf{E} = 0, \qquad (3) \qquad \qquad \text{div } \mathbf{E} = \frac{\rho}{\varepsilon}. \qquad (3a)$$

Eq. (3) states that $\mathbf{E}$ may be treated as gradient of a *scalar potential*

$$\mathbf{E} = -\text{ grad } \Psi, \qquad (4)$$

which evidently brings about a substantial simplification of the problem of integration.

According to (3a) this potential must satisfy the *Poisson equation*

$$\Delta\psi = -\frac{\rho}{\varepsilon}. \qquad (4a)$$

Lamellar field (curl **E** $= 0$) and *potential field* (**E** $= -$ grad Ψ) have the same meaning; the surfaces $\Psi =$ const. divide the field into layers (lamellae), to which the lines of force are orthogonal. The line integral of the field strength

$$\int_A^B E_s \, ds = \Psi_A - \Psi_B \tag{4b}$$

is independent of the path; carried out over any closed path $(B = A)$ it *vanishes*. The *voltage* V_{ab} is identical with the *potential difference*

$$\Psi_A - \Psi_B.$$

The summation problem mentioned above consists in the integration of Eq. (4a) and can be carried out directly with the aid of *Green's theorem*, for which we refer to Vol. II, §20, Nr. 1a. We obtain

$$4\pi\varepsilon\Psi = \int \frac{\rho}{r} \, d\tau, \qquad r = r_{PQ}. \tag{5}$$

P is the point at which Ψ is to be calculated, Q is the point of integration. The left side results from the integration over a small sphere surrounding the point $r = 0$, $Q = P$; the integral over the sphere bounding the region of integration externally vanishes provided that the total charge enclosed by this sphere is finite.

If the charge is not distributed in space, but concentrated on a surface or on a line, the mathematical method employed in (5) leads to

$$4\pi\varepsilon\Psi = \int \frac{\omega}{r} \, d\sigma, \tag{5a}$$

$$4\pi\varepsilon\Psi = \int \frac{\lambda}{r} \, ds; \tag{5b}$$

ω is the surface density, λ the line density (charge per unit length). A final step in this series leads us to the charge e concentrated in a point:

$$4\pi\varepsilon\Psi = \frac{e}{r}, \qquad (6) \qquad E = E_r = -\frac{\partial\Psi}{\partial r} = \frac{e}{4\pi\varepsilon r^2}. \tag{6a}$$

This is the *Coulomb field*. We could also have read it off directly from Eq. (3a), which, using Gauss's theorem, we could integrate over a sphere of radius r described about the charge e. We obtain then directly

$$\oint E_n \, d\sigma = \int \frac{\rho}{\varepsilon} \, d\tau = \frac{e}{\varepsilon}. \tag{6b}$$

In view of the spherical symmetry we must put $E_n = E_r =$ const. on the left, whereupon (6b) becomes in fact identical with (6a).

The Coulomb force F, with which two equal charges e at the distance r repel each other, follows from (6) according to our definition of field strength:

$$F = F_r = eE_r = \frac{e^2}{4\pi\varepsilon r^2}. \tag{7}$$

It has been customary in the past to write instead, for vacuum conditions,

$$F = f\cdot\frac{e^2}{r^2} \qquad \text{with} \qquad f = 1. \tag{8}$$

Here F is supposed to be measured in dynes, r in centimeters. However, in this manner the whole structure of our system of dimensions is cast aside; we pass from our former system of dimensions to the socalled *electrostatic*[1] *system* of cgs units. The charge e would then, according to (8), take on the unnatural and ungainly dimensions

$$e = \sqrt{\text{dyn cm}^2} = \text{cm}^{3/2}\,\text{g}^{1/2}\,\text{sec}^{-1}. \tag{8a}$$

Furthermore the charge e would be 1 if two equal charges e at a distance of 1 cm would repel each other in air with a force of 1 dyne.

We must reject, on dimensional grounds, Hertz's distinction between "true charge density" expressed by div **D** and "free charge density" expressed by div **E**. We shall designate the latter quantity correctly as "divergence of the lines of force"; in the preceding we have avoided it by writing ρ/ε for it.

B. *Magnetostatics*

Although in most treatments the analogy between electrostatics and magnetostatics is emphasized, our approach compels us to point out clearly the differences as well.

As we saw in (2), it is not the *intensity* **B**, but the *quantity* **H** that is lamellar. We again denote the corresponding scalar potential by Ψ, distinguishing where necessary between Ψ_e and Ψ_m, and find

$$\mathbf{H} = -\,\text{grad}\,\Psi. \tag{9}$$

We shall call the quantity ρ_m defined above in Eq. (2) simply "magnetic, density"; in view of the correspondence of **H** and **D** we can regard it as the direct analog of the electric density ρ.

How can we reconcile its existence, i.e. the Eq. div $\mathbf{H} \neq 0$ with the universally valid Eq. div $\mathbf{B} = 0$? This is only possible at points of local varia-

---

[1] The term *electrical system* seems to us to be preferable in principle to the customary term electrostatic system, since its application is not limited to equilibrium conditions, but may also be extended to electrodynamic processes. See §16*D*.

tion of permeability, as is shown by the following lines:

$$\operatorname{div} \mathbf{B} = \mu \operatorname{div} \mathbf{H} + \mathbf{H} \cdot \operatorname{grad} \mu = 0$$

$$\rho_m = \operatorname{div} \mathbf{H} = \mathbf{H} \cdot \operatorname{grad} \log (\mu_0/\mu).$$

(9a)

We will give a physical interpretation of this rather formal explanation of the concept of magnetic quantity and density by introducing the "magnetization" $\mathbf{M}$ in §12. We have no reason for a distinction between "true" and "free" magnetism, such as was also given by Hertz. For, since we have already interpreted div $\mathbf{H}$ as magnetic density, the differently dimensioned quantity div $\mathbf{B}$ is not a magnetic density. Furthermore it is everywhere equal to zero.

We now return to (9) and form the divergence of the vectors on the left and the right. We then obtain the Poisson equation of magnetostatics, i.e.

$$\Delta \Psi = - \rho_m$$

(9b)

If ρ_m is given throughout this is integrated, in analogy to (5), by

$$4\pi\Psi = \int \frac{\rho_m}{r} \, d\tau,$$

(10)

or, in analogy to (5a), for given surface charge ω_m, by

$$4\pi\Psi = \int \frac{\omega_m}{r} \, d\sigma.$$

(10a)

If the density is concentrated on a point pole and if we call

$$p = \int \rho_m \, d\tau$$

the pole strength (an opposite pole is imagined to lie at infinity), (10) leads to

$$4\pi\Psi = \frac{p}{r},$$

(10b)

$$H = H_r = -\frac{\partial \Psi}{\partial r} = \frac{p}{4\pi r^2}.$$

(10c)

This is the *Coulomb field* of the isolated magnetic pole. The Coulomb force, with which two poles of equal magnitude and the same sign repel each other, is, however, not $p\mathbf{H}$, but, according to our definition of the intensity $\mathbf{B}$

$$F = F_r = pB_r = p\mu H_r = \frac{p^2 \mu}{4\pi r^2}.$$

(11)

The fact that μ appears here in the numerator, although ε, in (7) appears in the denominator, results evidently from the inconsistency, pointed out in connection with Eq. (4.7), in the introduction of μ as compared with that of ε. (We would have liked to define the reciprocal of μ as the magnetic constant at that point.) In (11) μ evidently signifies the permeability of the surroundings of the magnetic pole p; for air (vacuum) we put $\mu = \mu_0$.

Just as in connection with (8) we took cognizance of an electrostatic system of units and a unit of charge corresponding to this system, so we can introduce, on the basis of (11), a magnetic system of units and a corresponding unit of pole strength. To this end (11) is replaced, for vacuum in particular, (we follow the pattern of Eq. (7) and what follows literally) by:

$$F = f \frac{p^2}{r^2} \tag{12}$$

and f is put equal to 1; F is supposed to be measured in dynes, r in centimeters. Our former system of units is once more cast aside, and we pass over to the *Gaussian magnetic cgs system*.[1] In this system the pole strength p has, according to (12), the same unsatisfactory dimension as the charge e in the electrical system (8a). Unity pole strength would correspond to a repulsion with a force of 1 dyne of two poles of equal sign and magnitude separated by 1 cm (in air as surrounding medium).

C. Rational and Conventional Units

We must now deal with the factor 4π in Coulomb's law. It is true that this is much less fundamental than the question of dimensions and bears to the latter only a historical relationship. Historically the forms (8) and (12) of Coulomb's law result from an effort to approach as closely as possible the customary form of Newton's law. We shall denote the suppression of the numerical factor 4π in Coulomb's law as *conventional*, our retention of it as *rational*. It is in fact evident that in a problem with spherical symmetry, such as the Coulomb problem, the factor 4π is appropriate (this follows in particular from our argument in (6b)). If we wish to avoid this factor, we must rewrite Poisson's equation (4a) as well as the second of Eqs. (1) as follows:

$$\Delta\Psi = -4\pi \frac{\rho}{\varepsilon}, \qquad \text{div } \mathbf{D} = 4\pi\rho. \tag{13}$$

The factor 4π would thus be improperly introduced into the fundamental equations of the Maxwell theory. Furthermore, the transparent expression (5.6) for the energy density would be distorted into

$$W_e = \frac{1}{8\pi} \mathbf{E} \cdot \mathbf{D}. \tag{14}$$

---

[1] The fact that Gauss employed mm instead of cm as unit of length is a superficial distinction.

Heaviside fought a life-long battle for the rational units. In this connection he pointed also to the expression for the capacity of a condenser (for details see §10, where the relationship with the expression for the energy density is also indicated): The *plate condenser* (area F, plate separation a) has, in rational and conventional units respectively, the capacity

$$K = \frac{F\varepsilon}{a} \quad \text{and} \quad \frac{F\,\varepsilon}{4\pi a}, \tag{15}$$

the *spherical condenser* (radius of sphere r, outer sphere imagined at infinity), the capacity

$$K = 4\pi\varepsilon r \quad \text{and} \quad \varepsilon r. \tag{15a}$$

We see that, with rational units, the factor 4π appears for the sphere, where it belongs; with conventional units it is missing for the sphere and appears for the plane condenser, where it does not belong.

Heaviside makes the following striking comparison: In passing from the measurement of distance to the measurement of area one might define as unit of area the area of a circle of radius 1. This would be logically possible. It would however lead to the strange result that a square with the side 1 would have the area $1/\pi$. Everyone would then say that π was at the wrong place. We said the same of the factor 4π in the formulas to the right in (15) and (15a).

D. *Final Determination of the Fundamental Constants* ε_0, μ_0 *in the MKSQ System*

The viewpoint of the rational units together with the requirement of meaningful dimensions and adaptation to the legal units leads to a quite definite choice of the fundamental constant μ_0 of vacuum. For we can obtain agreement between Eq. (11), which is dimensionally correct in our sense, and Eq. (12) by requiring

$$\left[\frac{\mu_0}{4\pi}\right] \frac{\text{joule S}^2}{\text{Q}^2\,\text{M}} = [f] \text{ cgs.} \tag{16}$$

$[f]$ is the numerical value of f in the cgs-system, which we wished to set equal to 1. The brackets on the left are the numerical value of the quantity $\mu_0/(4\pi)$ in our MKSQ-system; its dimension (see e.g. (4.7a)) is indicated. The conversion of these dimensions into the cgs-system follows from

$$Q = 1 \text{ Coulomb} = \tfrac{1}{10} \text{ cgs}, \quad M = 10^2 \text{ cm}, \quad \text{joule} = 10^7 \text{ erg.}$$

Accordingly

$$1 \, \frac{\text{joule S}^2}{\text{Q}^2\text{M}} = 10^7 \text{ cgs.}$$

With $[f] = 1$ we thus obtain, after cancelling the dimensional factor cgs on both sides of (16)

$$\left[\frac{\mu_0}{4\pi}\right] = 10^{-7}.$$

We obtain hence, entering the dimensions:

$$\mu_0 = 4\pi 10^{-7} \frac{\text{joule S}^2}{Q^2 M} = 4\pi 10^{-7} \frac{\Omega S}{M}. \tag{16a}$$

Regarding the unit of resistance, $\Omega =$ "ohm," here employed, see Eq. (4.5d).

We have thus determined one of the two fundamental constants of vacuum in such a manner that the demands stated above are satisfied. The numerical value of μ_0 so obtained, which is accurate to an arbitrary number of digits, shows clearly that our determination is not established by direct measurements, but by our choice of the unit of Q and is equivalent to the latter. *The other fundamental constant ε_0 of vacuum* follows then from the relation (6.6) which is supported by the sum total of the Hertzian experiments:

$$\varepsilon_0 = \frac{1}{\mu_0 c^2} = \frac{10^7}{4\pi c^2} \frac{M}{\Omega S}. \tag{17}$$

If we substitute for c the approximate value $c = 3 \cdot 10^8$ M/S we find

$$\varepsilon_0 \sim \frac{10^{-9}}{36\pi} \frac{S}{\Omega M}. \tag{18}$$

We can also write (17) in the form

$$4\pi c^2 \varepsilon_0 = 10^7 \frac{M}{\Omega S}. \tag{18a}$$

Division of (16a) by (18) and taking the square root leads to the following value for the "wave resistance of vacuum," introduced with (6.14), in terms of the unit Ω:

$$\sqrt{\frac{\mu_0}{\varepsilon_0}} \cong 120\pi\Omega \cong 377\Omega. \tag{19}$$

In the preceding we have disregarded the small differences between the velocity of light c and its approximate value $3 \cdot 10^8$ M/S, as well as the difference between the "international" and the "absolute," i.e., the ideal, ohm. These differences, which concern only the higher decimals, are of course of great importance in precision measurements and have called forth, in the determination of the Ω in relation to the old Siemens unit,[1]

---

[1] Resistance of a mercury thread 1 m long and 1 mm² in cross section at 0° C = 0.937 Ω.

the competitive efforts of the best experimenters (Kirchhoff, Lord Ray-leigh, F. Kohlrausch etc.). They play no role, however, in the general theory.

In summary: Our form of the Maxwell equations is adapted to the *rational choice of the units* MKSQ, with the value (16a) for μ_0 taken over from the *conventional Gaussian magnetic units*. Below we shall use the numerical values (16a), (18), (19) only in specific numerical computations and not introduce them, as often happens in engineering literature, into the general theory. Instead we shall always take account of the dimensions of all quantities, also those of ε_0, μ_0, and thus make ourselves independent of the particular choice of Q = 1 Coulomb.

§8. Four, Five, or Three Fundamental Units?

A. Supplementary Note on Our System of Four Units

Our four units MKSQ are simply intended to translate Giorgi's idea (introduction of a separate electrical unit) into a form which is particularly convenient for the theory. It is basically indifferent whether the unit of charge Q is employed or a standard resistance R, as Giorgi has occasionally advocated, for reasons of convenience of measurement. In view of the relationship Q = ampere-second we would of course also be content with the ampere as fourth unit. We take less kindly to the designation of Giorgi's system by the units MKSVA. In view of

$$VA = \text{watt} = \text{joule/sec}$$

these units are not independent of each other. We can well understand that the long-employed quantities V and A appear more convenient in use than our unit of charge Q. Nevertheless, of the two dimensions

$$E = \frac{\text{newton}}{Q} \qquad \text{and} \qquad E = \frac{\text{Volt}}{M}$$

the first appears to be the more natural one. Kalantaroff's system of the four units MSQO (magnetic flux) is selfconsistent, but seems, by the elimination of the unit of mass, somewhat too artificial for general use in physics.

It is to be welcomed, from our point of view, that, by international agreement, separate designations gauss and oersted have been introduced for the two magnetic vectors **B** and **H**. Historically, the name gauss also seems proper for **B**, since Gauss' methods of determining magnetic moment rest on measurements of force and hence refer to **B** and not to **H**. The unhappy term "magnetic field" for **H** should be avoided as far as possible. It seems to us that this term has led into error none less than Maxwell himself, who, in art. 625 of the Treatise puts the force exerted by the field on a magnetic pole m equal to $m\mathbf{H}$.

We have repeatedly stressed as an advantage of our system that it avoids the annoying powers of ten of the cgs-system. This applies to the electrical as well as to the mechanical quantities. The converse is true, however, for the *magnetic* quantities. The unit gauss of the magnetic induction **B** in the cgs-system is, by definition, equal to the unit in the cgs-system. Hence, transferred to our system of units, it acquires a power of ten. We determine the latter as follows:

Let $[B]$ be the numerical magnitude of a given field in our system of units, so that

$$B = [B] \frac{\text{newton S}}{\text{QM}} = [B] \frac{\text{joule S}}{\text{QM}^2} \tag{1}$$

We substitute again

$$\text{joule} = 10^7 \text{ erg}, \qquad M = 10^2 \text{ cm}, \qquad Q = \frac{1}{10} \text{ cgs-units.}$$

If, in particular, we set $[B] = 1$, we find from (1) as corresponding value of this quantity B in cgs-units:

$$B = 1 \frac{10^7}{10^3} \text{ cgs-units} = 10^4 \text{ gauss,} \tag{2}$$

and conversely

$$1 \text{ gauss} = 10^{-4} \frac{\text{joule S}}{\text{QM}^2} = 10^{-4} \frac{\text{VS}}{\text{M}^2}. \tag{3}$$

It may be mentioned in favor of this choice of unit that the gauss is inconveniently small for practical purposes, so that not only in engineering, but even in pure physics (except for terrestrial magnetism) the kilogauss must generally be employed (e.g. in the Zeeman effect). Hence our 10,000 times greater unit is to be preferred in practice.

In order to express the oersted in our system of units as well, we proceed from the relationship between **H** and **B**:

$$\mathbf{H} = \frac{\mathbf{B}}{\mu_0}. \tag{4}$$

We now set $H = 1$ oersted, $B = 1$ gauss, so that by (3)

$$B = 10^{-4} \frac{\text{joule S}}{\text{QM}^2}$$

and by (7.16a)

$$\mu_0 = 4\pi \cdot 10^{-7} \frac{\text{joule S}^2}{\text{Q}^2\text{M}}.$$

We then obtain from (4)

$$1 \text{ oersted} = 10^{-4} \frac{\text{joule S}}{\text{QM}^2} \bigg/ 4\pi \cdot 10^{-7} \frac{\text{joule S}^2}{\text{Q}^2\text{M}} ; \qquad (5)$$

so that

$$1 \text{ oersted} = \frac{1}{4\pi} 10^3 \frac{\text{Q}}{\text{SM}} = \frac{1}{4\pi} 10^3 \frac{\text{amp}}{\text{M}} \qquad (5a)$$

$$H_{\text{amp/M}} = 4\pi \cdot 10^{-3} H_{\text{oersted}}. \qquad (5b)$$

B. The Five Units MKSQP

It may be stated generally: A dimensional analysis will be more successful[1] in the degree in which more independent units are at its disposal. Our four units are hence more informative than the three units of the "absolute" system, in which the dimensional character of the fundamental electromagnetic vectors is obscured. The five independent units considered below are of even greater value from a general theoretical point of view.

We introduced the magnetic pole strength P as a dimension in §2, but expressed it immediately in (2.7) in terms of the charge Q, in accord with Ampère's hypothesis. Is this hypothesis binding even today, after the discovery of the neutron, a nuclear particle as basic and universal as the proton? The neutron has a magnetic moment which is not associated with any charge, unlike the electron and proton which, though endowed with *equal* charge of opposite sign, have magnetic moments of entirely *different* magnitude. Certainly an attempt to abandon Ampère's hypothesis and to introduce P as *independent fifth dimension* is justified and instructive. We shall, for the present, refrain from fixing the magnitude of P.

We write down the following sets of dimensional relations, which now show a complete correspondence:

$$E = \frac{\text{newton}}{\text{Q}} \qquad\qquad B = \frac{\text{newton}}{\text{P}}$$

$$D = \frac{\text{Q}}{\text{M}^2} \qquad\qquad H = \frac{\text{P}}{\text{M}^2}$$

$$\varepsilon = \frac{D}{E} = \frac{\text{Q}^2}{\text{joule M}} \qquad\qquad \frac{1}{\mu} = \frac{H}{B} = \frac{\text{P}^2}{\text{joule M}} \qquad (6)$$

$$\mathbf{E \cdot D} = \frac{\text{newton}}{\text{M}^2} = \frac{\text{joule}}{\text{M}^3} \qquad\qquad \mathbf{B \cdot H} = \frac{\text{newton}}{\text{M}^2} = \frac{\text{joule}}{\text{M}^3} .$$

[1] E. Fues, Z. Phys. *107*, 662, 1937, indicates an upper limit to the useful number of dimensions.

The entries for E, D, B, and H are identical with the original formulas in (2.1) to (2.9.) In accord with the note accompanying Eq. (4.7) we have entered the reciprocal of μ as analog of ε in our table.

The last line, which is *independent* of Q and P, has the dimension of energy density. On the other hand the (scalar or vector) product of **E** and **H** has a dimension which *depends* on P and Q:

$$\mathbf{EH} = \frac{P}{Q}\frac{newton}{M^2} = \frac{P}{Q}\frac{S}{M}\frac{joule}{M^2S}. \tag{7}$$

The last factor of the last expression has the dimension of energy flux (radiation vector). Let the factor multiplying it be $1/\Gamma$. We thus put

$$\Gamma = \frac{Q}{P}\frac{M}{S} \tag{8}$$

and write (7) in the form

$$\Gamma\mathbf{EH} = \frac{joule}{M^2S}. \tag{8a}$$

This dimensional equation suggests that the energy flux **S** is now to be defined as $\Gamma\mathbf{E} \times \mathbf{H}$. We furthermore compute the product $\varepsilon\mu$ from (6) and find

$$\varepsilon\mu = \frac{Q^2}{P^2} = \Gamma^2\left(\frac{S}{M}\right)^2. \tag{9}$$

We suspect from this that the velocity of light c is no longer given by $(\varepsilon_0\mu_0)^{-\frac{1}{2}}$, but by $\Gamma(\varepsilon_0\mu_0)^{-\frac{1}{2}}$.

The same factor Γ occurs now also in Maxwell's equations. We assert that these should be written:

$$\dot{\mathbf{B}} = -\Gamma \text{ curl } \mathbf{E}, \qquad \dot{\mathbf{D}} + \mathbf{J} = \Gamma \text{ curl } \mathbf{H}. \tag{10}$$

If, as in §5, we proceed to Poynting's theorem (scalar multiplication of the first equation with **H**, of the second equation with **E**), we obtain

$$\mathbf{H}\cdot\dot{\mathbf{B}} + \mathbf{E}\cdot\dot{\mathbf{D}} + \mathbf{E}\cdot\mathbf{J} + \Gamma \text{ div } \mathbf{E} \times \mathbf{H} = 0;$$

with the former definitions of the energy densities and of the Joule heat in (5.6) and (5.5) and with the definition of the energy flux suggested by (8a) this expresses the law of conservation of energy:

$$\dot{W}_m + \dot{W}_e + W_J + \text{div } \mathbf{S} = 0. \tag{11}$$

If, on the other hand, just as in §6, we integrate Eq. (10) for the case of the plane wave in vacuum propagated in the x-direction, we obtain the wave equation in the form

$$\varepsilon_0\mu_0 \ddot{\mathbf{E}} = -\Gamma^2 \text{ curl curl } \mathbf{E} = \Gamma^2 \Delta\mathbf{E}. \tag{12}$$

Since this is supposed to represent a process with the velocity of propagation c, our expectation suggested by (9) is confirmed:

$$\frac{\Gamma}{\sqrt{\varepsilon_0 \mu_0}} = c, \qquad \sqrt{\varepsilon_0 \mu_0} = \frac{\Gamma}{c}. \tag{12a}$$

The general form (10) of the Maxwell equations is not new. It was introduced by Emil Cohn, the friend and fellow student of Heinrich Hertz, and forms the basis of his important book[1] "Das elektromagnetische Feld." We have avoided Cohn's notation V, taking the place of our Γ, since we have otherwise disposed of V. It is true that Cohn does not work out the relationship of this constant with our unit P of pole strength, nor does he place dimensional considerations in the foreground as has been done here. Students of Cohn, in particular J. Zenneck, have used Cohn's system by preference.

H. A. Lorentz clearly recognized the advantages of Cohn's standpoint when, in 1902, he wrote his two great articles on Maxwell's theory and electron theory for the Enzyklopädie der mathematischen Wissenschaften. He wrote:[2] "Cohn's system has the advantage of easy transition to other systems, by specific choice of the values of V, ε_0, and μ_0. Eventual later advances in the understanding of the phenomena could be utilized for the ultimate determination of the units. On the other hand we could not bring ourselves to introduce indeterminate quantities into the formulas which are complex to begin with."

The "eventual later advances in the understanding of the phenomena" contemplated by Lorentz can only be expected when we have a *theory of the elementary particles* which now constitutes the greatest problem on the program of atomic physics; this would have to explain not only the magnetic moments, but also the possible masses and charges of the elementary particles. However, we can even now benefit by the flexibility of Cohn's system.

C. The Gaussian System of Only Three Units

We evidently return to our system with the four units MKSQ and our former form (4.4) of Maxwell's equations if we set

$$\Gamma = 1. \tag{13}$$

Then P has, according to (8), the dimension

$$P = Q \times \text{velocity}, \tag{13a}$$

---

[1] First Edition, Leipzig 1900, Second Edition, 1927.
[2] Vol. V, second part, p. 87.

in agreement with Ampère's hypothesis in (2.7). Furthermore, our specific choice of μ_0 and ε_0 in (7.16a) and (7.17) is evidently consistent with Eq. (12a) for these values of Γ and P.

We obtain another, also very simple, form of Maxwell's equations if we set

$$\Gamma = c. \tag{14}$$

Then, by (12a), the product $\varepsilon_0\mu_0$ must be a pure number. It is tempting to make ε_0 and μ_0 separately pure numbers and to set

$$\mu_0 = 1, \qquad \varepsilon_0 = 1. \tag{14a}$$

In this manner we pass over to the *Gaussian system of units*. In view of (14) the Maxwell equations then become (we confine ourselves first to *nonconductors*):

$$\frac{1}{c}\,\dot{\mathbf{B}} = -\text{ curl }\mathbf{E}, \qquad \frac{1}{c}\,\dot{\mathbf{D}} = \text{ curl }\mathbf{H}. \tag{15}$$

In view of (8) P and Q have now the same dimension. Hence, by Table (6), the dimensions of $\mathbf{E}$ and $\mathbf{B}$, as well as those of $\mathbf{D}$ and $\mathbf{H}$, also become mutually identical. (The same follows also from the form of Eqs. (15).) Furthermore the dimensions of the two pairs become the same, since now ε and μ, just as ε_0 and μ_0 in (14a), become pure numbers, equal to the pure numbers ε_{rel} and μ_{rel} introduced in (6.15). *The Gaussian system obscures the dimensional character of the four fundamental vectors* $\mathbf{E}$, $\mathbf{D}$, $\mathbf{B}$, $\mathbf{H}$ *completely, while Cohn's system expresses it most clearly.*

The two Coulomb laws (7.8) and (7.12) were written in the *conventional form* (with the factor 4π *suppressed*). This has the result that the 4π do not occur in the Maxwell Eqs. (15) for nonconductors but *arise once more* in their integration. For, taking the divergence and integrating Eqs. (15) with respect to t leads to:

$$\text{div }\mathbf{B} = \text{const.}, \qquad \text{div }\mathbf{D} = \text{const.}$$

The first constant is, of course, equal to zero; the second must now not be set equal to ρ, but equal to $4\pi\rho$:

$$\text{div }\mathbf{D} = 4\pi\rho, \tag{15a}$$

in order that the factors 4π cancel each other on the two sides of the equation if it is applied to a point charge $e = \int\rho\,d\tau$ and is integrated over a sphere about e. Only in this manner is the field strength $E_r = e/(\varepsilon r^2)$, obtained corresponding to the conventional form of the Coulomb force $\mathbf{F}$. From this follows also as the expression for the corresponding electrostatic potential $\Psi = e/(\varepsilon r)$, unlike Eq. (7.6), where the appropriate factor

4π appears on the left. Hence also (7.5) and (7.4a) must now be replaced by the less appropriate expressions

$$\varepsilon\Psi = \int \frac{\rho}{r}\, d\tau, \qquad \Delta\Psi = -\,4\pi\rho/\varepsilon. \qquad (15b)$$

The same follows from the form (7.12) of Coulomb's law for the magnetic density ρ_m and the magnetic potential Ψ_m:

$$\operatorname{div} \mathbf{H} = 4\pi\rho_m, \qquad \Psi_m = \int \frac{\rho_m}{r}\, d\tau, \qquad \Delta\Psi_m = -4\pi\rho_m. \qquad (15c)$$

We now extend Eq. (15) to the case of a *conductor*. Here we must temporarily multiply the conduction current $\mathbf{J}$, which is to be added to $\dot{\mathbf{D}}$, with a numerical factor γ which we shall determine in a moment. Hence we write in place of (15)

$$\frac{1}{c}\,\dot{\mathbf{B}} = -\operatorname{curl}\mathbf{E}, \qquad \frac{1}{c}\,(\dot{\mathbf{D}} + \gamma\mathbf{J}) = \operatorname{curl}\mathbf{H}. \qquad (16)$$

Taking the divergence of the second of these equations, as in (15a), and utilizing the definition of ρ given there leads to

$$4\pi\,\frac{\partial\rho}{\partial t} + \gamma\operatorname{div}\mathbf{J} = 0. \qquad (16a)$$

We must set $\gamma = 4\pi$ in order that this equation may express the conservation of charge, or in other words, the absence of sources of the total current $\mathbf{C}$; only then does (16a) become the analogue (4.4c) or (3.6b) of the hydrodynamic equation of continuity.

Having entered this value of γ in (16), we seek the expression for the *Poynting theorem* by the procedure followed at the beginning of §5. Multiplying the two equations (16) scalarly with $\mathbf{H}$ and $\mathbf{E}$ respectively, and utilizing the transformation (5.2) we obtain

$$\frac{1}{c}\,\mathbf{H}\cdot\dot{\mathbf{B}} + \frac{1}{c}\,\mathbf{E}\cdot\dot{\mathbf{D}} + \frac{4\pi}{c}\,\mathbf{E}\cdot\mathbf{J} + \operatorname{div}\mathbf{E}\times\mathbf{H} = 0. \qquad (17)$$

We compare this with the earlier form (5.7) of the same theorem:

$$\dot{W}_m + \dot{W}_e + \operatorname{div}\mathbf{S} = -\,W_J. \qquad (17a)$$

Since we cannot disturb Ohm's law W_J is still given by the product $\mathbf{E}\cdot\mathbf{J}$. We must hence divide (17) by $4\pi/c$ in order that (17) may correspond with (17a). Then a comparison of the terms of (17) and (17a) leads to

$$\dot{W}_m = \frac{1}{4\pi}\,\mathbf{H}\cdot\dot{\mathbf{B}}, \qquad \dot{W}_e = \frac{1}{4\pi}\,\mathbf{E}\cdot\dot{\mathbf{D}}, \qquad (17b)$$

$$\mathbf{S} = \frac{c}{4\pi}\,\mathbf{E}\times\mathbf{H}, \quad \text{and} \quad W_J = \mathbf{E}\cdot\mathbf{J} = \sigma E^2 \text{ as before.} \qquad (17c)$$

Integration of (17b) with respect to t, as on p. 27, yields (for isotropic and anisotropic media):

$$W_m = \frac{1}{8\pi}\, \mathbf{H}\cdot\mathbf{B}, \qquad W_e = \frac{1}{8\pi}\, \mathbf{E}\cdot\mathbf{D}; \tag{17d}$$

They express the localization of energy in conventional units. Already in connection with Eq. (7.14) we pointed out the unsuitable form of the denominator 8π, as compared with the denominator 2 in our rational notation (5.6). The same applies for the factor $c/(4\pi)$ in the present expression (17c) for the energy flux. Even in the Maxwell equations (16) the suppression of 4π, carried out at the wrong place, avenges itself: These equations, in the form appropriate for both conductors and nonconductors, become:

$$\frac{1}{c}\,\dot{\mathbf{B}} = -\operatorname{curl}\mathbf{E}, \qquad \frac{1}{c}\,\dot{\mathbf{D}} + \frac{4\pi}{c}\,\mathbf{J} = \operatorname{curl}\mathbf{H} \tag{18}$$

with the supplementary conditions, applying specifically for isotropic media

$$\mathbf{D} = \varepsilon\mathbf{E}, \qquad \mathbf{B} = \mu\mathbf{H}, \qquad \mathbf{J} = \sigma\mathbf{E}. \tag{18a}$$

We hope that by this summary we have facilitated for the reader the laborious transition between our two systems of units

$$\text{MKSQ (rational)} \rightleftarrows \text{cgs (Gauss, conventional)}$$

as far as possible. We have discussed the historical source of this annoyance at the end of §7. It is unavoidable in view of the present status of the question of units in electrical engineering, experimental physics, and theoretical physics. The following remarks may serve to clarify the situation.

H. A. Lorentz, when writing his articles for the *Enzyklopädie* in 1902, like Hertz, utilized the Gaussian system, postulating: *Electrical quantities* (including the electric current) *are measured electrically* (electrostatically), *magnetic quantities, magnetically.* Contrary to his original intention he decided, in the course of composing the articles, to convert the Gaussian system (unlike Gauss and Hertz) into *rational* units. In this manner the theoretical relationships became clearer and the 4π's were eliminated from the Maxwell equations. Lorentz set for vacuum $\varepsilon_0 = \mu_0 = 1$, as in our Eqs. (14a). In order to retain the rational form of the Coulomb force law he then had to introduce the 4π's appearing in it into the definition of the unit charge and the unit pole strength, respectively. This somewhat artificial conversion of units[1] has not found wide acceptance, in spite of the authority of Lorentz.

[1] See table on p. 87 of Vol. 5, part 2, of the *Enzyklopädie*.

We have here—also against our original intention—arrived at the decision to write the Gaussian system, insofar as we shall use it, in *conventional units*. The reason is the following: Since the year 1902 *atomic physics* has come to be the most important branch of our science. It deals always with conventional units, e.g. with the electron charge $e = 4.80 \cdot 10^{-10}$ (electrostatic cgs-units) and with the electric potential, e.g. in the hydrogen atom, $\Psi = e/r$ (not $\Psi = e/(4\pi r)$). We consider it inadvisable to overturn this whole formalism anew by passing over to the rational form of the Gaussian system or even to our system of four units.

On the other hand Giorgi's system of the units MKSQ, freed of 4π's, is most suitable for the macrophysical problems of this lecture. We are here in agreement with the international conventions, with the practice of engineering, and, in particular, with the textbooks of Mie (quoted on p. 10) and Pohl.[1] We regard the dogma of the scientific superiority of the *three* purely mechanical units cm, g, sec, which for example is supported in Kohlrausch, *Praktische Physik*, as outmoded.

D. Supplement Regarding Other Systems of Units

In the restriction to these two systems of units, the Gaussian system in conventional form and the MKSQ system, we follow the practice of the excellent textbook of Joos.[2] The Gaussian system (whether in rational or conventional form) is a *mixed* system, consisting of electrical (electrostatic) and magnetic cgs-units. There are however, as is well known, also a *purely electrical* and a *purely magnetic* system of units, of which the latter is particularly important, since the legal units volt, ampere, ohm, etc. are based on it.

The reason for setting up a separate *electrical system of units* rests on a certain quantitative difference between electrostatics and electrokinetics: Electrostatics deals with large voltages and small quantities of electricity, electrokinetics, with moderate voltages and large quantities of electricity. To give a comparison from hydrodynamics, the electric spark of a condenser discharge corresponds to a waterfall (great height, small quantity), the electric current, to a river (small grade, great flow), as is indicated in Fig. 7. Thus, for electrostatics, a small unit of charge and a large unit of field strength are suitable. The *electrostatic system* based on the electrical Coulomb law provides such units. In terms of this small unit of charge the charge of the electron (see above) has the relatively large value $4.80 \cdot 10^{-10}$ cgs-unit. The unit of charge in the *electromagnetic system* (equal to 10 coulombs), on the other hand, is larger by the factor c; in terms of it, the charge of the electron appears smaller by a factor c, i.e. equal to $1.60 \cdot 10^{-20}$ cgs $= 1.60 \cdot 10^{-19}$ coulomb (see p. 43). On the other hand, the unit of *field strength* in the electromagnetic system, according to the definition

[1] R. W. Pohl, *Elektrizitätslehre*, Springer. 8th and 9th Edition, 1943.
[2] G. Joos, Theoretical Physics, 2nd Ed., G. E. Stechert and Co., New York, 1950.

of the volt, is 10^{-8} volts/cm; that in the electrostatic system is c times as large, or 300 volts/cm. We shall return to this in §16D.

We have frightened generations of students with these two sets of values for charge and field strength (their number would be increased to 4 if, in addition to the usual conventional units, rational units would also be considered). It is, in our opinion, a special advantage of the introduction of our fourth unit of charge, Q, which is independent of all other units, that we need deal only with quite definite charges, expressed as multiples of Q.

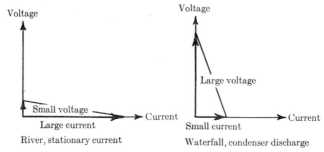

FIG. 7. The hydrodynamic representation of a stationary electric current and of a condenser discharge.

We quote finally an informative analog to the double (electrostatic and electromagnetic) measure of charge which, like so many other clarifications in the question of units, we owe to J. Wallot:[1] Suppose that someone had decided to describe mechanical processes in terms of only *two* independent units, cm and sec. He eliminates the gram as unit by setting either the density δ or the modulus of elasticity E of some standard material such as copper arbitrarily equal to 1. He can then express the mass m of a given copper rod in two ways, either by a measurement of volume according to the formula

$$\delta = \frac{m}{V}, \text{ which, because of } \delta = 1, \text{ leads to: } m = m_1 = V$$

or by a vibration experiment with longitudinal waves according to the formula

$$c^2 = \frac{E}{\delta} = \frac{EV}{m}, \text{ which, because of } E = 1, \text{ leads to: } m = m_2 = V/c^2.$$

If he now divides one of the two values of m so found by the other he obtains—perhaps to his surprise—the square of the velocity of propagation c of elastic waves in copper. The analogy to electrodynamics is striking and requires no further explanation.

[1] J. Wallot, Elektrotechn. Z., Vol. 43, Nr. 44 (1922), section 28 of the paper "Physical and Engineering Units." The latest relevant publication is Phys. Z. *44*, p. 17, 1943.

Part II

DERIVATION OF THE PHENOMENA FROM THE MAXWELL EQUATIONS

§9. *The Simplest Boundary-Value Problems of Electrostatics*

We have set up the fundamental equations of electrostatics in the beginning of §7 and have dealt with the resulting *summation problem* for a uniform medium in Eq. (7.5). We now turn to the *boundary-value problems* arising from the presence of *conductors* or *nonconductors of different dielectric constant*.

We think of the simplest electrostatic experiments: Let a metallic conductor of arbitrary shape, originally insulated, A, be connected to a source of potential V (with respect to ground) or B, be given a known charge (e.g. by a piezoquartz, see p. 78). We wish to know the field outside of the conductor. We describe this field by the potential Ψ associated with the field strength $\mathbf{E} = -\operatorname{grad}\Psi$. Let Ψ be set equal to zero at infinity in both cases, A and B. In both cases $\Delta\Psi = 0$ outside of the conductor; on the surface, as well as in the interior of the conductor, we have $\Psi = \Psi_L = \text{const.}$

A. Charging Problems

In case A, $\Psi_L = V$ is given, in case B, Ψ_L must be found. According to (3.12a) the surface charge density at any point $d\sigma$ of the surface of L is given by

$$\omega = D_n = \varepsilon E_n = -\varepsilon\left(\frac{\partial\Psi}{\partial n}\right)_L. \tag{1}$$

ε is the dielectric constant outside of the conductor, n the outward normal to the surface of L. According to (1) the total charge on L is

$$q = \int \omega \, d\sigma = -\varepsilon \int \frac{\partial\Psi}{\partial n} \, d\sigma. \tag{2}$$

In case A, q is sought: in case B, where q is given, (2) determines Ψ_L.

For the case of a *sphere*, of radius a, the appropriate solution of the differential equation $\Delta\Psi = 0$ may be written down immediately, in the form

$$\Psi = \frac{a}{r}\Psi_L. \tag{3}$$

This yields for A, since $\Psi_L = V$,

$$\Psi = \frac{a}{r} V. \tag{3a}$$

In case B, (2) and (3) lead to

$$q = 4\pi a^2 \frac{\varepsilon \Psi_L}{a}, \ \Psi_L = \frac{q}{4\pi\varepsilon a}, \ 4\pi\varepsilon\Psi = \frac{q}{r}. \tag{3b}$$

The charge q, uniformly distributed over the spherical surface, thus acts at a distance like a point charge concentrated at the center.

It is also possible to guess the field of a conductor of the shape of a prolate spheroid (ellipsoid of revolution with long axis as axis of symmetry). For this it is merely necessary to stretch, so to speak, the center of the sphere, which appears in the last Eq. (3b) as locus of the total charge, into the connecting line of the two focal points of the generating ellipse and to distribute the charge q uniformly over this line. If we call the distance of the two focal points from the center of the ellipsoid c, the linearly distributed charge density becomes $q/(2c)$ and its potential becomes, by Eq. (7.5b):

$$
\begin{aligned}
4\pi\varepsilon\Psi &= \frac{q}{2c} \int_{-c}^{+c} \frac{d\zeta}{\sqrt{x^2 + y^2 + (z - \zeta)^2}} \\
&= \frac{q}{2c} \log \frac{z + c + \sqrt{x^2 + y^2 + (z + c)^2}}{z - c + \sqrt{x^2 + y^2 + (z - c)^2}}
\end{aligned}
\tag{4}
$$

In Problem II.1 we will show that this expression assumes a constant value $\Psi = \Psi_L$ on each of the confocal ellipsoids pertaining to the given separation of foci and that therefore it solves our potential problem for each one of these ellipsoids. Since only the separation c of the focal points occurs in (4), this formula applies for all confocal ellipsoids of the family, in the sense that all of the equipotential surfaces of the confocal ellipsoids with the semiaxes $a_2 > a_1$, $b_2 > b_1$ are included among the equipotential surfaces of the ellipsoid with the semiaxes a_1, b_1. The ellipsoid $a = c$, $b = 0$, which degenerates to a straight line of length $2c$, also belongs to this family. In II.2 the limiting case of a paraboloid of revolution, and its degeneration, the field of a semi-infinite glass rod uniformly charged by friction, are studied from this point of view.

B. Induction Problems and Method of Reciprocal Radii

The "induction problem," which we shall specialize to an inducing point charge, is more complex than the "charging problem" treated thus far. Here also we can distinguish between two cases: A. The (otherwise arbitrarily shaped) conductor is grounded and B., it is insulated. The generally accepted meaning of "grounding" is a conducting connection with an infinitely distant surface at the potential $\Psi = 0$. "Insulation" signifies, for an originally uncharged conductor, that even after induction the total charge q continues to be zero.

Problem A is solved by *Green's function* $G(P, Q)$—more exactly, "Green's function of the potential equation for the exterior of the conductor L." Q is the "source point," which will be assumed to represent a "unit source," P, the "reference point." G is defined by the following conditions:

$$
\begin{aligned}
&\Delta G = 0 \text{ for all } P \neq Q \text{ outside of } L, \\
&G \to 1/(4\pi r_{PQ}) \text{ for } P \to Q \text{ (definition of unit source)}, \\
&G = 0 \text{ on the surface of } L, \\
&G \to 0 \text{ for } P \to \infty.
\end{aligned}
\tag{5}
$$

Green's function plays a central role not only in potential theory, but generally in the theory of linear differential equations, to be treated in Vol. VI. Here we shall merely point out its significance for our special problem. If Q represents the position of the inducing charge e, the solution of problem A is given by

$$
\Psi(P, Q) = \frac{e}{\varepsilon} G(P, Q)
\tag{6}
$$

and that of B by

$$
\Psi(P, Q) = \frac{e}{\varepsilon} G(P, Q) + \alpha \Psi;
\tag{6a}
$$

here Ψ is the solution of our "charging problem" A for the same conductor L, α, a parameter which, according to Eq. (2), is determined by the condition

$$
\int \frac{\partial \Psi(P, Q)}{\partial n} \, d\sigma = 0.
\tag{6b}
$$

As special case we consider once more a sphere of radius a. Its Green's function can be written down in closed form by the ingenious method of the young William Thomson, later Lord Kelvin, which will be treated in detail in Vol. VI, §23. With Q as source point and Q' as "electrical image of Q with reference to the sphere of radius a" this solution is:

$$
4\pi G(P, Q) = \frac{1}{r_{PQ}} - \frac{a}{\rho} \frac{1}{r_{PQ'}}.
\tag{7}
$$

$\rho = OQ$ is the separation of the source point Q from the center of the sphere, $\rho' = OQ'$, that of its image point Q'. They are related by the contion of "reciprocal radii":

$$
\rho\rho' = a^2,
\tag{8}
$$

from which Thomson's procedure has been given the name "method of reciprocal radii." It is immediately evident that our formula (7) satisfies the first, second, and last condition (5); the fulfilment of the third condition (5) can be demonstrated by elementary geometry.

According to (6), Eq. (7) yields for induction on the grounded sphere:

$$4\pi\varepsilon\Psi(P,Q) = \frac{e}{r_{PQ}} - \frac{e'}{r_{PQ'}}, e' = \frac{a}{\rho}e \tag{8a}$$

and for the insulated sphere, by (6a, b) and (3b):

$$4\pi\varepsilon\Psi(P,Q) = \frac{e}{r_{PQ}} - \frac{e'}{r_{PQ'}} + \frac{e'}{r_{PO}}, e' = \frac{a}{\rho}e. \tag{8b}$$

The last term in this formula corresponds to the added term $\alpha\Psi$ in (6a). It has the effect of raising the potential of our insulated sphere to the

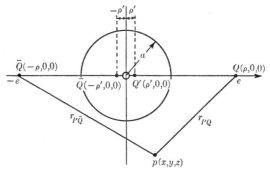

FIG. 8. Two charges $\pm e$ moving toward infinity and their electric images at a conducting sphere of radius a produce a uniform electric field and an electric dipole at the center of the sphere.

value $V = e'/(4\pi\varepsilon a)$ and of making the total charge on its surface $q = 0$, as it should be.

C. Conducting Sphere in a Uniform Field

For the practically unlimited possibilities of application of spherical images in potential theory we refer to the portion of Vol. VI cited above. Here we shall treat only the simple case of the sphere in *a uniform field*, whose lines of force may, for example, be parallel to the x-axis. In the absence of the conducting sphere the uniform field is given by

$$\Psi = -Fx, E_x = -\frac{\partial\Psi}{\partial x} = F, E_y = E_z = 0. \tag{9}$$

We imagine this field as resulting from the superposition of two fields, originating in the infinitely distant source points $Q, \bar{Q}$ (see Fig. 8) on the

x-axis, with the charges $\pm e$; their coordinates are $x = \pm \rho (\rho \to \infty), y = z = 0$. The superposition leads to the potential

$$4\pi\varepsilon\Psi = \frac{e}{r_{PQ}} - \frac{e}{r_{P\bar{Q}}} = e\{(\rho - x)^2 + y^2 + z^2\}^{-\frac{1}{2}}$$

$$- e\{(\rho + x)^2 + y^2 + z^2\}^{-\frac{1}{2}} = e\{\rho^2 + x^2 + y^2 + z^2\}^{-\frac{1}{2}} \qquad (9a)$$

$$\cdot \left(1 + \frac{\rho x}{\rho^2 + \cdots} - 1 + \frac{\rho x}{\rho^2 + \cdots}\right) \to \frac{2ex}{\rho^2}.$$

We have thus in fact a uniform field of the same form as (9) provided that we let e become infinite as ρ^2. To obtain quantitative agreement with (9) we must put

$$\frac{2e}{\rho^2} \to -4\pi\varepsilon F. \qquad (9b)$$

In Fig. 8 we have also shown the sphere of radius a and the appropriate spherical images of the source points Q, $\bar{Q}$ constructed for it:

$$Q' = \rho', 0, 0 \text{ and } \bar{Q}' = -\rho', 0, 0.$$

They approach each other as the charges $\pm e$ move apart and form in the limit an electric dipole with the moment

$$M = 2\rho'e'. \qquad (10)$$

Here we substitute from Eq. (8) $\rho' = a^2/\rho$ and from Eq. (8a, b) $e' = ea/\rho$. In view of (9b), Eq. (10) then states that the moment M assumes in the limit for $\rho \to \infty$ the finite value

$$M = 2\frac{a^2}{\rho}\frac{ea}{\rho} = -4\pi\varepsilon Fa^3. \qquad (10a)$$

We conclude therefore that the boundary value problem for the homogeneous field is solved by placing a virtual electric dipole of the finite moment M at the center of the sphere. The homogeneous field (9) is then replaced by the inhomogeneous field[1]

$$\Psi = -Fx + \frac{M}{4\pi\varepsilon}\frac{\partial}{\partial x}\frac{1}{r}, \qquad (11)$$

which contains the distortion created by the dipole. Since, now, r represents the distance from the center of the sphere,

$$r = (x^2 + y^2 + z^2)^{\frac{1}{2}}, \text{ so that } \frac{\partial}{\partial x}\frac{1}{r} = -\frac{x}{r^3}.$$

Eq. (11) hence becomes

$$\Psi = -Fx\left(1 + \frac{M}{4\pi\varepsilon F}\frac{1}{r^3}\right). \qquad (11a)$$

[1] The denominator $4\pi\varepsilon$ is to be added here for the same reason as the factor $4\pi\varepsilon$ on the left side of (9a).

If we substitute here the value of M from (10a), we obtain

$$\Psi = -Fx\left(1 - \frac{a^3}{r^3}\right). \tag{11b}$$

On the surface of the sphere, $r = a$, Ψ assumes the constant value $\Psi_L = 0$. The value of M found in (10a) by the method of reciprocal radii is thus confirmed for the *conducting sphere*.

D. Dielectric Sphere in a Uniform Field

We shall now show that the formula (11), with the value of M undetermined, has a much greater range of validity, i.e. will also fulfill the boundary conditions for a *non-conducting sphere of arbitrary dielectric constant*. If we distinguish the exterior and the interior of the sphere by the indices 1 and 2 (see Fig. 9), these conditions require that

$$\Psi_1 = \Psi_2, \; \varepsilon_1 \frac{\partial \Psi_1}{\partial n} = \varepsilon_2 \frac{\partial \Psi_2}{\partial n} \text{ for } r = a. \tag{12}$$

The first of these guarantees the continuity of the *tangential* components of $\mathbf{E}$; the second indicates the continuity of the *normal* component of $\mathbf{D}$, which, for an originally uncharged, nonconducting sphere, is equivalent to the absence of surface charge, $\omega = 0$ (Eq. (3.11)). We assert that both equations may be satisfied if for Ψ_1 (exterior of sphere, $r > a$) we use formula (11a) and for Ψ_2 (interior of sphere, $r < a$) we assume a *homogeneous field*, in the same direction as, but differing in strength from, the exterior primary field.

With $x = r \cos \theta$ (θ = geographic latitude on the sphere, measured from the field direction) we write tentatively

$$\Psi_1 = -F\left(r + \frac{M}{4\pi\varepsilon_1 F}\frac{1}{r^2}\right)\cos\theta, \; \Psi_2 = -F_2 \, r \cos \theta \tag{13}$$

and, according to (12), we must demand that for $r = a$ (cos θ cancels in both Eqs. (12)):

$$F\left(1 + \frac{M}{4\pi\varepsilon_1 F}\frac{1}{a^3}\right) = F_2,$$

$$F\left(1 - \frac{2M}{4\pi\varepsilon_1 F}\frac{1}{a^3}\right) = \frac{\varepsilon_2}{\varepsilon_1} F_2. \tag{13a}$$

From this we find, with $\varepsilon = \varepsilon_2/\varepsilon_1$ as *relative* dielectric constant:

$$\frac{F_2}{F} = \frac{3}{\varepsilon + 2}, \; \frac{M}{4\pi\varepsilon_1 F} = -\frac{\varepsilon - 1}{\varepsilon + 2}\, a^3. \tag{14}$$

Thus our assumption of a homogeneous field within the sphere has proved adequate. This field F_2 is weaker than the primary external field F if

$\varepsilon > 1$, as, for example, if the sphere is in air. The lines of force penetrate the interior of the sphere (see Fig. 9); although they are curved outside by the action of the (virtual) dipole moment they are straight and parallel to the x-axis within the sphere.

To understand this figure properly it should be noted that it does not represent the lines of *force* **E**, but the lines of *induction* **D**. The two systems of lines have the same direction but different density both inside and out-side of the sphere (see the remarks on magnetic lines of force or better tubes of force on p. 11), and hence behave differently at the surface. The **D**-lines are *source-free* not only within and outside of the sphere but also at its surface, because of the vanishing surface divergence (D_n is continuous); this does not apply to the **E**-lines (E_n is discontinuous). The fact that Fig. 9 represents the **D**-lines is evident from the fact that just one line passes through every point of the spherical surface. In the

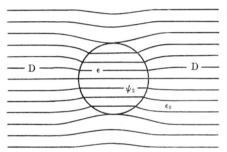

Fig. 9. A dielectric sphere in a uniform electric field. The excitation lines within and outside of the sphere.

case of the **E**-lines, more lines would arrive at the surface on the outside than leave it on the inside.

We consider two limiting cases, $\varepsilon \to \infty$ and $\varepsilon \to 0$. The first proves to be identical with that of the *conducting* sphere. Eqs. (14) yields then

$$M = -4\pi\varepsilon_1 F a^3, \qquad F_2 = 0, \qquad (14a)$$

in agreement with Eq. (10a) and with the fact that the interior of the sphere is field-free. This appears to contradict Fig. 9a, which shows a finite field inside of the sphere. We must note again, however, that this figure, as limiting case of Fig. 9, represents the **D**-field and that "$D_1 = \varepsilon E = $ finite" is consistent with passing to the limit $\varepsilon \to \infty$, $E \to 0$.

The limiting case $\varepsilon \to 0$ cannot be realized electrostatically,[1] but only

---

[1] Or only by assuming $\varepsilon_2 \ll \varepsilon_1$, i.e. considering a spherical cavity in a medium of very high dielectric constant. Then indeed the interior of the sphere is *relatively* free of **D**-lines, as is shown in Fig. 9b.

by placing a nonconductor in the stationary current field of a conductor. The magnetic analog would be a superconductor; hydrodynamically it represents the case of a rigid sphere immersed in a liquid whose flow is nonturbulent, incompressible and in parallel lines at infinity. In contrast with (14a) we have now

$$M = 2\pi\varepsilon_1 F a^3, \; F_2 = \frac{3}{2} F. \qquad (14b)$$

In spite of the finite value of F_2 the magnetic induction lines in the case of the superconductor and the hydrodynamic flow lines do not penetrate into

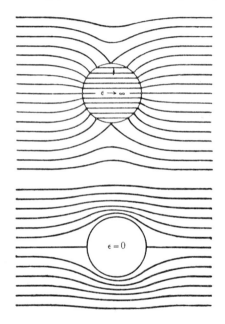

Fɪɢ. 9a. A conducting sphere in a uniform electric field. Representation of the excitation lines.

Fɪɢ. 9b. Stream lines about a rigid sphere. At the same time, magnetic lines of force about a superconducting sphere.

the interior of the sphere, as shown in Fig. 9b; they are pushed out of the interior since they must run tangential to the surface.

In the other limiting case, $\varepsilon \to \infty$, the lines of force are perpendicular to the surface,[1] as for a conductor. In Problem II.3 we shall indicate the close relationship between Figs. 9 and 9a. We shall return to these important formulas and figures in §11.

---

[1] The force lines at the upper and lower pole of the sphere (in three dimensions at the diametral plane passing through the poles) form an exception. They make an angle of 45° with the surface of the sphere (see figure), an angle of 90° with each other. This may for example be seen from the fact that the Taylor expansion of Ψ (Eq. (11b)) begins with a term of the second order. Maxwell calls such a point a "point of equilibrium" (see art. 112 of the Treatise).

E. Reflection and Refraction of Lines of Force at the Boundary of a Semi-infinite Dielectric

For the sake of completeness a rather trivial problem will be dealt with here, namely induction in a dielectric bounded by a plane (see Fig. 10): Let a unit electric charge Q be at the point $x = a$ in the right half-space, $x > 0$; it brings about a state of induction in the left halfspace, $x < 0$; ε is the *relative* dielectric constant of the left halfspace referred to the right halfspace. As in (12), the boundary conditions are

$$\Psi_1 = \Psi_2, \frac{\partial \Psi_1}{\partial x} = \varepsilon \frac{\partial \Psi_2}{\partial x} \quad \text{for} \quad x = 0. \tag{15}$$

A solution may be obtained with the aid of the simple reflection method familiar from optics: a virtual charge Q' of opposite sign is imagined at the point $x = -a$, which has an effect in medium 1, but has no effect in medium

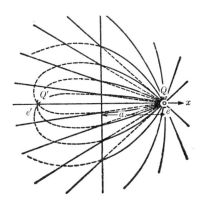

FIG. 10. In the right-hand halfspace (air) at point Q is a point charge producing induction in the left halfspace (dielectric). Representation of the excitation lines. In the right halfspace they are curved; their continuations (dotted in the figure) pass through the image point Q' of Q. In the left halfspace they are straight lines whose (dotted) continuations pass through Q.

2, in view of the finiteness of the field throughout this medium; here all effects appear to proceed from the primary charge. Introducing the two disposable parameters e'/e and e''/e we write tentatively:

$$4\pi\varepsilon_0 \Psi_1 = \frac{e}{r} + \frac{e'}{r'}, \qquad 4\pi\varepsilon_0 \Psi_2 = \frac{e''}{r}. \tag{16}$$

In the boundary plane $x = 0$, $r = r_{PQ}$ and $r' = r_{PQ'}$ are identical; at the same time

$$\frac{\partial}{\partial x} \frac{1}{r} = -\frac{x - a}{r^3}, \quad \text{and} \quad \frac{\partial}{\partial x} \frac{1}{r'} = -\frac{x + a}{r^3}$$

are equal and opposite, that is, are respectively equal to $\pm a/r^3$. Hence conditions (15) require

$$e + e' = e''$$

$$e - e' = \varepsilon e''$$

$$e' = \frac{1 - \varepsilon}{1 + \varepsilon} e, \qquad e'' = \frac{2}{1 + \varepsilon} e. \tag{16a}$$

Employing optical terminology, e'' might be called the "refracted" charge; the "reflected" charge e' of course becomes zero for $\varepsilon = 1$ (uniform dielectric, no discontinuity at $x = 0$). In the limiting case $\varepsilon \to \infty$ (conductor) $e' = -e$, $e'' = 0$, so that $\Psi_2 = 0$; this corresponds to the requirement of a constant potential for $x < 0$. For $\varepsilon \to 0$ the second condition (15) becomes $\partial \Psi_1 / \partial x = 0$ and leads to $e' = +e$. See in this connection Fig. 10 (arbitrary $\varepsilon > 1$), Fig. 10a ($\varepsilon \to \infty$), and Fig. 10b ($\varepsilon \to 0$, i.e. $\varepsilon_{\text{right}} \gg \varepsilon_{\text{left}}$).

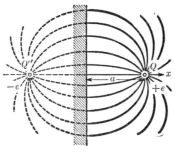

FIG. 10a. Limiting case $\varepsilon \to \infty$; the left halfspace is a conductor.

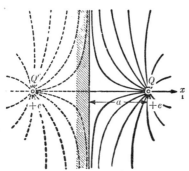

FIG. 10b. Limiting case $\varepsilon \to 0$; dielectric constant of the right halfspace very large compared with that of the left halfspace.

§10. Capacity and its Connection with Field Energy

We consider two conductors of arbitrary shape L_1 and L_2 and give them charges $+q$ and $-q$ respectively. Such a system is called a *condenser* because the field between them is concentrated and limited to their neighborhood. The lines of force pass from L_1 to L_2 without diverging to infinity. Let the potential Ψ have the constant values Ψ_1 and Ψ_2 on L_1 and L_2. The potential difference between them is then

$$V = \Psi_1 - \Psi_2 = \int_{L_1}^{L_2} \mathbf{E} \cdot d\mathbf{s}. \tag{1}$$

In this line integral we are permitted to leave the shape of the path indefinite; every path from L_1 to L_2 (it need by no means be a line of force) yields, as we know, the same value of the potential difference for a lamellar electrostatic field.

The ratio q/V is called the *capacity*, i.e. the ability of the system to take up charge. We employ for it the letter K (instead of the often-employed symbol C, which we use so often with the meaning "constant"):

$$K = \frac{E}{V}. \tag{2}$$

The unit of capacity is the "farad":

$$1 \text{ farad} = 1 \frac{\text{coulomb}}{\text{volt}} = 1 \frac{Q^2}{\text{joule}}; \tag{3}$$

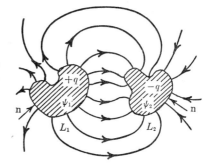

FIG. 11. The pair of conductors L_1, L_2 with the charges $\pm q$ and the potentials Ψ_1, Ψ_2 form an electric condenser.

which is thus defined in our system without the appearance of inconvenient powers of ten. On the other hand, in electromagnetic cgs-units, according to (3),

$$1 \text{ farad} = \frac{10^{-1}}{10^8} \text{ cgs} = 10^{-9} \text{ cgs}; \tag{3a}$$

thus the socalled absolute cgs-unit of capacity would be $= 10^9$ farad. The microfarad $(= 10^{-6}$ farad$)$ is employed more often than the farad.

According to (3) and (4.6a) the dielectric constant of vacuum has the dimension farad/M and, according to (7.18), it has the magnitude

$$\varepsilon_0 = \frac{10^{-9}}{36\pi} \frac{\text{farad}}{M} = \frac{10^{-3}}{36\pi} \frac{\text{microfarad}}{M}. \tag{3b}$$

However, if one chooses to give the capacity in "absolute electrostatic units" not only is the numerical value of the capacity different but also its dimensions (here cm!) are changed and a very confusing state of affairs results.

We shall consider, as simplest example,

A. The Plate Condenser

This condenser is to consist of two conducting plates with (large) area F, which are placed parallel and facing each other with (small) separation *a.* We imagine their charges $\pm q$ to be distributed over F with uniform surface density ω. Hence we have for $x = \pm a/2$

$$\omega = \mp \frac{q}{F}, \qquad \text{and also } \omega = D_n \text{ (Eq. (3.12a))}. \tag{4}$$

We regard the field between the plates as homogeneous, neglecting as we already have in (4), the perturbation at the edge zones. The lines of force are then normal to the plates *throughout*, and we find, with the value of D given by (4)

$$E_x = \frac{\omega}{\varepsilon} = \frac{q}{F\varepsilon}. \tag{4a}$$

It follows hence that

$$V = \int_{-a/2}^{+a/2} E_x \, dx = \frac{qa}{F\varepsilon} \tag{4b}$$

and thus, from (2)

$$K = \frac{F\varepsilon}{a}. \tag{5}$$

This is the "rational" value of our capacity, previously given in (7.15).

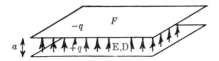

FIG. 12. The plate condenser with the field $\mathbf{E}$, $\mathbf{D}$ between plates considered as uniform.

If, for example, we consider $F = 20 \cdot 20 \text{ cm}^2$, $a = 1$ mm, and $\varepsilon = 2\varepsilon_0$ (paraffin filling) we find, with the value (3b) of ε_0 in our system of units,

$$K = \frac{80}{36\pi} \, 10^{-9} \text{ farad} = \frac{2}{9\pi} \cdot 10^{-8} \text{ farad} = \frac{2}{9\pi} \cdot 10^{-2} \text{ microfarad}. \tag{5a}$$

Thus, to obtain a capacity comparable with a microfarad, a very large number of condensers of the type considered have to be added together.

The neglected edge correction and the resulting inhomogeneity of the field will be treated in Problem II.4.

B. Spherical Condenser

This is to consist of an inner sphere of radius r_1 and an outer sphere of radius r_2. The spheres need not be conductors throughout; it suffices if the outer surface of the inner sphere, and the inner surface of the outer sphere are "coated with tin foil." (The same remark applies for the plate condenser.) Let the inner sphere have the charge $+q$, the outer sphere, the charge $-q$. The field is spherically symmetrical—**E** and **D** depend only on r and are directed radially. Hence for every r between r_1 and r_2

$$\int D_r \, d\sigma = 4\pi r^2 D = q,$$

$$E_r = \frac{D}{\varepsilon} = \frac{q}{4\pi\varepsilon r^2} \cdot$$

(6)

From E_r we obtain

$$V = \int_{r_1}^{r_2} E_r \, dr = \frac{q}{4\pi\varepsilon} \left(\frac{1}{r_1} - \frac{1}{r_2} \right) = \frac{q}{4\pi\varepsilon} \frac{r_2 - r_1}{r_1 r_2} \cdot$$

(7)

This leads to

$$K = \frac{q}{V} = 4\pi\varepsilon \frac{r_1 r_2}{r_2 - r_1} \cdot$$

(8)

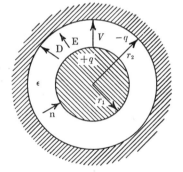

Fig. 13. The spherical condenser with its radial field **E**, **D** and a voltage $V = \Psi_1 - \Psi_2$.

If we allow r_2 to become infinite we obtain

$$K = 4\pi\varepsilon r_1 ,$$

(8a)

i.e. the value for the capacity for this limiting form of spherical condenser which was designated as "rational" in Eq. (7.15a). If, on the other hand, both radii in (8) are permitted to approach infinity while yet their difference $r_2 - r_1 = a$ remains finite, and if only a finite segment F of the two spherical surfaces $4\pi r_1^2$ and $4\pi r_2^2$ is considered, the capacity of this segment becomes according to (8)

$$K = \frac{F\varepsilon}{a}$$

(8b)

i.e., identical, as it should be, with that of the plate condenser in Eq. (5).

The cylindrical condenser (Leyden jar[1]) will be treated in Problem II.5.

C. Capacity of an Ellipsoid of Revolution and of a Straight Piece of Wire

In Eq. (9.4) we have given the potential Ψ of an ellipsoid with the charge q. If we substitute for x, y, z the coordinates of any point of its surface, e.g. one of the two endpoints of its major axis $x = y = 0$, $z = a$, Ψ assumes the value V, where V is its potential relative to an infinitely distant ground ($\Psi = 0$). If, at the same time, we introduce the minor axis b in place of half the focal separation c, we obtain $c = \sqrt{a^2 - b^2}$. Eq. (9.4) thus becomes[2]

$$\frac{V}{q} = \frac{1}{K} = \frac{1}{8\pi\varepsilon\sqrt{a^2 - b^2}} \log \frac{a + \sqrt{a^2 - b^2}}{a - \sqrt{a^2 - b^2}}$$

$$= \frac{1}{4\pi\varepsilon\sqrt{a^2 - b^2}} \log \frac{a + \sqrt{a^2 - b^2}}{b}. \tag{9}$$

For $b \to a$, in the limit, the capacity (8a) of the sphere is obtained. In a similar manner the capacity of the axially symmetric *ellipsoidal condenser*, consisting of an inner and an outer ellipsoid of the confocal family, may be derived.

On the other hand, if we let $b \to 0$, our ellipsoid degenerates into a *straight segment* of length $l = 2c$ (focal separation). This may be thought of as representing a straight wire[3] of radius $b \to 0$. Its capacity becomes, according to (9)

$$K = 2\pi\varepsilon l / \log \frac{l}{b}. \tag{9a}$$

We will obtain a similar logarithmic formula for the selfinduction of a straight piece of wire (see §15).

D. Energetic Definition of Capacity

The usual elementary definition (2) of capacity may appear rather arbitrary and formal. We attain a physically more significant understanding of it by considering the *energy of the electrostatic field*.

---

[1] Also known as "Kleist jar," since it was built by the pastor of Kleist in Cammin (Pomerania) and first demonstrated at the Danzig Scientific Society in December 1745. At the beginning of the next year it was demonstrated in Leyden and has hence become internationally famous. In this manner it acquired its present customary name, which is thus purely accidental in origin.

[2] Compare Kohlrausch, *Praktische Physik*, 12th Ed., p. 631. The factor 4π is missing here (conventional system of units). Obviously the last formula in Eq. (9) results from the preceding one by rationalization of the denominator.

[3] Not, it is true, a cylindrical wire, but one getting thinner toward the ends.

We utilize Green's theorem in the form (3.16) of Vol. II:

$$\int \operatorname{grad} U \cdot \operatorname{grad} V \, d\tau + \int U \Delta V \, d\tau = \int U \frac{\partial V}{\partial n} \, d\sigma. \qquad (10)$$

U and V are two arbitrary continuous functions. The integration on the left side is extended over an arbitrary region of space, the integration on the right, over its surface; n is the normal to the surface of the space, pointing outward. We put $U = V = \Psi$ and obtain, because of $\Delta \Psi = 0$, $\mathbf{E} = -\operatorname{grad} \Psi$, from (10)

$$\int \operatorname{grad} \Psi \cdot \operatorname{grad} \Psi \, d\tau = \int \mathbf{E}^2 \, d\tau = \int \Psi \frac{\partial \Psi}{\partial n} \, d\sigma. \qquad (10a)$$

Let the volume integration on the left be carried out through the region exterior to the two conductors L_1 and L_2 in Fig. 11, the surface integration on the right, over the two conductor surfaces and a sphere K with the very large radius $r = R$. The integral over K vanishes.[1] (10a) then leads to

$$\int \mathbf{E}^2 \, d\tau = \Psi_1 \int_{L_1} \frac{\partial \Psi}{\partial n} \, d\sigma + \Psi_2 \int_{L_2} \frac{\partial \Psi}{\partial n} \, d\sigma. \qquad (10b)$$

After multiplication with $\varepsilon/2$ we can write instead

$$\frac{1}{2} \int \mathbf{E} \cdot \mathbf{D} \, d\tau = -\frac{1}{2} \left(\Psi_1 \int_{L_1} D_n \, d\sigma + \Psi_2 \int_{L_2} D_n \, d\sigma \right) = \frac{q}{2} (\Psi_1 - \Psi_2) = \frac{1}{2} qV.$$

At the left end of this multiple equation we have the total energy of the volume considered, which we shall call W:

$$W = \int W_e \, d\tau.$$

We then find, with the definition (2) of K,

$$W = \frac{1}{2} qV = \frac{K}{2} V^2 = \frac{q^2}{2K}. \qquad (11)$$

---

[1] We have, by Gauss's theorem, for an arbitrary system of n conductors L_1, L_2, $\cdots$ L_n, since $\operatorname{div} \mathbf{D} = 0$,

$$0 = \int_K D_n \, d\sigma + \int_{L_1} D_n \, d\sigma + \cdots + \int_{L_n} D_n \, d\sigma = \int_K D_n \, d\sigma - \sum_{i=1}^{n} q_i.$$

If Ψ on K is put equal to Ψ_∞, it follows that

$$\int_K \Psi \frac{\partial \Psi}{\partial n} \, d\sigma = -\frac{\Psi_\infty}{\varepsilon} \int_K D_n \, d\sigma = -\frac{\Psi_\infty}{\varepsilon} \Sigma q_i.$$

In the present case this vanishes because of $q_1 = q$, $q_2 = -q$. In general, with $\Sigma q_i \neq 0$, it is only necessary to put $\Psi_\infty = 0$, i.e. to refer the potentials Ψ_i to the zero level at infinity, to make it possible to carry over the following formulas also to unneutral systems. This is to be remembered for section E.

This fundamental relation is reminiscent of the expression for the kinetic energy of a particle in rectilinear motion in terms of its velocity v and its momentum $p = mv$:

$$W = \frac{1}{2} vp = \frac{p^2}{2m} = \frac{m}{2} v^2. \tag{11a}$$

In both cases the energy is factored (see p. 11) into the product of an entity of quantity and one of intensity or expressed by the square of one of these two quantities. The quantity is in one case q, in the other, v, the intensity, V and p, respectively. Comparison of (11) and (11a) shows that the capacity K corresponds to the reciprocal mass $1/m$, which we might call "compliance," in contrast with the "inertia" m. However, the analogy is not very profound and will have to be modified in §33.

We can call (11) an energetic definition of capacity, just as (11a) may serve as energetic definition of inertia.

E. The Capacities of an Arbitrary System of Conductors

If we pass from two conductors L_1, L_2 with charges $\pm q$ to an arbitrary number of conductors L_1, L_2, $\cdots$ L_n with charges q_1, q_2, $\cdots$ q_n, where once more the total charge $\Sigma_{i=1}^n q_i$ is assumed to be zero, Green's theorem (10a, b) shows directly that the total energy W of the system is the sum of n terms, according to the formula

$$W = \frac{1}{2} \sum_{i=1}^{n} \Psi_i q_i. \tag{12}$$

Here Ψ_i denotes the constant value of the potential on the conductor L_i. Now, however, Ψ_i depends not only on q_i, but depends *linearly* on all the q_j as well. This follows from the general representation (7.5) of the potential.

In order to perceive this, we rewrite (7.5) in terms of the surface charge ω_j, whose distribution on every conductor L_j we can assume as known, in place of the volume charge ρ, and put $\omega_j = q_j \omega_j'$, where ω_j' is the distribution of unit charge on L_j (in the presence of the remaining conductors!). Then (7.5) becomes

$$\Psi_i = \sum_{j=1}^{n} q_j H_{ij}. \tag{13}$$

The coefficients which appear here,

$$H_{ij} = \frac{1}{4\pi\varepsilon} \int \frac{\omega_j'}{r_{ij}} d\sigma_j, \tag{13a}$$

are purely geometrical quantities, which depend only on the location of the L_j relative to each other and relative to L_i; they are independent of the choice of the origin $r_{ij} = 0$ on L_i, since Ψ_i has the same value for every choice of this point. Eqs. (13) and (13a) thus confirm the linear relation between the Ψ_i and q_j.

The solution of the system (13) of n equations for the n charges q_j yields

$$q_j = \sum_{i=1}^{n} K_{ij} \Psi_i \quad \text{with} \quad K_{ij} = \frac{\Delta_{ij}}{\Delta}. \tag{14}$$

Δ_{ij} is the sub-determinant of the $n \cdot n$ row determinant Δ of the H_{ij} associated with the term i, j. Maxwell calls the H_{ij} "potential coefficients" of the system in art. 87 of his Treatise and the K_{ij}, "capacity coefficients."

Substituting the relations (13) and (14) in (12) yields the *multiple equation*

$$W = \tfrac{1}{2}\Sigma\Psi_i q_i = \tfrac{1}{2}\Sigma\Sigma H_{ij} q_i q_j = \tfrac{1}{2}\Sigma\Sigma K_{ij}\Psi_i\Psi_j . \tag{15}$$

which generalizes our earlier Eq. (11). Since W is a quantity determined by the state of the system (the work done in charging the system must be independent of the "path," i.e. the sequence of the individual processes), the K and H fulfill the reciprocity relations

$$K_{ij} = K_{ji}, \qquad H_{ij} = H_{ji} . \tag{15a}$$

Capacity and potential coefficients play a role in communications, where complicated systems of interacting conductors are of frequent occurrence. Their theoretical calculation is difficult since it presupposes the solution of the potential problem of the multiconductor system in question. As we have seen, even for two conductors the solution is possible only for particularly simple shapes of the conductors (plane, sphere, ellipsoid). In general approximations are required.

In Problem II.6 we shall discuss the (somewhat complicated) relation between these coefficients and the elementary definition of capacity in Eq. (2).

§11. General Considerations on the Electric Field

The following statements and concepts apply not only to electrostatic, but also to arbitrarily varying fields.

A. The Law of Refraction for the Lines of Force

The boundary conditions applying at the interface between two insulators of different dielectric constant,

$$E_{\text{tang}} \text{ continuous and } D_{\text{norm}} \text{ continuous} \tag{1}$$

(the latter in the absence of surface charge, Eq. (3.11)) show directly that the "angle of incidence" α_1 and the "angle of refraction" α_2, both measured with respect to the normal to the interface and given by

$$\tan \alpha_1 = \left(\frac{E_{\text{tang}}}{E_{\text{norm}}}\right)_1, \qquad \tan \alpha_2 = \left(\frac{E_{\text{tang}}}{E_{\text{norm}}}\right)_2,$$

are related by

$$\frac{\tan \alpha_1}{\varepsilon_1} = \frac{\tan \alpha_2}{\varepsilon_2}. \tag{2}$$

This "law of refraction of the electric lines of force" deviates from the optical law of refraction not only in the appearance of the tangent instead of the sine, but also in the direction of refraction: *In entering into the electrically denser medium a line of force is refracted away from the normal to the interface.* We here describe the medium with the higher dielectric constant as "electrically denser." If this is medium 2, then it follows from Eq. (2) that

$$\tan \alpha_2 > \tan \alpha_1.$$

Examples of this phenomenon are shown in Figs. 10 (refraction at a plane) and 9 (refraction at a sphere). The conductor (limiting case $\varepsilon_2/\varepsilon_1 = \infty$) satisfies this law of refraction inasmuch as here generally $\alpha_1 = 0$ (lines of force normal to surface of conductor).

B. On the Definition of the Vectors $\mathbf{E}$ and $\mathbf{D}$

Now we do not accept the "positivistic" standpoint, according to which only observables may be employed in theoretical physics, but instead are of the opinion that the introduction of not directly observable quantities is justified whenever the resulting conclusions agree with experiment (as in the kinetic theory of gases). Nevertheless we demand that the concepts introduced in a hypothesis may be based at least on an imaginary experiment, i.e. an observational method, even if it cannot be carried out in practice.

In §2 we defined the electric field strength dynamically as the force on a unit test charge. This force can, however, only be measured in air (more generally, in a fluid) by the motion produced by it. Thus the definition fails in the solid body.

In order to define the component of $\mathbf{E}$ at a given point in a given direction s within the solid body we proceed as follows: We drill a *tube* with the direction s in the body at the point in question. The tube is so narrow and so short that it does not appreciably disturb the field elsewhere; it remains empty or is filled with air. According to the boundary condition (1) the field-strength component E_s is the same within it as in the surrounding solid body and as it was originally in the tube. Thus $\mathbf{E}_s$ can be measured within the tube on a test body which has been introduced, and by varying the direction of the tube all three components of the vector $\mathbf{E}$ can be obtained. If the field is not stationary, but variable, we must measure more rapidly than the field changes.

Our definition of **D** in §2 requires supplementation in even greater degree. We described, in Eq. (4), **D** as the quantity of electricity which, at a given point, has passed through an area F during the excitation of the field, divided by the magnitude of F. More precisely, we obtain in this manner the component D_n of **D** in the direction of the normal n to F. This explanation is unsatisfying since it does not contain specific directions for measurement. We can, however, obtain such directions by the following imaginary experiment:

We place at the point in question a plate condenser,[1] whose surfaces F are made normal to the n-direction; the space between the plates is to be filled with the surrounding dielectric (if we are dealing with a solid body a slit must be cut into it into which the condenser fits exactly). In view of the boundary condition (1) the value of D_n in the condenser is equal to that in its surroundings and hence also equal to the value of D_n which prevailed before the introduction of the condenser at the point in question. We can now measure D_n on our condenser directly as the surface charge ω on that coating toward which the prescribed direction n points.

In this manner the displacement **D** also becomes, in a sense, an "observable quantity."

C. The Concept of Electric Polarization; the Clausius-Mossotti Formula

We give up temporarily the purely phenomenological point of view of Maxwell's theory and attempt to construct a molecular model of the dielectric. A molecule consists of positive and negative charges (protons and electrons), but acts as a neutral entity in the absence of a field. With the application of the field the charges are separated and form a dipole.[2] The induced moment **m** is proportional to the external field and is a characteristic of the molecule.

Such a moment has the dimension charge·lever arm = QM. If we pass from the single molecule to the sum of the molecular moments "per unit volume" the dimension

$$\frac{QM}{M^3} = \frac{Q}{M^2}$$

---

[1] If the field is inhomogeneous the condenser must be made adequately small. Its metallic coatings distort the field, but do not disturb the measurement of D_n between the plates.

[2] We think here of *nonpolar molecules*. The *polar molecules*, which have been studied with great success, both experimentally and theoretically, by Debye, have a dipole even in the absence of a field. The formulas of the text would have to be altered for polar molecules and would show a dependence on temperature. For details see Debye: *Polar Molecules*, Dover, New York, 1945. In analogy to paramagnetism, polar molecules may be called *paraelectric*.

is obtained. It corresponds to the dimension (2.4) of $\mathbf{D}$. We shall designate this sum $\Sigma \mathbf{m}$ divided by the volume with $\mathbf{P}$ and call it the *polarization*.[1]

We shall divide $\mathbf{D}$ into one part which is present even in the absence of the molecules, and another part which is produced by the molecules. The first corresponds to the case of vacuum and is $\mathbf{D}_0 = \varepsilon_0 \mathbf{E}$, where $\mathbf{E}$ is the applied, macroscopically measurable field; the second is our polarization $\mathbf{P}$. Hence we write

$$\mathbf{D} = \mathbf{D}_0 + \mathbf{P} = \varepsilon_0 \mathbf{E} + \mathbf{P}. \tag{3}$$

$\mathbf{D}$ is the macroscopically measurable excitation and is hence equal to $\varepsilon \mathbf{E}$. Accordingly (3) leads to

$$\mathbf{P} = (\varepsilon - \varepsilon_0)\mathbf{E}. \tag{4}$$

We also wish to determine $\mathbf{P}$ from the behavior of the molecules in the electric field. In agreement with the notation of Eqs. (9.9)ff. we call this field $\mathbf{F}$ and indicate by this that it differs from the macroscopic field $\mathbf{E}$. The difference between them results from the effect of the polarized molecules according to the formula

$$\mathbf{F} = \mathbf{E} + \frac{1}{3}\frac{\mathbf{P}}{\varepsilon_0}. \tag{5}$$

In order not to interrupt our train of thought we defer proof of this expression to section D. The moment $\mathbf{m}$ acquired by the individual molecule is proportional to this $\mathbf{F}$. We put

$$\mathbf{m} = \alpha \varepsilon_0 \mathbf{F}, \tag{6}$$

where α is a constant characteristic of the molecule. Here the molecule is assumed to be isotropic; otherwise $\mathbf{m}$ and $\mathbf{F}$ would not have to have the same direction.

If N is the number of molecules per unit volume, we obtain from (6) and (5):

$$\mathbf{P} = \underset{\substack{\text{unit}\\\text{volume}}}{\Sigma \mathbf{m}} = N\alpha\varepsilon_0 \mathbf{F} = N\alpha \left(\varepsilon_0 \mathbf{E} + \frac{1}{3}\mathbf{P} \right). \tag{7}$$

If we substitute here expression (4) for $\mathbf{P}$ and cancel the common factor $\mathbf{E}$, we find

$$\varepsilon - \varepsilon_0 = N\alpha \left(\varepsilon_0 + \frac{\varepsilon - \varepsilon_0}{3} \right) = \frac{N\alpha}{3}(\varepsilon + 2\varepsilon_0)$$

or, if we pass to the relative dielectric constant $\varepsilon/\varepsilon_0$:

[1] Every individual moment has the direction of its lever arm as axis. Hence the direction of $\mathbf{P}$ is obtained by the geometric addition of all the $\mathbf{m}$ in the volume considered and passing to the limit of a sufficiently small volume.

$$\frac{\varepsilon_{rel} - 1}{\varepsilon_{rel} + 2} = \frac{N\alpha}{3} \tag{8}$$

This is the Clausius-Mossotti formula. In optics, where ε_{rel} is the square of the refractive index, it is known as the Lorenz-Lorentz formula.

To clarify the physical content of Eq. (8), we multiply numerator and denominator with m, the mass of the individual molecule. Thus we obtain, in the product Nm, the mass of unit volume of the dielectric or its *density* and, at the same time, in the quotient α/m, a new constant characteristic of the molecule. Then in Eq. (8) the left side is *proportional to the density*. This assertion can be tested directly on compressed gases for which ε_{rel} differs appreciably from 1. For highly diluted gases, where $\varepsilon_{rel} \sim 1$, $\varepsilon_{rel} + 2 \sim 3$, Eq. (8) leads to

$$\varepsilon_{rel} - 1 = N\alpha = \rho \frac{\alpha}{m}. \tag{8a}$$

Historically it may be mentioned that Mossotti, in his paper dating as far back as 1850, treated the molecules as *conducting* spheres, which were assumed to be distributed in some fashion in the imponderable "ether." We know from §9 that an external field $\mathbf{F}$ induces a moment $\mathbf{M}$ in such a sphere whose magnitude is given by (Eq. (9.10a))

$$\mathbf{M} = 4\pi\varepsilon_0 a^3 \mathbf{F}.$$

Our molecular constant α has then, according to the definition (6), the value

$$\alpha = 4\pi a^3. \tag{9}$$

If this expression is substituted in (8) the right side becomes

$$\frac{4\pi}{3} a^3 N. \tag{9a}$$

This is simply the ratio of the volume of the spheres contained in unit volume to unit volume (dimensionless, as it should be).

We summarize what we have learned about the concept of dielectric displacement which was inadequately explained in §2. $\mathbf{D}$ is composed of two parts, a vacuum portion $\mathbf{D}_0 = \varepsilon_0\mathbf{E}$ and a portion arising from matter $\mathbf{P}$:

$$\mathbf{D} = \mathbf{D}_0 + \mathbf{P}. \tag{10}$$

We call $\mathbf{P}$ the polarization of the matter; it is also the electric moment per unit volume of the dielectric. Similarly, the dielectric constant is made up of two parts, its vacuum component ε_0 and its component arising from matter $\varepsilon_0\eta$:

$$\varepsilon = \varepsilon_0(1 + \eta). \tag{11}$$

The material constant η, which is defined as a pure number, is called the *electric susceptibility*. It is pleasant to note that in both Eqs. (10) and (11) the factor 4π, which otherwise occurs in $\mathbf{P}$ and η, is absent as the result of the rational character of our MKSQ-system. In determining $\mathbf{P}$ and η we had to differentiate between the field strength $\mathbf{F}$ acting on the molecule and the macroscopically defined field strength $\mathbf{E}$; the difference between them arises from the field of the neighboring molecules.

As already noted on p. 9, the designation "dielectric displacement" for $\mathbf{D}$ really fits only its polarization component $\mathbf{P}$. In vacuum there is no charge which can be displaced and $\mathbf{D}$ is nevertheless by no means equal to zero. This is the reason why we have preferred (see p. 8) the term "excitation" for $\mathbf{D}$.

D. Supplement to the Calculation of the Polarization

We are here concerned with the proof of Eq. (5). We consider an arbitrary molecule, surround it with a small sphere whose radius b is nevertheless very great compared to the molecular radius, and remove from the sphere all matter except the one molecule at the center; this molecule is therefore in vacuum. It is thus acted upon by the field strength $\mathbf{E}$, corresponding to the first term on the right side of (5). The removal of the molecules from the interior of the sphere does not result in a change of the field resulting from the matter present, provided that the molecules were distributed randomly, i.e. that they were not oriented in any way by the molecule under consideration. If we limit ourselves to isotropic dielectrics we can assume this.[1]

After exclusion of this sphere we can treat the remaining dielectric as a continuous medium, i.e. neglect its molecular structure and proceed according to Maxwell's phenomenological theory. Hence we shall replace the action of the residual dielectric by charge densities ω on the elements $d\sigma$ of the inner bounding sphere of the cavity of radius b; here we shall have to determine ω not from the complete $\mathbf{D}$, but only from its molecular component $\mathbf{P}$ in Eq. (3). Since $\mathbf{P}$ differs from zero only for $r > b$ (in vacuum, for $r < b$, $\mathbf{P} = 0$), ω is not given by the difference of two $\mathbf{P}$-values (as in (3.11) by that of two $\mathbf{D}$-values), but directly by $\omega = P_n$. In view of the fact that $\mathbf{P}$ has the same direction as the primary field $\mathbf{E}$, which shall have the x-direction, we find

$$\omega = P_n = P_x \cos \theta \text{ with } \theta = \text{angle of } n \text{ with respect to } x. \quad (12)$$

According to Coulomb's law (7.7) the contribution of $\omega \, d\sigma$ to the field strength acting on our molecule at $r = 0$ in the direction of the radius vector is

---

[1] H. A. Lorentz has proved this also for crystals of cubic structure; for other symmetries, as well as for associating liquids, the assumption in the text is unproved.

$$dF = \frac{\omega \, d\sigma}{4\pi\varepsilon_0 b^2} = \frac{P_x \cos\theta}{4\pi\varepsilon_0 b^2} \, d\sigma \tag{13}$$

and its x-component, with which alone we are concerned,

$$dF_x = \frac{P_x \cos^2\theta}{4\pi\varepsilon_0 b^2} \, d\sigma. \tag{13a}$$

Integration over the whole sphere, with $d\sigma = b^2 \sin\theta \, d\theta \, d\varphi$ leads to

$$F_x = \frac{P_x}{4\pi\varepsilon_0} \cdot 2\pi \int_0^\pi \cos^2\theta \sin\theta \, d\theta = \frac{1}{3}\frac{P}{\varepsilon_0}. \tag{14}$$

This corresponds to the second term of the right side of Eq. (5), which is proved herewith.

E. Permanent Polarization

We have assumed so far that the polarization is caused by an external field and vanishes with it. That is not the case in general. We have already noted (p. 73, footnote 2) that permanent electric moments exist on a *molecular* scale. It is true that they compensate each other, particularly in the liquid or gaseous state, because of the thermal disorder *in any finite volume*, so that here also the resulting polarization vanishes with the external field **E**. However, if a substance made up of such polar molecules (a wax or resin) is liquefied by heating and exposed to a strong electric field, the latter forces the molecular moments largely into its direction. After solidification the substance retains its polarization for a time even if the field is subsequently removed. If the environment could be made completely insulating a substance would be obtained with a macroscopically *permanent electric field*.

Heaviside has christened a substance treated in this manner with the rather forced name "electret," in view of its analogy to the permanent magnet.

The assumption of a completely insulating environment is, however, never satisfied. Even pure, highly diluted air is, because of radioactive emanation and, in particular, because of cosmic radiation, always somewhat ionized and hence conducting. An electret hence tends to lose its effectiveness to the outside in the course of hours or days.

There are however also natural substances with similar properties. We find these among crystals which are asymmetric in structure (crystals with a polar axis). The most familiar example is *tourmalin*. A crystal is in general made up of positively and negatively charged ions which, if there is an imperfect symmetry of structure, have an electric moment in any elementary domain. Depending on the lattice of the crystal, the elementary moments may combine to form a macroscopic moment, which then produces an electric field in its neighborhood.

Such a field can in fact be detected on *fresh fragments* of tourmalin. Since the environment, as noted above, never insulates perfectly, the field decays in the course of a few hours. Surface charges are built up by the conduction currents at the entrance and exit points of the lines of force which then compensate the external field of the interior electric moment. The difference between the electric and magnetic permanent moment consists merely in the fact that there are no such conduction currents in the magnetic field. Hence a steel magnet shows no appreciable change in its field in the course of decades. This difference between electret and magnet is not fundamental, but only quantitative.

The polar asymmetry existing in tourmalin may be produced artificially in other, less asymmetric[1] crystals by subjecting them to a deformation. This distorts the crystal lattice and impresses an electric moment proportional to the deformation. The crystal thus becomes *piezoelectric*. Quartz is the typical representative of this class of substances. It was used by Pierre Curie, as "piezo-quartz," to produce well-defined quantities of electric charge. It is true that here also, because of imperfect insulation, the electric charge decays with a certain finite relaxation time. Of even greater importance in more recent times has been the role of the quartz crystal when vibrating with its characteristic frequency and producing a corresponding oscillatory electric field. Inversely, by applying an alternating field of this frequency it is possible to maintain the characteristic vibration of the quartz at a constant amplitude. In this manner one obtains (Cady) an ideal microscale of time, which plays its well known role in present-day radio engineering.

In all these cases (electret, tourmalin, piezoquartz) the external electric field may be calculated from the inner moment which is assumed to be known. We will omit this, however, since the calculation is quite similar to that of the external field of a permanent magnet, which is carried out below.

§12. The Field of the Permanent Bar Magnet

The forces which emanate from certain forms of iron have excited the popular imagination since the earliest times. The Greeks called the carriers of such effects *magnets*.[2] The Chinese were the first to utilize their interaction with the great magnet "Earth" for geographic orientation on the

---

[1] The degree of the required asymmetry may be predicted exactly by means of the general rules of Voigt. See Vol. II, §40.

[2] Apart from *steel* the metals *cobalt* and *nickel*, which are related to iron, show permanent magnetism, similarly the *Heusler alloys*, containing manganese, which adjoins iron in the periodic system. The iron ore $Fe_2O_3 \cdot FeO$ which, crystallizes in a cubic lattice, is known as *magnetite*, the hexagonal FeS (with admixture of Fe_2S), as pyrrhotin or magnetic gravel; both are characterized by permanent magnetism and magnetic anisotropy.

broad expanses of their country. In the 18th Century it was fashionable to attribute all mysterious processes in the human body to "animal magnetism" (Mesmer). Today the importance of the natural or *permanent magnets* is outdistanced by that of the *electromagnets*. In spite of this we shall begin with some consideration of permanent magnets; we shall then be in a position to cover briefly the general properties of the magnetostatic field, in analogy to those of the electrostatic field in §13.

It is true that the nature of the permanent magnets lies outside the range of the Maxwell theory and can be understood only with the aid of atomic physics. It is based on the spin of the electron and its magnetic moment, of which the Maxwell theory is, of course, ignorant. The same remark applies eventually to the electromagnet: the electric currents which produce the electromagnet, unless they are generated electrodynamically by induction, have their origin in electrochemical processes, which are foreign to the Maxwell theory; only the magnetic fields proceeding from the electromagnet are described by the latter. In similar manner, the fields proceeding from permanent magnets fit into the framework of the Maxwell theory.

We commence with the *magnetization*, which we shall call $\mathbf{M}$ or, to begin with, $\mathbf{M}^*$, as counterpart to the electric polarization $\mathbf{P}$; the analog of our equation of definition (11.3) for $\mathbf{P}$,

$$\mathbf{D} = \varepsilon_0 \, \mathbf{E} + \mathbf{P}, \tag{1}$$

would be, from our point of view,

$$\mathbf{H} = \frac{1}{\mu_0} \mathbf{B} + \mathbf{M}^*. \tag{1a}$$

(Here $\mathbf{H}$ and $\mathbf{M}^*$ are "quantities" like $\mathbf{D}$ and $\mathbf{P}$, $\mathbf{B}$ is an "intensity" like $\mathbf{E}$; $1/\mu_0$ corresponds to ε_0, as emphasized in Eq. (4.7)). Solved for $\mathbf{B}$, Eq. (1a) yields

$$\mathbf{B} = \mu_0(\mathbf{H} - \mathbf{M}^*). \tag{1b}$$

The customary definition of magnetization is, on the other hand, contained in the equation

$$\mathbf{B} = \mu_0(\mathbf{H} + \mathbf{M}), \tag{2}$$

which we shall utilize from here on. We shall return to Eq. (1b) in §13D, in discussing diamagnetism. As defined by (2), $\mathbf{M}$ signifies the *part of the excitation derived from matter* and is at the same time *the sum*, referred to unit volume, *of the moments of elementary magnets*, just as $\mathbf{P}$ was a corresponding sum of electrical elementary moments. In the following we shall imagine the distribution of $\mathbf{M}$ within the magnet to be given arbitrarily and shall calculate from the Maxwell equations the corresponding fields of the vectors $\mathbf{B}$ and $\mathbf{H}$.

We know from §7 that $\mathbf{H}$ is throughout *lamellar*, $\mathbf{B}$ throughout *solenoidal* (free of sources). In view of the absence of sources of $\mathbf{B}$ Eq. (2) leads to the condition

$$\operatorname{div} \mathbf{H} = -\operatorname{div} \mathbf{M}. \tag{3}$$

At a surface of discontinuity of $\mathbf{M}$ the Eq. $\operatorname{div} \mathbf{B} = 0$, which presupposes that $\mathbf{B}$ is continuous and differentiable, is replaced by the condition that the "surface divergence" of $\mathbf{B}$ vanishes; i.e.

$$B_n + B_{n'} = 0, \tag{3a}$$

where n and n', as in (3.7), denote the normals of the surface of discontinuity pointing toward opposite sides. Applied to the surface of a magnet, at which $\mathbf{M}$ jumps from the external value $\mathbf{M} = 0$ to a value of $\mathbf{M}$ which, in general, differs from zero, (3a) yields in view of (2)

$$H_n + H_{n'} = -M_n. \tag{3b}$$

If, now, $\mathbf{H} = -\operatorname{grad} \Psi$ is substituted in Eqs. (3) and (3b), the differential equation of the problem

$$\Delta \Psi = \operatorname{div} \mathbf{M}, \tag{4}$$

and the surface condition

$$\frac{\partial \Psi}{\partial n} + \frac{\partial \Psi}{\partial n'} = M_n \tag{4a}$$

are obtained. We add, as boundary condition at infinity

$$\Psi = 0. \tag{4b}$$

Since, in the two Eqs. (4) and (4a), the right sides can be assumed to be known, we are dealing here not with a boundary value problem, but, in the sense of §7, Eqs. (10) and (10a), with a simple summation problem. The solution is

$$4\pi\Psi = -\int \frac{\operatorname{div} \mathbf{M}}{r} \, d\tau - \int \frac{M_n}{r} \, d\sigma. \tag{5}$$

The first term sums all magnetic volume densities ρ_m in the *interior* of the magnet, the second all surface densities ω_m on its *boundary*. The negative signs result from the fact that, according to (4) and (4a)

$$\rho_m = -\operatorname{div} \mathbf{M}, \qquad \omega_m = -M_n.$$

(If we had continued to employ $\mathbf{M}^* = -\mathbf{M}$ in our calculation, the signs would have been positive, as in electrostatics.) We consider two special cases: *a.* homogeneous magnetization parallel to the bar axis, *b.* magnetization increasing from zero toward the center, also parallel to the bar axis.

For a the first integral on the right in Eq. (5) vanishes because of div $\mathbf{M} = 0$, for b, the second one because of $M_n = 0$.

a. Only the two end surfaces contribute to the surface integral in (5) since the normal component of $\mathbf{M}$ vanishes, by assumption, on the sides. The pole strengths of the magnet are thus, in a sense, uniformly distributed over the end surfaces; if the cross-section area is F the two total pole strengths are $\pm P = \pm FM$.

Approximate integration of (5) for *large* distance of the reference point from the magnet (see Fig. 14: p, r_1, r_2 distance of reference point from bar axis and from the centers of the end surfaces, respectively, z, coordinate

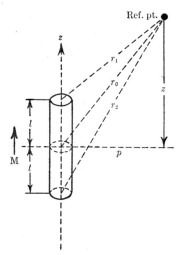

FIG. 14. Bar magnet, magnetized longitudinally. Point of reference on the outside.

of the reference point parallel to the bar axis, measured from its center, $2l$, length of bar) leads to

$$4\pi\Psi = -MF\left(\frac{1}{r_1} - \frac{1}{r_2}\right) \text{ with } \begin{array}{l} r_1^2 = (z - l)^2 + p^2 \\ r_2^2 = (z + l)^2 + p^2 \end{array}. \tag{6}$$

Series expansion yields

$$\frac{1}{r_1} = \frac{1}{r_0}\left(1 + \frac{1}{2}\frac{2zl + l^2}{r_0^2} + \cdots\right),$$

$$\frac{1}{r_2} = \frac{1}{r_0}\left(1 - \frac{1}{2}\frac{2zl + l^2}{r_0^2} + \cdots\right),$$

with $r_0 = \sqrt{z^2 + p^2}$ = distance of reference point from center of bar. Hence, according to (6),

$$4\pi\Psi = -2lP\frac{z}{r_0^3} = 2lP\frac{\partial}{\partial z}\frac{1}{r_0}. \tag{6a}$$

As was to be expected, the external action of the magnet for distances $\gg 2l$ is that of a dipole with the lever arm $2l$ and pole strength $P = MF$.

The same representation by the surface integral in Eq. (5) applies for the interior of the bar, but this requires a more careful evaluation. For the sake of brevity we limit ourselves to the center line of the bar and

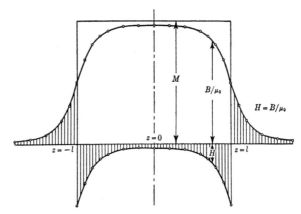

FIG. 15. Demagnetization of a uniformly magnetized bar. In the drawing it is assumed that $a = l/4$.

assume the cross section to be circular (a = radius, ρ = distance of the point of integration from the center of the end surfaces). We then find

$$4\pi\Psi = 2\pi M \left\{ \int_0^a \frac{\rho \, d\rho}{r_1} - \int_0^a \frac{\rho \, d\rho}{r_2} \right\}, \tag{7}$$

$$r_1^2 = (l - z)^2 + \rho^2, \qquad r_2^2 = (l + z)^2 + \rho^2.$$

The evaluation yields, since $0 \leqq |z| \leqq l$:

$$\Psi = \frac{M}{2} [\{(l - z)^2 + a^2\}^{\frac{1}{2}} - \{(l + z)^2 + a^2\}^{\frac{1}{2}} + 2z], \tag{7a}$$

$$H = -\frac{\partial\Psi}{\partial z} = \frac{M}{2} \left[\frac{l - z}{\{(l - z)^2 + a^2\}^{\frac{1}{2}}} + \frac{l + z}{\{(l + z)^2 + a^2\}^{\frac{1}{2}}} - 2 \right]. \tag{7b}$$

Since l is in any case many times larger than a, we obtain

$$H \sim 0, \qquad \frac{\partial H}{\partial z} \sim 0 \quad \text{for} \quad z = 0$$

$$H = -\frac{M}{2}, \qquad \frac{\partial H}{\partial z} = \mp \frac{M}{2} \frac{1}{a} \quad \text{for} \quad z = \pm l;$$

as shown in Fig. 15, there is a sharp decrease of $-H$ at the two ends of the bar and a vanishing of a high order at the center of the bar. The sign of $\mathbf{H}$

is opposite to that of $\mathbf{M}$; $\mathbf{H}$ has what is commonly expressed as a "demagnetizing action." This is seen also in the pattern of $\mathbf{B}$. Though $\mathbf{B}$ attains almost the full magnitude $\mu_0\mathbf{M}$ at the center of the bar, it is only half as large at the ends, the same of course, inside and outside of the bar.

b. Let $\mathbf{M}$ be constant in every cross section of the bar, but be dependent on z in such fashion that $\mathbf{M}$ vanishes at the ends $z = \pm l$ and increases parabolically toward the center:

$$M = \frac{C}{2}\left(1 - \frac{z^2}{l^2}\right), \qquad \operatorname{div} \mathbf{M} = -\frac{Cz}{l^2}. \tag{8}$$

As already mentioned, the surface integral in (5) vanishes here, and only the volume integral remains to be calculated.

With ζ, ρ, ψ for the coordinates of the point of integration, and z, p, φ for the coordinates of the reference point Eq. (5) yields:

$$4\pi\Psi = \frac{C}{l^2}\iiint \frac{\zeta}{r}\, d\zeta\, d\sigma, \tag{9}$$

$$r^2 = (z - \zeta)^2 + p^2 + \rho^2 - 2p\rho\cos(\varphi - \psi), \qquad d\sigma = \rho\, d\rho\, d\psi.$$

The field at a distance here, just as in case a, is that of a dipole. If the average magnetization $\bar{M}$ is computed for the length of the bar and if as in a, we put $\bar{M}F = P$, the moment of the dipole field becomes, as in (6a), $2lP$.

Within the bar, in particular on the bar axis $p = 0$, we obtain from (9), by carrying out the integrations with respect to ρ and ψ

$$\Psi = \frac{C}{2l^2}\int_{-l}^{+l}\zeta\, d\zeta[\{(z - \zeta)^2 + a^2\}^{\frac{1}{2}} - |z - \zeta|]$$

$$= \frac{C}{2l^2}\int_{-l}^{+l}\zeta\, d\zeta\{(z - \zeta)^2 + a^2\}^{\frac{1}{2}} - \frac{C}{2l^2}\left(\frac{1}{3}z^3 - zl^2\right) \tag{10}$$

and

$$H = -\frac{\partial\Psi}{\partial z} = \frac{C}{2l^2}\int_{-l}^{+l}\zeta\, d\zeta\, \frac{\partial}{\partial\zeta}\{(z - \zeta)^2 + a^2\}^{\frac{1}{2}} + \frac{C}{2l^2}(z^2 - l^2)$$

$$= \frac{C}{2l}\left[\{(z - l)^2 + a^2\}^{\frac{1}{2}} + \{(z + l)^2 + a^2\}^{\frac{1}{2}} + \frac{z^2 - l^2}{l}\right] \tag{11}$$

$$- \frac{C}{2l^2}\int_{-l}^{+l}|(z - \zeta)^2 + a^2|^{\frac{1}{2}}\, d\zeta.$$

The symbol $|\ |$ in the last integral indicates that the sign of this square root, just as that of the preceding ones, is to be positive.

We will show that, just as in the case of uniform magnetization, $\mathbf{H}$ is *nearly zero* (of the order a/l) everywhere except at the ends of the bar. Exclusion of the bar ends signifies $a \ll l - |z|$. We may then neglect a in (11) and obtain

$$H = \frac{C}{2l}\left[l - z + l + z + \frac{z^2 - l^2}{l} - \frac{z^2 + l^2}{l} \right],\tag{11a}$$

i.e. zero (more exactly, vanishing to the order Ca/l). For $z = \pm l$, i.e. at the bar ends, H is small to the same order of magnitude. This does not apply, however, to dH/dz. This is given by[1]

$$\frac{dH}{dz} = \frac{C}{2l}\left(1 + \cdots\right).\tag{12}$$

Thus H has *positive* values, appreciably differing from zero, only in a small region near the two ends of the bar. This is illustrated by Fig. 16.

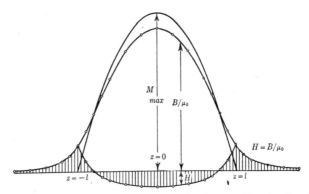

Fig. 16. Demagnetization of a bar magnetized according to the formula $M = \frac{C}{2}\left(1 - \frac{z^2}{l^2}\right)$. As in Fig. 15, $a = l/4$.

Also here $\mathbf{H}$ has a "demagnetizing action," i.e. is opposite to the impressed moment $\mathbf{M}$. B/μ_0 approximates the full value of $\mathbf{M}$ in the whole middle portion of the bar and deviates from it slightly only at the ends.

The fact that $\mathbf{H}$ always acts in the direction of demagnetization (for arbitrary distribution of the magnetization and for any shape of the magnet) may be recognized from the following: The lines of force ($\mathbf{B}$-lines) are closed

[1] Differentiation of (11) leads to the exact elementary formula

$$\frac{dH}{dz} = \frac{C}{2l^2}\left\{ \frac{z(z - l) + a^2}{\sqrt{(z - l)^2 + a^2}} - \frac{z(z + l) + a^2}{\sqrt{(z + l)^2 + a^2}} + 2z \right\}$$

from which (12) is readily derived. The dots in (12) indicate terms of the order a/l. The upper positive sign in (12) refers to $z = +l$ and positive dz; the gradient of H toward the *interior* of the bar (*negative* dz) is thus *negative*, just as at the other end of the bar.

because of div $\mathbf{B} = 0$ and part of their path lies in the interior, and part lies in the region outside of the magnet. We carry out a line integral of $\mathbf{H}$ over such a closed line of force in the positive $\mathbf{B}$-direction. Then we obtain, because of the lamellar character of $\mathbf{H}$, as for any closed path,

$$\oint H_s \, ds = 0. \tag{12a}$$

The part of the integral over the path *outside* of the magnet, where the directions of $\mathbf{H}$ and $\mathbf{B}$ coincide, is *positive*; hence the part of the integral over the path *within* the magnet must be negative:

$$\int_{\text{inside}} H_s \, ds < 0 \tag{12b}$$

On the other hand, the integral of $\mathbf{B}$ over the same part is, by assumption, positive. Eq. (2) shows that this applies even more to the integral of $\mathbf{M}$:

$$\int_{\text{inside}} M_s \, ds = \frac{1}{\mu_0} \int_{\text{inside}} B_s \, ds - \int_{\text{inside}} H_s \, ds > 0. \tag{12c}$$

The two inequalities (12b, c) for $\mathbf{H}$ and $\mathbf{M}$ show together that $\mathbf{H}$ has within the magnet, along any line of force, on the *average* the *opposite* direction of $\mathbf{M}$.

We have dealt with the preceding rather arbitrarily selected problem in such detail because most textbooks contain little of a quantitative nature regarding the vectors $\mathbf{B}$, $\mathbf{H}$, and $\mathbf{M}$ in the interior of a magnet. We have seen that if $\mathbf{M}$ is known $\mathbf{H}$ may in principle be evaluated, in accord with the rules of potential theory, by a simple summation, whereupon $\mathbf{B}$ is also known. It is true that the assumption of a known distribution of $\mathbf{M}$ is not fulfilled in practice. Bar magnets are therefore unsuited for a practical study of ferromagnetism: we shall return to this later.

Although our calculation was limited to the center line of the bar magnet, Figs. 17 and 18 give quantitative information[1] regarding the shape of the lines of force and lines of excitation throughout the interior of a uniformly magnetized bar magnet: The $\mathbf{B}$-lines are drawn into the interior, the $\mathbf{H}$-lines pushed out of it. On the outside the two sets of lines coincide, of course, since $\mathbf{B} = \mu_0 \mathbf{H}$.

The *ring magnet*, provided with a narrow gap, is both simpler and of greater practical importance than the bar magnet. Because of the equivalence of all cross sections the magnetization may here be regarded as uniform, so that div $\mathbf{M} = 0$, and $\mathbf{M}$ is everywhere parallel to the center line.

---

[1] These figures were kindly prepared by Prof. J. Jaumann, by a graphical method which was developed by Maxwell for the numerous line-of-force patterns at the end of his Treatise and which is widely employed by electrical engineers; see also art. 123 of the Treatise.

A "magnetic coating" exists only on the gap faces; between them a magnetic field is formed which is similar in geometry to the electric field in a plate condenser. This applies not only to a permanent ring magnet, but almost identically to the ring-shaped electromagnet with an iron core.

Referring once more to Eq. (12a), we consider the line integral of the excitation **H** carried out over the center line of the ring. It may be divided into two parts, the short section through the air gap of thickness a, which we shall traverse in the positive direction of the magnetic condenser field

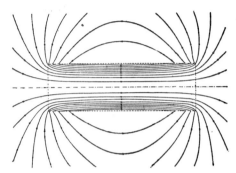

FIG. 17. Lines of force of a uniformly magnetized bar magnet; they are drawn into the interior.

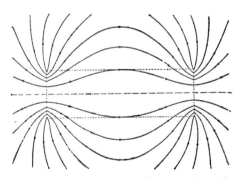

FIG. 18. Lines of excitation of a uniformly magnetized bar magnet; they are pushed out of the interior.

which exists there, and the long part through the ring-shaped iron core of length l, on which the integration is to be carried out in the same sense. We put $H = H_G$ in the gap and $H = H_I$ in the iron core, respectively. Eq. (12a) yields

$$l \cdot H_I = - a \cdot H_G .$$

Thus a "demagnetization" of the iron core goes with the magnetization of the air gap. We have here the same state of affairs as in Figs. 15 and 16, only in a much simpler and more obvious form.

We return once more to the bar magnet and to the definition of its pole strength P. From our point of view it is a *quantity* of magnitude

$$P = \oint H_n \, d\sigma, \tag{13}$$

where the closed surface σ envelops, starting from the center of the bar, the one or the other half of the bar in arbitrary manner. This definition of P corresponds to the definition (7.2) of the magnetic volume density $\rho_m = \operatorname{div} \mathbf{H}$ and states that P is equal to the sum of all magnetic quantities $\rho_m d\tau$ which are present in the half of the bar in question.

In contrast to this the pole strength is often defined as *intensity* in the literature and described in terms of the magnetic flux. We shall call the pole strength so defined $\bar{P}$ and write

$$\bar{P} = \int B_n \, d\sigma \tag{14}$$

The surface σ cannot now be closed since otherwise, in view of $\operatorname{div} \mathbf{B} = 0$, $\bar{P} = 0$. Rather, the cross section q passing through the center of the bar must be *excluded* from the integration; or, as an alternative, the integration is carried out *only* over this cross section with reversed sign of the normal n:

$$\bar{P} = \int B_n \, dq. \tag{14a}$$

It may readily be shown with the aid of (13) and (14) that then, very nearly,

$$\bar{P} = \mu_0 P \tag{15}$$

where μ_0 is the permeability of the surroundings.[1] In fact, we convinced ourselves above that $\mathbf{H}$ very nearly vanishes in the central cross section, so that the closed integration in (13) may be replaced by the open integration in (14), where we may put $H_n = B_n/\mu_0$. The definitions (13) and (14) would thus be practically equivalent in the presence of a plane of symmetry (which, incidentally, exists also for the horseshoe magnet). However, for asymmetric shape or asymmetric magnetization relation (14) fails and only definition (13) remains meaningful. It is also recommended by the fact that it exactly corresponds to the definition of charge:

$$e = \oint D_n \, d\sigma.$$

---

[1] While P, according to (13), may be regarded as an *internal* property of the magnet, $\bar{P}$ also depends, according to (15), on its surroundings. This evidently arises from the fact that an environment differing from vacuum contains itself magnetic moments which, quite understandably, are included in $\bar{P}$. See in this respect Phys. Z. 1935, p. 424.

§*13. General Considerations on Magnetostatics and Its Boundary-Value Problems*

While, in §12, we have considered only the proper field of permanent magnets, we shall now approach the behavior of *arbitrary bodies* in an *outside field*, arising from either permanent magnets or electromagnets. The laws which apply here closely parallel those of electrostatics; however, contrary to our classification of the field vectors into entities of quantity and intensity, **H** here corresponds to **E**, **B** to **D**. This follows from the familiar fundamental equations, e.g. (7.2), (7.9), and (7.9a)

$$\text{curl } \mathbf{H} = 0, \qquad \mathbf{H} = - \text{ grad } \Psi, \tag{1}$$

$$\text{div } \mathbf{B} = 0, \qquad \text{div } \mathbf{H} = \rho_m \tag{2}$$

and the boundary conditions at an interface:

$$\text{continuity of the tang. comp. of } \mathbf{H} \text{ and the normal comp. of } \mathbf{B}. \tag{3}$$

The line integral of the magnetic excitation, which, by (1), is independent of the path, i.e. depends only on the endpoints A, B, we shall call magneto-motive force and designate by U_{AB}, in analogy to V_{AB} in §2, U_{AB} bears following relation to the magnetic potential:

$$U_{AB} = \int_A^B H_s \, ds = \Psi_A - \Psi_B. \tag{4}$$

The "loop magnetomotive force" is given by

$$U_{A=B} = \oint H_s \, ds = 0, \tag{5}$$

irrespective of the manner in which the closed path is traversed, whether it passes through magnetized material or through air.

We designate the "induction" or the "flux" through an arbitrarily shaped surface σ by the usual symbol Φ:

$$\Phi = \int B_n \, d\sigma. \tag{6}$$

In view of (2), Φ depends only on the boundary curve of the surface σ, i.e. is identical for all surfaces passing through the same boundary curve. For a closed surface it vanishes, of course:

$$\Phi = \oint B_n \, d\sigma = 0. \tag{6a}$$

We shall now discuss briefly the magnetic analogs of the topics dealt with in §§11, 10, and 9.

A. The Law of Refraction of the Lines of Magnetic Excitation

The law of refraction (11.2) of the magnetic lines of *force* may be carried over to the lines of magnetic *excitation*, but applies also to the magnetic lines of force because of the identical direction of **H** and **B**. If the angles α_1 and α_2 have the same meaning as before,

$$\frac{\tan \alpha_1}{\mu_1} = \frac{\tan \alpha_2}{\mu_2}. \tag{7}$$

We prefer to speak here of the lines of force because they, unlike the lines of excitation, are individually continued into the second medium. We may therefore say: Every individual **B**-line is refracted away from the normal in entering the more permeable medium (e.g. $\mu_2 > \mu_1$).

B. Definition of the Vectors **H** and **B**, Particularly in Solid Bodies

To measure the component of **H** in a given direction at a given point by means of an imaginary experiment a short narrow tube must be drilled, to measure **B**, a thin slit must be cut. Within the cavity (vacuum or filled with air) so prepared a deflection experiment can be carried out and **B** or $\mathbf{H} = \mathbf{B}/\mu_0$, respectively, be determined from the force, so measured, acting on a test body.[1]

C. The Magnetization **M** in Any Non-Ferromagnetic Substance

As in Eq. (12.2) we define **M** by

$$\mathbf{B} = \mu_0(\mathbf{H} + \mathbf{M}) \tag{8}$$

and set

$$\mathbf{M} = \kappa\mathbf{H}. \tag{8a}$$

κ is a material constant, the *magnetic susceptibility* of the substance. Physically more significant is the *molar susceptibility*

$$\chi = \kappa M/\rho, \tag{8b}$$

where M is the mass of a mole, the so-called molecular weight of the substance, and ρ is its density. (As we know, the proportionality between **M** and **H** does not in general apply for ferromagnetic materials). $\mathbf{B} = \mu\mathbf{H}$ and Eqs. (8) and (8a) lead to

$$\frac{\mu}{\mu_0} = 1 + \kappa, \qquad \kappa = \frac{\mu - \mu_0}{\mu_0}. \tag{9}$$

---

[1] A magnet needle, a wire traversed by current, or, eventually, a bismuth spiral; from our standpoint such an experiment yields the intensity **B** directly; from it is determined, in the tube experiment, the proportional quantity **H**. For details see also §11B.

The conventional form of Eqs. (8) and (9) is marred by the appearance of 4π as factor of $\mathbf{M}$ and κ (see our corresponding remark regarding $\mathbf{P}$ and η in connection with Eq. (11.3)). We must hence note that in the use of experimental data employing conventional notation the factor 4π must be added. That is indicated in the small table in the following section D.

D. Dia- and Paramagnetism

Diamagnetism corresponds to *dielectricity*; like the latter it is a general property of ponderable matter and *independent of temperature*. Both owe their origin to the electronic (and nuclear) structure of matter. *Paramagnetism* occurs only for magnetically polar molecules, i.e. molecules which have a magnetic moment of their own. (Electrically polar molecules were discussed on p. 73.) Paramagnetism is temperature-dependent and hence has a statistical origin. This phenomenon, like ferromagnetism, lies outside of Maxwell's theory. The paramagnetic susceptibility obeys the *law of Curie and Langevin*:

$$\chi = \frac{C}{T}, \qquad \begin{aligned} C &= \text{Curie constant} \\ T &= \text{absolute temperature} \end{aligned} \qquad (10)$$

This dependence on the temperature indicates that increasing thermal agitation interferes with the alignment of the magnetic moments in the field direction, decreasing thermal agitation favors it.

In the diamagnetic case we have

$$\mu < \mu_0, \qquad \kappa < 0 \qquad (11)$$

This apparent difference from the dielectric case

$$\varepsilon > \varepsilon_0, \qquad \eta > 0 \qquad (11a)$$

is explained in the manner already indicated following Eq. (4.7): the true analog of ε is not μ, but $1/\mu$. The magnetic parallel of the statement $\varepsilon > \varepsilon_0$ is thus

$$\frac{1}{\mu} > \frac{1}{\mu_0}, \qquad \mu < \mu_0, \qquad \kappa < 0$$

corresponding to (11). The negative sign of $\mathbf{M}^*$ in Eq. (12.1b) is also related to this. For, if we put $\mathbf{M}^* = \kappa^* \, \mathbf{H}$, the susceptibility κ^* so defined becomes, in view of $\mathbf{M}^* = -\mathbf{M} = -\kappa\mathbf{H}$,

$$\kappa^* = -\kappa > 0$$

in the diamagnetic case corresponding to $\eta > 0$ in (11a). The introduction of $\mathbf{M}^*$, which was suggested previously but immediately given up above, thus corresponds in fact to the inner relationship of dielectricity and diamagnetism.

For paramagnetism, we have, in contrast with (11),

$$\mu > \mu_0 , \qquad \kappa > 0. \tag{11b}$$

The numerical value of κ is very small for both paramagnetic and diamagnetic materials. The following represent extreme values:

Paramagnetism	Diamagnetism
$\kappa = + 4\pi \cdot 1.8 \cdot 10^{-6}$ for O_2	$\kappa = - 4\pi \cdot 0.007 \cdot 10^{-6}$ for N_2
$\kappa = + 4\pi \cdot 782 \cdot 10^{-6}$ for Pd	$\kappa = - 4\pi \cdot 160 \cdot 10^{-6}$ for Bi

The paramagnetic values refer here to 18°C.

The Clausius-Mossotti law, Eq. (11.8), with ε replaced by μ, gives the dependence on density.

E. Soft Iron as Analog to the Electric Conductor

With certain restrictions soft iron may be classified with the paramagnetic substances. The initial value of the permeability (referred to the value for vacuum) is several thousandfold, according to the variety of iron; for increasing $\mathbf{H}$, $\mathbf{B}$ approaches a saturation value in the neighborhood of 21,000 gauss. The equation $\mathbf{B} = \mu \mathbf{H}$ must hence be replaced by the functional relationship $\mathbf{B} = \mathbf{B}(\mathbf{H})$, as was mentioned already on p. 21.

Just as we noted on p. 61, that the electric conductor corresponds, with respect to the electrostatic boundary conditions and boundary-value problems, to the limiting case $\varepsilon \to \infty$ of a dielectric, so we may regard soft iron ($\mu \to \infty$) as the magnetostatic analog of the electric conductor. Eq. (7) shows, in fact, that the magnetic lines of force are perpendicular to the surface of soft iron ($\alpha_1 \to 0$ follows from $\mu_2 \to \infty$). If two such pieces of soft iron Fe_1 and Fe_2 are placed at different magnetic potentials (e.g. if they are placed on the poles of a horseshoe magnet), the $\mathbf{H}$-lines between them are similar to the $\mathbf{E}$-lines in Fig. 11. Thus, in a sense, a magnetic condenser is obtained.

In engineering applications it is also convenient to introduce the concept of "magnetic resistance" and to employ a "magnetic Ohm's law."

F. Specific Boundary-Value Problems

The methods of solution developed in §9 may be taken over directly into magnetostatics; an example is the imaging at a plane, where it makes no difference whether the induction in halfspace 2 in Fig. 10 is produced by a single pole or a dipole in halfspace 1. The same remark applies for the method of reciprocal radii as applied to a sphere (previously regarded as a conductor, here as consisting of soft iron).

We refer in particular to the sphere in a uniform magnetic field. Within there is a uniform field F_2, while outside the field becomes nonuniform, through the superposition on the original field F of the field of a virtual magnetic moment M at the center of the sphere, with its axis in the field direction. ("Field" here denotes "excitation field".) The values of F_2 and M are, by (9.14),

$$F_2 = \frac{3}{\mu + 2} F, \qquad M = -\frac{\mu - 1}{\mu + 2} a^3 \cdot 4\pi\mu_0 F. \qquad (12)$$

Here a = radius of sphere, $\mu = \mu_2/\mu_0$ = relative permeability of the sphere referred to its surroundings (air). F_2 is stronger than F for diamagnetic substances, weaker for paramagnetic materials. In the interior of soft iron $\mathbf{H} \cong 0$, just as in the electric conductor.

G. The Uniform Field within an Ellipsoid of Revolution

The solution (9.13) for the sphere, translated into magnetic terms, becomes, after substitution of the values (12)

$$\Psi_1 = -F\left(r - \frac{\mu - 1}{\mu + 2}\frac{a^3}{r^2}\right)\cos\theta, \qquad \Psi_2 = -\frac{3}{\mu + 2} Fr \cos\theta. \qquad (13)$$

This solution satisfied the boundary conditions

$$\Psi_1 = \Psi_2, \qquad \frac{\partial\Psi_1}{\partial r} = \mu\frac{\partial\Psi_2}{\partial r} \qquad (13a)$$

because the factor $\cos\theta$, which varies over the sphere, factors out of these equations. We will show that a solution of the same form is valid for the ellipsoid.

In passing from the sphere to the ellipsoid we must first replace the spherical polar coordinates r, θ by corresponding elliptical coordinates, which we shall call u, v. We proceed here from the well-known parametric representation of the ellipse (instructions for Problem II.1), in which we write for the principal axes a and b

$$a = c \cosh u, \qquad b = c \sinh u,$$
$$c = \sqrt{a^2 - b^2} = \text{independent of } u. \qquad (14)$$

Rotation about the long axis (z-axis, angle of rotation φ) produces a family of elongated confocal ellipsoids of revolution, corresponding to Eq. (19.17a) in Vol. II:

$$\frac{z^2}{c^2 \cosh^2 u} + \frac{x^2 + y^2}{c^2 \sinh^2 u} = 1. \qquad (14a)$$

Let the ellipsoid considered by us be one of these, namely that with the parameter $u = u_0$. The relation between x, y, z and the elliptical coordinates u, v, φ is the following:

$$z = c \cosh u \cos v,$$

$$x = c \sinh u \sin v \cos \varphi, \tag{15}$$

$$y = c \sinh u \sin v \sin \varphi.$$

The expression for the line element ds consequently is given by

$$\frac{ds^2}{c^2} = (\cosh^2 u - \cos^2 v)(du^2 + dv^2) + \sinh^2 u \sin^2 v \, d\varphi^2. \tag{15a}$$

According to the rule (3.9b) in Vol. II the potential equation in these coordinates, becomes:

$$\frac{\partial}{\partial u}\left(\sinh u \sin v \frac{\partial \Psi}{\partial u}\right) + \frac{\partial}{\partial v}\left(\sinh u \sin v \frac{\partial \Psi}{\partial v}\right) = 0 \tag{16}$$

if the potential is independent of the cyclic coordinate φ.
One solution is the uniform field, parallel to the major axis,

$$\Psi = \frac{z}{c} = \cosh u \cos v. \tag{17}$$

We seek a second solution of such form that the factor $\cos v$ which varies over the surface of the ellipsoid corresponding to the factor $\cos \theta$ giving geographic latitude on the sphere, is cancelled out in the boundary conditions. We write for this second solution

$$\Psi = f(u) \cos v \tag{17a}$$

and obtain from (16) for the differential equation

$$\frac{d}{du}(\sinh u \, f'(u)) - 2 \sinh u \, f(u) = 0 \tag{17b}$$

which is evidently satisfied by the uniform field $f = \cosh u$. Following a general rule,[1] we place the desired second solution equal to the product of the known first solution and an unknown function $U(u)$:

$$f(u) = \cosh u \, U(u). \tag{18}$$

The resulting differential equation for $U(u)$

$$U'' + \frac{3 \sinh^2 u + 1}{\sinh u \cosh u} U' = 0$$

can be integrated directly and yields, with A and B as integration constants,

---

[1] It corresponds to the method of solving an algebraic equation with one known root.

$$U'(u) = \frac{A}{\sinh u \cosh^2 u}, \qquad U(u) = \frac{A}{2} \log \frac{\cosh u - 1}{\cosh u + 1} + \frac{A}{\cosh u} + B,$$

$$f(u) = A \left(1 + \frac{\cosh u}{2} \log \frac{\cosh u - 1}{\cosh u + 1} \right). \qquad (18a)$$

In the last formula we have omitted the term multiplied with B, since this corresponds to our known solution representing a uniform field.

We now complete our expression (17) for the external field by the addition of (18a) and retain the uniform-field expression (17) for the internal field:

$$\Psi_1 = F \cosh u \cos v + A \left(1 + \frac{\cosh u}{2} \log \frac{\cosh u - 1}{\cosh u + 1} \right) \cos v, \qquad (19)$$

$$\Psi_2 = F_2 \cosh u \cos v.$$

We regard the constant F of the external field, which has been added, as known; the two constants A and F_2 are to be determined from the boundary conditions. These are identical with (13a), with dr replaced by du (more exactly, by ds_u, the line element in the direction of the normal to the ellipsoid,

$$ds_u = c\sqrt{\cosh^2 u - \cos^2 v} \, du \, ,$$

where, however, the square root cancels out in the second Eq. (13a)). The boundary conditions demand hence for $u = u_0$

$$F + A \left(\frac{1}{\cosh u_0} + \frac{1}{2} \log \frac{\cosh u_0 - 1}{\cosh u_0 + 1} \right) = F_2, \qquad (19a)$$

$$F + A \left(\frac{\cosh u_0}{\sinh^2 u_0} + \frac{1}{2} \log \frac{\cosh u_0 - 1}{\cosh u_0 + 1} \right) = \mu F_2. \qquad (19b)$$

Subtraction yields the somewhat simpler relation

$$A = \cosh u_0 \sinh^2 u_0 (\mu - 1) F_2$$

and substitution thereof in (19a)

$$F_2 \left\{ 1 - (\mu - 1) \sinh^2 u_0 \left(1 + \frac{1}{2} \cosh u_0 \log \frac{\cosh u_0 - 1}{\cosh u_0 + 1} \right) \right\} = F. \qquad (20)$$

The field strength F_2 inside is thus expressed in terms of the known strength of the original external, homogeneous field.

If we utilize Eqs. (14) and denote by a, b, c the principal axes and focal distance from the center of our ellipsoid $u = u_0$, we may write instead of (20)

$$F_2 \left\{ 1 - (\mu - 1) \frac{b^2}{c^2} \left(1 + \frac{1}{2} \frac{a}{c} \log \frac{a - c}{a + c} \right) \right\} = F \qquad (20a)$$

or, in terms of the numerical eccentricity $\varepsilon = c/a$ and the magnetic susceptibility $\kappa = \mu - 1$:

$$F_2 \left\{ 1 + \kappa \frac{1 - \varepsilon^2}{\varepsilon^3} \left(\frac{1}{2} \log \frac{1 + \varepsilon}{1 - \varepsilon} - \varepsilon \right) \right\} = F. \tag{21}$$

We note here that:

a. for $\kappa = 0$ we have of course

$$F_2 = F; \tag{21a}$$

b. for $\varepsilon \to 0$ series expansion of (21) leads to

$$F_2 = F \bigg/ \left(1 + \frac{\kappa}{3} \right). \tag{21b}$$

The field inside is *weaker* than outside for paramagnetic bodies, stronger for diamagnetic bodies.

c. The same applies for $\varepsilon \to 1$. If we put $\eta = 1 - \varepsilon$, (21) yields

$$F_2 = F \bigg/ \left\{ 1 + \kappa\eta \left(\log \frac{2}{\eta} - 2 \right) \right\}. \tag{21c}$$

In the limiting case b the ellipsoid becomes nearly *spherical*; in fact, (21b) is identical with Eq. (12) for the sphere. In the limiting case c the ellipsoid degenerates to a *thin rod*.

The problem treated here is usually related to a famous formula of Dirichlet for the gravitational potential of a triaxial ellipsoid uniformly filled with matter. This procedure is mathematically more elegant than ours, but is rather indirect. We have preferred the direct method of the magnetic boundary-value problem because it appears to give us more profound insight into the physical conditions.

Our solution of the magnetic problem is of course transferable without change to the corresponding electrostatic problem.

H. The So-Called Demagnetization Factor

The ellipsoid and its limiting forms (sphere, rod) is the *standard shape* of the magnetic test body because it alone possesses a uniform and easily calculable internal field when introduced into an originally uniform external field. For other shapes the determination of the internal field leads to a practically insoluble *boundary-value problem*; the internal field is by no means uniform, but varies from point to point.

It is clear, however, that all questions concerned with the magnetic properties of the material depend on the internal field F_i. This field interacts with the molecular components of the material directly, while the *external* field F has no direct effect on them. Accordingly we may express our Eq. (8a) more precisely by

$$M = \kappa F_i. \tag{22}$$

Such questions become particularly important with *ferromagnetic* materials, where the differences between the external and the internal field (F and F_i in our present notation) are extremely large; for para- and diamagnetic materials they are negligible because of the smallness of κ. Since for ferromagnetic, as for paramagnetic, materials $F_i < F$, we write, with P denoting a numerical factor,

$$F_i = F - PM \tag{22a}$$

or, in view of (22),

$$F_i = F - \kappa P F_i, \tag{22b}$$

$$F_i(1 + \kappa P) = F. \tag{22c}$$

The numerical factor P is a measure for the attenuation of the external field by the presence of the magnetizable material and is hence called the *demagnetization factor*. A comparison of (22c) and (21) yields for its value

$$P = \frac{1 - \varepsilon^2}{\varepsilon^3} \left(\frac{1}{2} \log \frac{1 + \varepsilon}{1 - \varepsilon} - \varepsilon \right). \tag{23}$$

It is purely geometric in character. For the two limiting cases of (21b) and (21c) we have

$$P = 1/3 \text{ for } \varepsilon \to 0, \text{ sphere,}$$

$$P = \eta \left(\log \frac{2}{\eta} - 2 \right) \to 0 \quad \text{for} \quad \eta \to 0, \text{ rod.}$$

Intermediate values are readily calculated from (23) (log always signifies in this book the natural logarithm) and are tabulated for example, by Kohlrausch.[1]

It is clear that this factor P has a legitimate meaning only for the ellipsoid and its degenerate forms, since only here we are dealing with the ratio of one uniform field to another. Clearly the boundary-value problem for other body-shapes cannot be circumvented by the employment of a numerical factor which is guessed in some manner. In practice the procedure consists of measuring the value of **B** experimentally with an induction coil at some characteristic points (e.g. the center of the test body).

§14. Some Remarks on Ferromagnetism

This section does not pretend to be an introduction to the broad field of ferromagnetism, but merely intends to mention certain important features

---

[1] F. Kohlrausch, *Praktische Physik*, 12th Ed., p. 540. It should be noted that the factor 4π, by which Kohlrausch's Eq. (4) differs from our Eq. (23), is, with us, included in the definition of κ. See the remark at Eq. (9) above.

which, since they lie outside our subject, will be indicated rather than logically derived. As has already been noted at the beginning of §12, ferromagnetism is not based on Maxwell's phenomenological theory, but on the more profound laws of atomic physics and on the statistical behavior of electrons.

A. The Weiss Domains

The sign of the ferromagnetic susceptibility and its temperature dependence indicate that it, like the paramagnetic susceptibility, results from the alignment of elementary magnets in a magnetic field. The fact that it differs from the paramagnetic susceptibility in order of magnitude shows, however, that we are here dealing not with individual, freely mobile, magnets, but with whole groups of them which, perfectly aligned within the group, have different preferential directions. The individual group is "saturated" internally even in the absence of an external field, while in a macroscopic block of ferromagnetic material saturation occurs only at a field excitation of 10 to 1000 oersted.

This concept of the ferromagnetic state is illustrated by a model constructed by Ewing. Magnet needles are arranged in a lattice on a board. With the earth's field compensated by a current loop, they arrange themselves in groups or rows in which they are parallel. This state is stable against external disturbances, such as shaking of the board. A weak external field produces only a slight deflection from the equilibrium position since the internal aligning field is much stronger than the external field. Complete alignment in the direction of the external field, i.e. saturation of the entire system, takes place only at a very much higher field strength.

Pierre Weiss has elaborated this interpretation of ferromagnetism in all directions, both experimentally and theoretically, and has, together with Langevin, given it a thermodynamic basis. The individual groups are known as *Weiss domains*. Their size is estimated at about 10^{-5} cm in linear dimension, corresponding to $5 \cdot 10^{8}$ Fe-atoms. The smallest ferromagnetically active domains are however certainly very much smaller and contain fewer than 100 Fe-atoms.[1] It is tempting to identify them with the single crystals of which a polycrystal of the material is composed on a microscopic scale. However, it is necessary to assume such a subdivision into Weiss domains even for the macroscopic single crystal, since its behavior is qualitatively similar to that of the polycrystal. (It is true that quantitatively the shape of the hysteresis loop, discussed below, differs from that for the polycrystal; it has a rectangular shape.)

---

[1] See H. König, Naturwiss. 1946, p. 1.

B. The Electron Spin as Elementary Magnet

All ferromagnetic materials are conductors of electricity, i.e. contain *free electrons*.[1] We have very definite reasons for regarding these electrons as the elements whose alignment causes ferromagnetism. These are the so-called "gyromagnetic effects": *magnetization by rotation* of a bar of a material of the iron group (J. S. Barnett 1914) and *rotation by magnetization* of a small ferromagnetic rod suspended on a torsion fiber (Einstein and de Haas 1915). In both cases measurements gave as ratio of the mechanical to the magnetic moment *half the* value to be expected if the effects resulted from the electron *orbits* in the atom. It must therefore be concluded that it is not the charge of the revolving electron and the magnetic field produced by it that are responsible, but the inner structure of the electron itself. The electron possesses, apart from its charge, an inner mechanical moment, a *"spin,"* and a *magnetic moment* which is *twice as large* as the magnetic moment which would be assigned classically to its spin. This *magnetic anomaly* of the electron is the general result which may be deduced from the total observed data on the anomalous Zeeman effects (Goudsmit and Uhlenbeck 1925). It explains directly the observed results of the two gyromagnetic effects and proves at the same time that the electron spin plays the role of the magnetic needles in Ewing's model.

On the basis of this discovery Heisenberg, in 1928, with the aid of modern electron statistics, was able to proceed toward a true physical understanding of ferromagnetism and to calculate qualitatively the extraordinary magnitude of the inner magnetic field in a Weiss domain. We see from this how long the road is from the Maxwell theory to the actual theory of ferromagnetism and it becomes evident that we cannot travel this road.

C. Hysteresis Loop and Reversible Magnetization

The figure which represents the magnetization M of a ferromagnetic material as function of the excitation H with increasing and decreasing H is well known. If the material is originally unmagnetized, the "virginal curve" is first traversed, beginning in the origin $H = 0$, $M = 0$ and passing over into the horizontal asymptote $M = M_S$ of *saturation* for sufficiently large H. If, from this point, H is permitted to decrease, the characteristic lies above the virginal curve and cuts the ordinate axis in a point $H = 0$, $M = M_R$, which indicates the *remanent magnetization*. If H is permitted to decrease still further, i.e. is reversed in direction, a region is entered in which **B** and **H** have opposite directions. The iron specimen has then become a "permanent magnet". With further decrease of H the hysteresis loop cuts the axis of abscissas in a point $H = -H_c$, $M = 0$, where the

---

[1] Atomic theory has not demonstrated fully why just the atoms of the iron group are ferromagnetically active.

remanent magnetization is just nullified. H_C is called the "coercive force". If H is decreased further, the negative saturation $M = -M_S$ is approached. If, now, H is once more increased, the gradually rising characteristic remains below the descending branch and below the virginal curve. It does not pass through the origin, but cuts the axis of abscissas in a point $H = +H_C$, $M = 0$. The ascending branch is symmetrical through the origin to the descending branch and approaches finally once more the positive saturation $M = +M_S$.

As a rule M_R is approximately $\frac{1}{2}M_S$. In order that the magnet may retain its remanence for all opposing fields which occur it is important that H_C may be as large as possible. This is the case for hard steel (tungsten steel has $H_C \sim 70$ oersteds).

The ascending and descending branches form together the *hysteresis loop*; its area is the magnetic work $\oint \mathbf{H} \cdot d\mathbf{B}$, which is performed on the material in a complete cycle. If the increase of H is stopped, in traversing the virginal curve, before reaching saturation, e.g. for $H_1 < H_S$, and H is then permitted to decrease to $H = -H_1$, then to increase to $H = +H_1$, a smaller hysteresis loop is obtained, which lies inside of the one previously described. For very small $H_1 = \delta H$ the loop degenerates into a twice traversed line; its area becomes zero and the process is *reversible*. The ratio $\delta M/\delta H$ defines the *initial susceptibility* κ_0. It is possible to carry out such a reversible process not only at the origin, but at any arbitrary point of the cycle and to define for every such point a *reversible susceptibility* κ_{rev}.

With reversible magnetization the elementary magnets are *deflected* only slightly out of their original position in the direction of the external field. With irreversible magnetization some *reorientations* take place as well. In both cases changes occur in the boundaries of the Weiss domains, which are described as wall displacements; they are small for reversible, large for irreversible processes. In an induction coil with telephone connection they become acoustically noticeable by noises (clicks) and can be made visible on an oscilloscope as *Barkhausen jumps*. In fact, for sufficient oscilloscopic magnification, the apparently continuous course of the hysteresis loop resolves itself, particularly in the descending branch, into a sequence of small steps.

The individual processes which go to make up the magnetization curves are thus of varied nature. They depend on the composition of the iron specimen and on its microcrystalline structure; even for the single crystal they depend on the orientation relative to the magnetic field. It is the problem of the metallurgist to find the alloy (permalloy, perminvar, cobalt steel) suited for each purpose (transformer laminations, communications engineering).

D. Thermodynamics

Ferromagnetism is even more dependent on temperature than para-magnetism. Above a certain critical temperature ferromagnetism ceases and passes over into ordinary paramagnetism. This critical temperature is designated with θ and is called the *Curie point*. On the centigrade scale $(\theta = \theta_{cent} + 273)$ we have:

	for iron	cobalt	nickel
$\theta_{cent} =$	770	1120	358

For $T > \theta$ we have in place of Curie's law (13.10) the Curie-Weiss law

$$\chi = \frac{C}{T - \theta}. \tag{1}$$

This suggests that there may be no difference in principle between para-magnetism and ferromagnetism, that, in other words, the Curie point, in the former case, lies close to absolute zero. With this assumption a region of ferromagnetic behavior near $T = 0$ is to be expected also for ordinary paramagnetic materials. It then seems reasonable to transfer Langevin's statistical and thermodynamic theory of paramagnetic materials to the conditions of ferromagnetism.

In fact Weiss in this manner arrived at a representation of the whole complex of ferromagnetic phenomena which, in its main features, is satis-factory. However, this representation utilizes concepts with which we will only be able to deal in Volume V. Also detailed questions of atomic physics play here a role, such as the question of the atomic unit of magnetic moment (Bohr's magneton as compared with the Weiss magneton, which is smaller by a factor of five) and the question to what extent, in addition to the spin moment of the free electrons, the orbital moment of the electrons bound in the atom, which is twice as large, must be considered. The standard textbook[1] of R. Becker and W. Döring gives complete information on all relevant questions.

§15. Stationary Currents and Their Magnetic Field. Method of the Vector Potential

Since the assumption of stationary fields demands $\partial/\partial t = 0$ throughout, the Maxwell equations (4.4) reduce to

$$\text{curl } \mathbf{E} = 0, \qquad \mathbf{J} = \text{curl } \mathbf{H}. \tag{1}$$

---

[1] *Ferromagnetismus*, Springer, Berlin 1939. For a less detailed treatment, see F. Bitter, *Introduction to Ferromagnetism*, McGraw-Hill, New York, 1937.

The second of these leads to

$$\text{div } \mathbf{J} = 0. \tag{2}$$

For the *surface* of a conductor carrying current we have therefore

$$J_n = 0. \tag{2a}$$

(2) states that the electricity within a conductor behaves like an incompressible fluid (increase and decrease of current density for a narrowing and broadening of the conductor, respectively). Also *Kirchhoff's branching laws for linear conductors*, which Kirchhoff worked out as solution to a seminar problem given by F. Neumann (see p. 1), rest in the final analysis on (2) and the existence of the electrical potential (see below).

Eqs. (1) state that $\mathbf{E}$ has *everywhere* a potential, $\mathbf{E} = -\text{ grad } \Psi_e$, while $\mathbf{H}$ has a scalar potential, $\mathbf{H} = -\text{grad } \Psi_m$, only *outside* of the current-carrying conductors. We will deal with this scalar potential in §16. Here we shall give a representation of $\mathbf{B}$ (and hence also one of $\mathbf{H}$) which is valid both *inside* and *outside* of the conductors. We recall here Helmholtz's representation of the velocity field $\mathbf{v}$ for given distribution of turbulence ω in Vol. II, Eq. (20.13). In Helmholtz's analogy the electric current density $\mathbf{J}$ corresponded[1] to the second, the magnetic excitation $\mathbf{H}$, to the first. Just as there, we introduce a *vector potential* $\mathbf{A}$, by setting:[2]

$$\mathbf{B} = \text{curl } \mathbf{A}. \tag{3}$$

The second Eq. (1) then becomes

$$\text{curl } \frac{1}{\mu} \text{ curl } \mathbf{A} = \mathbf{J}. \tag{4}$$

For constant μ we can write instead

$$\text{curl curl } \mathbf{A} = \mu \mathbf{J}. \tag{4a}$$

---

[1] But for a factor $\frac{1}{2}$ whose suppression was justified in Vol. II, p. 15 by the requirements of electrodynamics.

[2] It is customary to write instead $\mathbf{H} = \text{curl } \mathbf{A}$, which, however, assumes the complete absence of sources of $\mathbf{H}$, a condition which is not fulfilled in regions of nonvanishing magnetic density ρ_m. Our formula (3) is more satisfactory, since div $\mathbf{B} = 0$ throughout; in addition, it will generally simplify our formulas, particularly in Part III. Incidentally, both formulas amount to much the same thing if the assumption is made that μ is constant everywhere, which occurs already in Eq. (4a) of the text. For nonconstant μ the summation problem to be solved in (7) would have to be supplemented by a magnetostatic *boundary-value problem* (determination of the discontinuity of B_{tang} at the boundary between media of different permeability for our expression for $\mathbf{A}$ and determination of the magnetic surface densities appearing there for the usual expression for $\mathbf{A}$, respectively).

We here employ the general transformation (6.2) with the restriction to Cartesian coordinates which is there emphasized and obtain instead of (4a) the form which is more convenient for integration:

$$\Delta \mathbf{A} - \text{grad div } \mathbf{A} = - \mu \mathbf{J}. \tag{4b}$$

This may be simplified by the supplementary condition

$$\text{div } \mathbf{A} = 0, \tag{5}$$

which transforms (4b) into

$$\Delta \mathbf{A} = - \mu \mathbf{J}. \tag{6}$$

The condition (5) may be added since $\mathbf{A}$, for given $\mathbf{B}$, is determined by Eq. (3) with the exception of the gradient of a scalar function. The latter may be utilized to satisfy (5). For, if $\mathbf{A}_1$ is *any* solution of (3),

$$\mathbf{A} = \mathbf{A}_1 + \text{grad } f \tag{6a}$$

is similarly a solution; if we now write

$$\text{div grad } f = \Delta f = - \text{div } \mathbf{A}_1 , \tag{6b}$$

which, according to the well-known integration procedure of Poisson's equation is always possible, we find div $\mathbf{A} = 0$.

This method of integration yields at the same time as the solution of (6):

$$4\pi \mathbf{A}_P = \int \frac{\mu \mathbf{J}_Q \, d\tau_Q}{r_{PQ}} . \tag{7}$$

Here the point of integration $Q = \xi, \eta, \zeta$ traverses the entire interior of the conductors; $P = x, y, z$ is the reference point for which the Cartesian components A_x, A_y, A_z are to be calculated. It was shown in Vol. II, §20, Nr. 2a that this representation satisfies (5) provided that $\mathbf{J}$ is solenoidal, in accord with (2), and that μ is constant. The integration in Eq. 7 is to be extended over the *closed* current field $\mathbf{J}$ (just as in Vol. II over the *closed* vortex rings).

The current density $\mathbf{J}$ appearing in (7) may be obtained as solution of a potential problem. Since $\mathbf{J} = \sigma \mathbf{E}$ the potential equation applies, for constant σ, just as much for $\mathbf{J}$ as for $\mathbf{E}$:

$$\Delta \mathbf{J} = 0. \tag{8}$$

For varying σ Eq. (8) takes on a somewhat more complicated form. The total current I is obtained from $\mathbf{J}$ by integration over any cross section of the conductor:

$$I = \int J_n \, d\sigma. \tag{9}$$

The familiar fact that I has a fixed value independent of position and shape of the cross section follows from integration of Eq. (2) over a segment of the conductor bounded by two arbitrary cross sections as for the analogous spatial law of conservation of vortex theory (see Vol. II, p. 136, Fig. 24).

We shall now give some applications of our representation (7).

A. The Law of Biot-Savart

We subdivide the three-dimensional conductor into current tubes with the cross section dq , normal to the tube axis, and the element of length ds; the current $J_n\, dq$ in such a tube element, which has the same dimension as the total current I, we shall also call I, for the present. We then can set

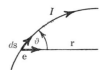

FIG. 19. The law of Biot-Savart, derived from the vector potential of an element of current.

$\mathbf{J}\, d\tau = I\, d\mathbf{s}$, where the direction of the current flow $\mathbf{J}$ is indicated by the vectorial character of $d\mathbf{s}$. By (7) the contribution of our tube element to $\mathbf{A}$ then becomes

$$4\pi\, d\mathbf{A} = \frac{\mu I\, d\mathbf{s}}{r}$$

and, by (3), the corresponding contribution to $\mathbf{B}$ is,

$$4\pi\, d\mathbf{B} = \operatorname{curl} \frac{\mu I\, d\mathbf{s}}{r}. \tag{10}$$

We shorten the remaining calculation by employing the symbolic vector ∇ :

$$\operatorname{curl} \frac{\mu I\, d\mathbf{s}}{r} = \nabla \times \frac{\mu I\, d\mathbf{s}}{r} = \left(\operatorname{grad} \frac{1}{r}\right) \times \mu I\, d\mathbf{s}. \tag{11}$$

For $I\, d\mathbf{s}$ is dependent on x, y, z in direction, but not in magnitude; $1/r$ depends on x, y, z in magnitude, but not in direction (being a scalar). We see furthermore from Fig. 19 that

$$\operatorname{grad} \frac{1}{r} = -\frac{\mathbf{r}}{r^3} = -\frac{\mathbf{e}}{r^2}, \tag{11a}$$

where $\mathbf{r}$ denotes the radius vector from the current element to the reference point and $\mathbf{e}$ the corresponding unit vector. Substitution of (11) and (11a) in (10) yields

$$4\pi \, d\mathbf{B} \;=\; \frac{-\mu I}{r^2} \, \mathbf{e} \times d\mathbf{s}, \qquad 4\pi \mid d\mathbf{B} \mid \;=\; \frac{\mu I \, ds}{r^2} \sin \vartheta. \tag{12}$$

ϑ signifies the angle between the vectors $\mathbf{e}$ and $d\mathbf{s}$; the direction of $d\mathbf{B}$, in view of the negative sign in (12) and the meaning of the vector product, is that of a left-hand screw for the direction of rotation $\mathbf{e} \to d\mathbf{s}$. Let us imagine a magnetic unit pole at the reference point and let the vector $d\mathbf{B}$ act on it; $d\mathbf{B}$ then represents the *Biot-Savart force* exerted on the unit pole by the current element $I \, d\mathbf{s}$. The corresponding line of force then surrounds the current element, in a *right-hand* screw direction, as shown in the figure and as expected. Evidently the first Eq. (12) is the more complete one, since it expresses the dependence on direction which is characteristic of the magnetic field in the simplest and most appropriate manner; we have added the second form merely because it is the historically more familiar one.

B. The Magnetic Energy of the Field of Two Conductors

If we denote the energy integrated over space by W we obtain, for the magnetic energy density W, from Eq. (5.6)

$$2\mathrm{W} \;=\; \int \mathbf{H} \cdot \mathbf{B} \, d\tau \;=\; \int \mathbf{H} \cdot \mathrm{curl} \, \mathbf{A} \, d\tau. \tag{13}$$

For the evaluation we utilize the vector formula (5.2), previously derived in connection with the Poynting theorem, which we rewrite in terms of our present symbols ($\mathbf{A}$, $\mathbf{H}$ in place of $\mathbf{U}$, $\mathbf{V}$) as follows:

$$\mathbf{H} \cdot \mathrm{curl} \, \mathbf{A} \;=\; \mathbf{A} \cdot \mathrm{curl} \, \mathbf{H} + \mathrm{div} \, \mathbf{A} \times \mathbf{H}. \tag{14}$$

We assert that the second term on the right vanishes in the integration over infinite space. According to Gauss's theorem this term yields

$$\int \mathrm{div} \, \mathbf{A} \times \mathbf{H} \, d\tau \;=\; \int (\mathbf{A} \times \mathbf{H})_n \, d\sigma, \tag{14a}$$

where the integration on the right is to be carried out over a surface bounding the region at a great distance, e.g. a sphere of radius R. Let the two conductors, whose total magnetic energy is to be determined, be entirely confined to a finite region. The distance of all their points from the infinitely distant element of area $d\sigma$ of the bounding sphere may then be set equal to the constant value R. By Eq. (7) $\mathbf{A}$ approaches zero on $d\sigma$ as $1/R$ and, by the law of Biot-Savart, $\mathbf{H}$ approaches zero as $1/R^2$. Since $d\sigma = R^2 \, d\Omega$ ($d\Omega =$ solid angle intercepted by $d\sigma$), the right side of (14a) approaches zero as R^2/R^3.

In view of (1), (14) and (13) lead to

$$2W = \int \mathbf{A} \cdot \operatorname{curl} \mathbf{H} \, d\tau = \int \mathbf{A} \cdot \mathbf{J} \, d\tau. \tag{15}$$

The integration is now to be carried out only over the conductors 1 and 2, since $\mathbf{J}$ is zero everywhere outside of them. If we designate the point of integration in (15) by P ($d\tau_P$ instead of $d\tau$) and if we substitute $\mathbf{A} = \mathbf{A}_P$ from (7), the simple volume integral is replaced by a double volume integral:

$$\frac{2W}{\mu} = \frac{1}{4\pi} \iint \mathbf{J}_P \cdot \mathbf{J}_Q \, \frac{d\tau_P \, d\tau_Q}{r_{PQ}}. \tag{16}$$

In evaluating this integral we have to distinguish four cases, depending on the position of P and Q on conductors 1 and 2:

      a. P and Q on 1,   b. P and Q on 2,
      c. P on 1, Q on 2,   d. P on 2, Q on 1;

the cases c and d are however alike in view of the symmetry of (16) with respect to P and Q. We can write the result in the form

$$W = \tfrac{1}{2}(L_{11} I_1^2 + L_{22} I_2^2 + 2L_{12} I_1 I_2), \tag{17}$$

$$L_{11} = \frac{\mu}{4\pi} \iint \mathbf{j}_1 \cdot \mathbf{j}_1' \, \frac{d\tau_1 \, d\tau_1'}{r_{11}'}, \qquad L_{22} = \frac{\mu}{4\pi} \iint \mathbf{j}_2 \cdot \mathbf{j}_2' \, \frac{d\tau_2 \, d\tau_2'}{r_{22}'}, \tag{17a}$$

$$L_{12} = \frac{\mu}{4\pi} \iint \mathbf{j}_1 \cdot \mathbf{j}_2 \, \frac{d\tau_1 \, d\tau_2}{r_{12}}. \tag{17b}$$

The factor 2 in the product term in (17) results from the equality of cases c and d, which causes the coefficient L_{21} also to be given by (17b), i.e. $L_{21} = L_{12}$. I_1, I_2 are the total currents in conductors 1 and 2, which, as noted at (9), are independent of the place in the conductor. 1, 1' are two points on the conductor 1; 2, 2' two points on conductor 2. Division of the current densities $\mathbf{J}_1$, $\mathbf{J}_2$ by I_1, I_2 leads to the purely geometrically defined "current-line density vectors"

$$\mathbf{j}_1 = \frac{\mathbf{J}_1}{I_1}, \qquad \mathbf{j}_2 = \frac{\mathbf{J}_2}{I_2}. \tag{17c}$$

The L's are called *induction coefficients*: L_{11}, L_{22} are the *coefficients of self-induction*, L_{12} is the *coefficient of mutual induction*. Maxwell uses the letter M in place of L_{12}.

The unit of the coefficients L (or M) is the *henry*.[1] In accord with (17)

---

[1] Joseph Henry, 1792–1878, American physicist, discovered almost simultaneously with Faraday the appearance of an electromotive force in a coil when the magnetic field in its interior is changed.

this unit is fixed in value and dimension by the statement

$$1 \text{ henry} = 1 \frac{\text{joule}}{\text{I}^2} = 1 \frac{\text{joule S}^2}{\text{Q}^2} . \tag{18}$$

Converted into electromagnetic cgs-units we find, since

$$Q = 10^{-1} \text{ cm}^{\frac{1}{2}} \text{g}^{\frac{1}{2}}, \quad 1 \text{ joule} = 10^7 \text{ cm}^2 \text{g sec}^{-2},$$
$$1 \text{ henry} = 10^9 \text{ cm} = 1 \text{ quadrant of the earth.} \tag{18a}$$

From our standpoint we can, however, attach no significance to this apparently so simple dimension and this relationship to the earth's circumference since it rests on the arbitrary assumptions of the electromagnetic system of units. At the same time we are glad to point out the following relation between the henry and the permeability in vacuum given by (7.16a):

$$\mu_0 = 4\pi \cdot 10^{-7} \frac{\text{joule} \cdot \text{S}^2}{\text{Q}^2 \text{M}} = 4\pi \cdot 10^{-7} \frac{\text{henry}}{\text{M}} . \tag{18b}$$

Compare with this the analogous relation between the dielectric constant of vacuum and the farad as given by Eq. (10.3b).

C. Neumann's Potential as the Coefficient of Mutual Induction

In (17b) it is possible to pass to the limiting condition of linear conductors, i.e. infinitely thin wires. In (17a) this is not permitted since r_1' (or r_{22}') would vanish as the two integration points approach each other and the convergence of the integrals would be destroyed.

We write in (17b)

$$d\tau_1 = dq_1 \, ds_1, \quad d\tau_2 = dq_2 \, ds_2 \tag{19}$$

and combine dq_1 and $\mathbf{j}_1$, dq_2 and $\mathbf{j}_2$. The products $\mathbf{j}_1 \, dq_1$ and $\mathbf{j}_2 \, dq_2$ then have by (17c), unit magnitude and their scalar product is equal to the cosine of the angle θ_{12} between the two directions of flow ds_1 and ds_2. Hence:

$$\frac{4\pi}{\mu} L_{12} = \int ds_1 \int ds_2 \frac{\cos \theta_{12}}{r_{12}} = \iint \frac{d\mathbf{s}_1 \cdot d\mathbf{s}_2}{r_{12}} . \tag{20}$$

This amazingly simple and beautiful representation was discovered[1] by Franz Neumann as early as 1845. It is known as *Neumann's potential* according to (17) it represents that portion of the magnetic energy which results from the interaction of the two circuits. A relative displacement or rotation of the two circuits with the currents I_1 and I_2 left unaltered hence is accompanied by a change in energy δW in the amount

$$\delta W = I_1 I_2 \, \delta L_{12} . \tag{20a}$$

[1] *Abhandl. Preuss. Akad.*, reprinted in Ostwald's *Klassiker*, Nr. 10.

The *work* which must be done in a displacement or rotation, as well as the *force* or *torque* which one circuit exerts on the other, are related to this.

In spite of the simplicity of expression (20) the actual calculation of the mutual induction coefficient is rather inconvenient. To begin with we work out a formal mathematical example, i.e. two straight parallel wires of length l separated by distance a. The condition of closed circuits stated in Eq. (7) is temporarily not fulfilled here. We shall take due account of it only

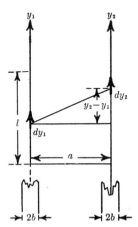

Fig. 20. The coefficient of mutual induction of two straight, parallel segments of wire of length l. The finite cross section of the wires is indicated at the bottom of the figure.

when we reach Eq. (24). Referring to Fig. 20, we have ($dy_1\, dy_2$ in place of $ds_1\, ds_2$, $\cos\theta_{12} = 1$):

$$\frac{4\pi}{\mu} L_{12} = \int_0^l dy_1 \int_0^l \frac{dy_2}{\sqrt{a^2 + (y_2 - y_1)^2}}. \tag{21}$$

The formula of integration already employed in (9.4) yields for the second integral

$$\int_0^l \frac{dy_2}{\sqrt{a^2 + (y_2 - y_1)^2}} \tag{22}$$
$$= \log(l - y_1 + \sqrt{a^2 + (l - y_1)^2}) - \log(-y_1 + \sqrt{a^2 + y_1^2}).$$

A further formula, which may readily be checked by differentiation,

$$\int^z \log(\zeta + \sqrt{a^2 + \zeta^2})\, d\zeta = z \log(z + \sqrt{a^2 + z^2}) - \sqrt{a^2 + z^2} + \text{const}$$

yields for the integration of the first term on the right of (22)

$$\int_0^l dy_1 \log(l - y_1 + \sqrt{a^2 + (l - y_1)^2})$$

$$= l \log(l + \sqrt{a^2 + l^2}) - \sqrt{a^2 + l^2} + a$$

and for the integration of the second term

$$- \int_0^l dy_1 \log(-y_1 + \sqrt{a^2 + y_1^2})$$

$$= -l \log(-l + \sqrt{a^2 + l^2}) - \sqrt{a^2 + l^2} + a.$$

A simple transformation under the logarithm sign yields for the sum of the two

$$\frac{4\pi}{\mu} L_{12} = 2l \log \frac{l + \sqrt{a^2 + l^2}}{a} - 2\sqrt{a^2 + l^2} + 2a. \tag{23}$$

We assume $l \gg a$, and obtain as a first approximation

$$\frac{4\pi}{\mu} L_{12} = 2l \left(\log \frac{2l}{a} - 1 \right). \tag{24}$$

It is worth noting in this result that we have not obtained simple proportionality to l, since the parenthesis depends on l logarithmically. Accordingly we can not speak of a mutual induction coefficient per unit length of our wires.

The reason for this is the following: Our derivation assumes, as has been stressed repeatedly, two *closed* circuits, while our example deals with two circuit segments. Our result (24) is nevertheless meaningful. For example, it is possible to determine with its aid the mutual inductance of two parallel rectangles such as occur in Ampere's basic experiments. For such a pair of rectangles (the second rectangle is supposed to be obtained from the first by a parallel displacement perpendicular to its plane) only parallel pairs of sides contribute; for mutually perpendicular sides the product $ds_1 \cdot ds_2$ occurring in Eq. (20) is equal to zero. The mutual inductance of two such rectangles becomes equal to the sum of four terms of the form of (24).

D. *The Coefficient of Selfinduction*

As already noted we cannot in this case pass to the limit of the linear conductor, but must return to the double volume integrals in (17a). We can readily convince ourselves, however, that then any convergence difficulty is avoided. For if, for an arbitrary position of 1, we employ polar coordinates r, ϑ, φ with this point as origin to locate the point $1'$, $d\tau_1' = r^2 dr \sin \vartheta \, d\vartheta \, d\varphi$ and the denominator $r_{11'} = r$ cancels one of the factors r in $d\tau_1'$. However the carrying out of the integrations becomes now, in general, even more awkward than in C.

We therefore limit ourselves to a simple mathematical example, namely a straight wire of the (great) length l and the (small, but finite) cross section q. We consider two current filaments parallel to the axis of the wire

(y-axis) and employ once again Fig. 20, where now the two linear currents are assumed to refer to the same wire of cross section q. Let dq_1, dq_2 be the cross sections of the two current filaments; their separation, formerly denoted by a, will now be called ρ since it is variable, depending on the position of the two current filaments within q. The volume elements are once again given by (19), with $ds_1 = dy_1$, $ds_2 = dy_2$, and the vectors defined in (17c) become

$$j_1 = j_2 = \frac{l}{q}.$$

The defining equation (17a) for the selfinductance then takes the form

$$\frac{4\pi}{\mu} L = \iint \frac{dq_1\, dq_2}{q^2} \int_0^l \int_0^l \frac{dy_1\, dy_2}{\sqrt{\rho^2 + (y_2 - y_1)^2}}. \tag{25}$$

The second double integral has exactly the same form as (21). We can utilize the approximate evaluation (24) here also and obtain

$$\frac{4\pi}{\mu} L = \frac{2l}{q^2} \iint dq_1 dq_2 \left(\log \frac{2l}{\rho} - 1 \right)$$

$$= \frac{2l}{q^2} \left\{ \iint dq_1 dq_2\, (\log 2l - 1) - \iint dq_1 dq_2 \log \rho \right\}.$$

In the first term on the right the integration with respect to dq_1 and dq_2 can be carried out easily; the second term requires more detailed discussion because of the variability of $\rho = \rho_{12} = $ separation of our two current filaments. We note the preliminary result

$$\frac{4\pi}{\mu} L = 2l\{\log 2l - 1 - \log \bar{\rho}\}, \qquad \log \bar{\rho} = \frac{1}{q^2} \int dq_1 \int dq_2 \log \rho_{12}. \tag{26}$$

Maxwell calls the quantity $\bar{\rho}$ here introduced the *mean geometric separation* of the elements dq_1, dq_2 within the cross section q.[1] It can be determined more elegantly by an electrostatic consideration than by direct calculation.

In terms of polar coordinates the two-dimensional potential equation becomes

$$\Delta\Phi = \frac{\partial^2\Phi}{\partial x^2} + \frac{\partial^2\Phi}{\partial y^2} = \frac{1}{\rho} \frac{d}{d\rho} \rho \frac{d\Phi}{d\rho} + \frac{1}{\rho^2} \frac{\partial^2\Phi}{\partial\varphi^2} = 0. \tag{27}$$

[1] Treatise, art. 691 ff. In explanation of the notation we remark: The integral to be evaluated in (26) is the arithmetic mean of all values of $\log \rho$ occurring on our surface q. In view of the relation

$$\Sigma \log \rho_i = \log \Pi \rho_i$$

this arithmetic mean of the logarithms is at the same time the logarithm of the *geometric* mean of all the ρ_i.

Apart from a multiplying and an additive constant the solution which is independent of φ is known to be the "logarithmic potential"

$$\Phi = \log \rho.$$

This signifies, in two dimensions, a negative charge concentrated at the point $\rho = 0$. If the charge is distributed over the area q with a positive surface density f, and with the surface element of q being designated with dq_2, Green's theorem yields for its potential at the reference point 1.

$$2\pi\Phi_1 = - \int_q f \log \rho_{12} \, dq_2. \tag{28}$$

This is the two-dimensional analog to the familiar Eq. (7.5). If, in (28), we put $f = -2\pi$, we obtain the inner integral in (26),

$$\Phi_1 = \int_q \log \rho_{12} \, dq_2 \tag{28a}$$

and our desired mean geometric separation may be written

$$\log \bar{\rho} = \frac{1}{q^2} \int \Phi_1 \, dq_1. \tag{28b}$$

The integral (28b) can be readily evaluated in the case that q is a circle, e.g. of radius b, and the point 1 coincides with the center of the circle. We then have $\rho_{12} = \rho$, i.e. equal to the polar coordinate employed previously, and $dq_2 = \rho \, d\rho \, d\varphi$. We denote by Φ_0 the corresponding specific value of Φ_1. Eq. (28a) then takes the form

$$\Phi_0 = \int_0^{2\pi} d\varphi \int_0^b \log \rho \, \rho \, d\rho. \tag{29}$$

The integral with respect to ρ can readily be evaluated by integration by parts; thus

$$\int_0^b \log \rho \, \rho \, d\rho = \frac{\rho^2}{2} \log \rho \, \Big|_0^b - \int_0^b \frac{\rho}{2} \, d\rho = \frac{b^2}{2} (\log b - \tfrac{1}{2}).$$

Hence (29) leads to

$$\Phi_0 = \pi b^2 (\log b - \tfrac{1}{2}). \tag{29a}$$

Furthermore Φ_1, as the potential for a known surface distribution, can be calculated for arbitrary location of point 1 directly from Poisson's equation, which, in analogy to (7.4a), in two dimensions takes the form

$$\Delta\Phi_1 = -f,$$

where f is the surface density. In our special case ($f = -2\pi$, circle q of radius b, Φ_1 a function of ρ only) it becomes

$$\frac{1}{\rho}\frac{d}{d\rho}\,\rho\,\frac{d\Phi_1}{d\rho} = 2\pi . \tag{30}$$

Integrated twice this yields

$$\rho\,\frac{d\Phi_1}{d\rho} = \pi\rho^2 + A, \qquad \Phi_1 = \frac{\pi\rho^2}{2} + A\log\rho + B.$$

In order that this expression for Φ_1 may pass over, for $\rho = 0$, into the expression (29a) for Φ_0, we must set

$$A = 0, \quad B = \Phi_0, \quad \text{hence} \quad \Phi_1 = \pi b^2\left(\log b - \frac{1}{2} + \frac{1}{2}\frac{\rho^2}{b^2}\right). \tag{31}$$

If we substitute this expression for Φ_1 in (28b) we obtain

$$\log\bar{\rho} = \frac{1}{q}\int dq_1(\log b - \tfrac{1}{2}) + \frac{1}{2q}\int dq_1\frac{\rho^2}{b^2}$$

$$= \log b - \tfrac{1}{2} + \frac{1}{b^4}\int_0^b \rho^3\,d\rho = \log b - \tfrac{1}{4}. \tag{32}$$

For $\bar{\rho}$ itself we obtain from this the peculiar value

$$\bar{\rho} = b/\sqrt[4]{e}. \tag{32a}$$

Here the proportionality with b is rather obvious in view of the definition of the geometric mean (see last footnote); however the numerical factor is to be found only by detailed analysis, which following Maxwell, we have here based on potential theory. It may be mentioned that Maxwell carried out these considerations even for cross sections of arbitrary shape.

To pursue our real goal, the calculation of the selfinductance L, we return to (26). We then find, utilizing (32)

$$\frac{4\pi}{\mu}L = 2l\left\{\log 2l - \log b - \frac{3}{4}\right\} = 2l\left\{\log\frac{2l}{b} - \frac{3}{4}\right\}. \tag{33}$$

This formula confirms our original expectation that the transition to the linear conductor ($b \to 0$) is not permissible for the selfinductance. Regarding the dependence on l we must make the same remark as at the end of section C: We cannot, by division of l, obtain the *selfinductance per unit length of an infinitely long wire*; however, we can, as for the mutual inductance, piece together the selfinductance of any *closed* circuit, made up of straight wires, by adding up terms of the form of (33).

We shall further answer a rather obvious objection which may be raised against the somewhat indirect derivation of (33). The magnetic field of an infinitely long wire is known from Fig. 4 and the corresponding equations (10) to (13) on p. 24. Cannot the energy and selfinduction of the straight wire be calculated much more directly from them?

With I as current we found (writing now b, c in place of the earlier a, b)

$$r < b : H = \frac{r}{b} \frac{I}{2\pi b}; \qquad r > b : H = \frac{I}{2\pi r}.$$

Hence the contribution of the interior of the wire to the energy per unit length is

$$\frac{\mu}{2} 2\pi \int_0^b H^2 r \, dr = \frac{\mu}{4\pi} \frac{I^2}{b^4} \int_0^b r^3 \, dr = \frac{\mu}{16\pi} I^2 \tag{33a}$$

and the contribution of the exterior:

$$\frac{\mu}{2} 2\pi \int_b^c H^2 r \, dr = \frac{\mu}{4\pi} I^2 \int_b^c \frac{dr}{r} = \frac{\mu}{4\pi} I^2 \log \frac{c}{b}. \tag{33b}$$

From this we find for the selfinductance per unit length, from the energetic formula of definition (17),

$$\frac{\mu}{2\pi} \left(\frac{1}{4} + \log \frac{c}{b} \right). \tag{33c}$$

This expression becomes logarithmically infinite if we let $c \to \infty$, i.e. pass over to the single wire without return conductor. Our intended simplified derivation hence fails—quite understandably—as the result of the unphysical assumptions of the problem, which must lead to an infinite energy content of space for any unit length of the wire as we pass to the limit $c \to \infty$.

E. Selfinductance of the Two-Wire Line

The system of two straight parallel wires traversed by current in opposite directions plays an important role in electric power transmission and is known, in Hertz's experiments with high frequency waves, as a Lecher system. We shall determine the selfinductance per unit length of such a system (with appropriate restriction to the direct-current case).

So as to be able to utilize Fig. 20, we shall call the (very great) length of the wires l, their separation a, and their radius b. We proceed from the energetic formula (17), where we put

$$I_1 = I, \qquad I_2 = -I, \qquad L_{11} = L_{22} = L.$$

We then obtain

$$W = \tfrac{1}{2} L_D I^2, \qquad L_D = 2(L - L_{12}). \tag{34}$$

We can designate L_D, introduced here, as inductance of the *line*, regarded as a uniform system.[1] Substitution from (33) and (24) yields:

$$L_D = \frac{\mu}{\pi} l \left(\log \frac{2l}{b} - \frac{3}{4} - \log \frac{2l}{a} + 1 \right).$$

In evaluating the logarithmic terms the two terms $\pm \log 2\, l$ drop out and the following simple formula remains:

$$\frac{L_D}{l} = \frac{\mu}{\pi} \left(\log \frac{a}{b} + \frac{1}{4} \right). \tag{35}$$

L_D/l is the desired selfinductance per unit length of the two-wire line, which evidently is independent of l. Hence the transition to $l \to \infty$, which, according to (33) and (24), was inappropriate in the expressions for L/l and L_{12}/l because of the term $\log 2l$, is now feasible. This is obviously related to the fact that the field of a two-wire line traversed by oppositely directed currents corresponds to that of a circuit closed at infinity. On the other hand the transition to the linear two-wire line ($b \to 0$), which was possible in the expression for the mutual inductance, cannot be carried out even now.

F. General Theorem Regarding Energy Transmission by Stationary Currents

We consider the section of an arbitrarily shaped wire between two cross sections F_1 and F_2. Let it contain a "load" in which electrical energy is translated into work or some other form of energy, e.g. a light bulb. We ask what *power* is supplied to the load (the Joule heat generated in our section of wire to be counted as part of the load).

We extend the cross sections F_1, F_2 to a closed surface F and calculate the power N as the inward-directed energy flux through this surface. According to Poynting's theorem (5.7a) and the meaning of Joule heat (5.5) we obtain under stationary conditions

$$N = \int_F S_n \, dF = \int \mathbf{E} \cdot \mathbf{J} \, dV. \tag{36}$$

V is the volume enclosed by F. Under stationary conditions we have everywhere within V

$$\text{curl } \mathbf{E} = 0, \qquad \mathbf{E} = -\text{grad } \Psi, \qquad \text{hence } \mathbf{E} \cdot \mathbf{J} = -\text{grad } \Psi \cdot \mathbf{J}.$$

We transform this with the aid of the obvious and universally valid identity

$$\text{div}(\Psi \mathbf{J}) = \text{grad } \Psi \cdot \mathbf{J} + \Psi \text{div } \mathbf{J}$$

---

[1] A similar definition is employed in electrical engineering for multiple conductor systems with the designation "operating selfinductance," if under the conditions of operation all circuit currents are determined by one of them.

where $\mathbf{J}$ is any vector, Ψ, any scalar. For our meaning of $\mathbf{J}$ the last term vanishes. We hence deduce from (36)

$$N = -\int \mathrm{div}(\Psi\mathbf{J})\,dV \tag{37}$$

and, by application of Gauss's theorem

$$N = \int_F \Psi J_n\,dF. \tag{38}$$

This integral need only be carried out over the surfaces of entry and exit, F_1 and F_2, of the current, since only here J_n differs from zero. It is most convenient to choose F_1 and F_2 as the equipotential surfaces $\Psi = \Psi_1$ and $\Psi = \Psi_2$. We then obtain from (38)

$$N = \Psi_1\int_{F_1} J_n\,dF_1 + \Psi_2\int_{F_2} J_n\,dF_2. \tag{39}$$

With the convention regarding the direction of n established above

$$\int J_n\,dF_1 = -\int J_n\,dF_2 = I$$

and hence, by (39),

$$N = (\Psi_1 - \Psi_2)I = VI. \tag{40}$$

V is the voltage drop between F_1 and F_2 (in volts) and I, the current entering and leaving (in amperes).

This fundamental formula, which expresses the power directly in watts (joules/S), has here been derived for stationary conditions; in §18 it will be found applicable also to "quasistationary" states. To avoid misunderstandings it should be noted that Eq. (40) makes no statement regarding the brightness of the light bulb or the energy radiated by it. This phenomenon lies outside of the domain of Maxwell's theory and rests on atomic processes which are made possible by the Joule heat supplied to the filament, but whose energy balance has nothing to do with Eq. (40). Our Eq. (40) follows the energy conversion only up to the generation of the Joule heat, not beyond it.

§16. *Ampère's Method of the Magnetic Double Layer*

In the last paragraph we had to introduce the concept of the *vector potential* in order to arrive at a representation of the magnetic field without and within the current-carrying conductor. If we content ourselves with a representation which applies only *outside* of the conductor we can get along with the ordinary *scalar* magnetic potential Ψ.

Outside of the conductors we have, by (15.1),

$$0 = \operatorname{curl} \mathbf{H}, \qquad \mathbf{H} = -\operatorname{grad} \Psi. \tag{1}$$

If we add the condition div $\mathbf{B} = 0$ and assume uniform permeability, e.g. $\mu = \mu_0$, outside, we have furthermore

$$\operatorname{div} \operatorname{grad} \Psi = \Delta \Psi = 0. \tag{2}$$

This potential Ψ is *not*, however, a *unique* function of position, as in the magnetostatic case. For every closed loop about a conductor carrying I it changes by the amount I, independently of the shape and length of the path:

$$\Psi_1 - \Psi_2 = \oint H_s \, ds = \pm I. \tag{3}$$

Here Ψ_1 and Ψ_2 are the values of Ψ at the starting point 1 and the coinciding endpoint 2 of the loop. The upper or lower sign of I applies depending on whether the loop forms a right or left screw with the direction of the current. On the other hand, for every closed path which does not link such a conductor:

$$\Psi_1 - \Psi_2 = \oint H_s \, ds = 0. \tag{3a}$$

The proof of both Eqs. (3) and (3a) follows again from (15.1). An arbitrary surface σ bounded by the path of integration cuts the conductor in question in the first instance; in the second it can always be placed so that it cuts no conductor. If the component of Eq. (15.1) normal to every surface element $d\sigma$ is formed and integrated over all $d\sigma$, we obtain

$$\int_\sigma J_n \, d\sigma = \int_\sigma \operatorname{curl}_n \mathbf{H} \, d\sigma.$$

By (15.9) the left side is the total current traversing the surface σ, i.e. I for Eq. (3) and 0 for (3a). The right side may be transformed by Stokes's theorem into the line integral over s.

For multiple loops about the conductors in one or the other direction a change in Ψ is obtained which is equal to the sum of the changes corresponding to individual circuits:

$$\Psi_1 - \Psi_2 = \sum n_k I_k .$$

Here n_k denotes the number of circuits about the kth conductor. Thus Ψ is infinitely multiplevalued. Any two "branches" of Ψ differ by a constant which is a sum of the currents I_k multiplied by integer coefficients n_k.

A. The Magnetic Shell for Linear Conductors

To give a prescription for the computation of Ψ which is unique in spite of this multiplicity we must confine ourselves to the limiting case of linear conductors (cross section $\rightarrow 0$); furthermore, it will suffice for the present to consider a single conductor. Let it be Λ. Through Λ as boundary we place an otherwise arbitrarily shaped "branch cut" surface S and forbid passage through S. In this manner we select, from the infinitely many-valued potential, a "function branch." Carrying over the already somewhat daring language of the *Riemannian surfaces* into three dimensions, we could also say: Of the "Riemannian space," whose infinite number of "leaves" have the branch line Λ in common and are joined in the branch cut S, we separate out one leaf as alone physically significant. This leaf has become "singly connected."

The calculation of Ψ may now be carried out by simple application of Green's theorem:

$$\int (u\Delta v - v\Delta u)\, d\tau = \int \left(u\,\frac{\partial v}{\partial n} - v\,\frac{\partial u}{\partial n} \right) d\sigma. \tag{4}$$

Here we put

$$u = \Psi, \qquad v = \frac{1}{r}, \qquad r = r_{PQ} \tag{4a}$$

and carry out the volume integral on the left over all points P of our physical leaf, excluding a sphere of radius $\rho \rightarrow 0$ about the source point Q and a sphere of radius $R \rightarrow \infty$ cutting off the infinite; the latter sphere may have an arbitrary point O as center, which may for example be situated on S. Correspondingly, the surface integral on the right is to be carried out over the two spherical surfaces K_ρ and K_R, as well as the two "sides" of the branch cut S. Since on the sphere K_ρ,

$$u\,\frac{\partial v}{\partial n} = -\Psi_Q \left(\frac{d}{dr}\frac{1}{r} \right)_{r=\rho} = \frac{1}{\rho^2}\,\Psi_Q$$

integration over $d\sigma$ here yields evidently

$$4\pi\Psi_Q\,. \tag{5}$$

The sphere K_R contributes

$$-\frac{1}{R^2}\int \Psi\, d\sigma - \frac{1}{R}\int \frac{\partial \Psi}{\partial n}\, d\sigma.$$

According to the law of Biot-Savart $\mathbf{H}$ decreases with increasing R as $1/R^2$. Hence the second of the above integrals is finite and, with its factor, vanishes as $1/R$. The first integral becomes infinite only in proportion to

R and yields similarly a vanishing contribution when multiplied with $1/R^2$. Thus the contribution of K_R is zero. We could confirm this result which is here based on the law of Biot-Savart, i.e. the method of the vector potential, also by the method of this paragraph.

Finally, we must consider the two sides 1 and 2 of the branch cut. In view of the opposite direction of n on the two sides we have

$$\left(\frac{\partial v}{\partial n}\right)_1 = -\left(\frac{\partial v}{\partial n}\right)_2. \tag{6}$$

Since

$$\frac{\partial u}{\partial n} = \frac{\partial \Psi}{\partial n} = -H_n$$

is a physical quantity which has nothing to do with the mathematical fiction of our branch cut we have, in addition to (6),

$$\left(\frac{\partial u}{\partial n}\right)_1 = -\left(\frac{\partial u}{\partial n}\right)_2. \tag{6a}$$

The sum of the contributions of the two sides of the branch cut to the right side of (4) may hence be written

$$\int \left\{ (u_1 - u_2)\left(\frac{\partial v}{\partial n}\right)_1 - (v_1 - v_2)\left(\frac{\partial u}{\partial n}\right)_1 \right\} d\sigma. \tag{6b}$$

Here $v_1 - v_2$ vanishes because of the meaning of $v = 1/r$, but, by Eqs. (3) and (4a), $u_1 - u_2 = \pm I$. Hence (6b) takes the form

$$\pm \int I \frac{\partial}{\partial n_1} \frac{1}{r} \, d\sigma. \tag{7}$$

To fix the sign, consider Fig. 21. Here 1 is to denote that side of S on which the normal n_1 directed toward S forms a *right screw* with the direction of current flow. The loop from 1 to 2 shown in the figure then forms a *left screw* with the direction of the current. Hence by prescription (3) we must choose the negative sign in formula (7). If we now write n for n_1 and place the factor I, which is a constant for all pairs of points 1, 2, ahead of the integral sign, this becomes

$$-I \int_S \frac{\partial}{\partial n} \frac{1}{r} \, d\sigma. \tag{7a}$$

Together with (5) we then obtain as the value of the right side of (4):

$$4\pi\Psi_Q - I \int_S \frac{\partial}{\partial n} \frac{1}{r} \, d\sigma.$$

Since the left side of (4) vanishes because $\Delta u = \Delta v = 0$, we obtain as the *final representation of* Ψ:

$$4\pi\Psi_Q = I \int_s \frac{\partial}{\partial n}\frac{1}{r}\,d\sigma. \tag{8}$$

The integral on the right has a very simple geometrical significance: It is the *solid angle* Ω intercepted by the circuit Λ at the point Q. In fact the integrand in Eq. (8)

$$\frac{\partial}{\partial n}\frac{1}{r}\,d\sigma = \frac{1}{r^2}\cos(n, r)\,d\sigma = \frac{d\sigma_n}{r^2}$$

is the surface element df of the unit sphere about Q cut out by the radii directed toward the boundary of $d\sigma$; $d\sigma_n$ is the corresponding surface element of a sphere of radius r. Hence

$$\Omega = \int_S df \tag{8a}$$

is the total area cut out on the unit sphere by the cone of radii directed toward the boundary of S, i.e. the above mentioned solid angle.

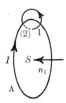

Fig. 21. The magnetic line integral about the conductor Λ, extended from side 1 to side 2 of the branch cut S.

The potential jump at the branch cut, $\Psi_1 - \Psi_2$, now also acquires a certain simple meaning. For if we place our point Q on side 1 of S the cone of radii degenerates into a flat fan and the solid angle Ω to 2π; on the other hand, if we place it on side 2, the solid angle becomes $\Omega = -2\pi$. If we form the difference of Eq. (8) for the two cases we find

$$4\pi(\Psi_1 - \Psi_2) = I(2\pi - [-2\pi]),$$

i.e. the potential jump demanded by (3).

We will supplement this *geometrical* interpretation of the expression (8) by a *magnetic* interpretation: We speak of a *double layer* on the branch cut S with the *magnetic surface densities* $\pm\,\omega_m$, which we think of as distributed parallel to S at a distance dn from each other. We regard the current I as the moment per unit area of this magnetic double layer:

$$I = \omega_m\, dn. \tag{9}$$

Eq. (8) may then be written

$$4\pi\Psi = \int_S \omega_m\, dn\, \frac{\partial}{\partial n}\frac{1}{r}\, d\sigma. \tag{9a}$$

If we designate by $r_\pm$ the distances of the point Q from the positive and the negative layer respectively we have

$$\frac{\partial}{\partial n}\frac{1}{r}\, dn = \frac{1}{r_+} - \frac{1}{r_-} \quad \text{and} \quad 4\pi\Psi = \int \frac{\omega_m\, d\sigma}{r_+} - \int \frac{\omega_m\, d\sigma}{r_-}. \tag{9b}$$

Accordingly Eq. (8) is in fact the potential of a *magnetic double layer* whose moment has the constant value I over its entire surface. Following Ampère we call the carrier of this double layer a *magnetic shell*; the linear conductor Λ forms the *boundary* of this shell.

In this connection we state a generally valid law of potential theory for simple and double layers: A simple layer of the type (7.5a)

$$4\pi\Psi = \int \frac{\omega}{r}\, d\sigma$$

leaves the *potential Ψ continuous* in passing through the carrier surface σ, but leads to a *discontinuous* normal component of the potential *gradient*, since

$$\left(\frac{\partial\Psi}{\partial n}\right)_1 - \left(\frac{\partial\Psi}{\partial n}\right)_2 = \omega;$$

on the other hand, a double layer of the type (8)

$$4\pi\Psi = \int I\, \frac{\partial}{\partial n}\frac{1}{r}\, d\sigma$$

makes the *potential discontinuous*, but leaves its *gradient continuous*. We have here

$$\Psi_1 - \Psi_2 = I, \quad \text{but } \mathbf{H}_1 - \mathbf{H}_2 = 0.$$

B. *Magnetic Energy and Magnetic Flux*

The calculation of the magnetic energy of a linear conductor in a medium of uniform permeability, i.e. the carrying out of the integration in the expression defining this energy:

$$W = \frac{1}{2}\int \mathbf{H}\cdot\mathbf{B}\, d\tau = \frac{\mu}{2}\int \mathbf{H}^2\, d\tau, \tag{10}$$

becomes particularly simple by the above method. We utilize here the so-called "second form of Green's theorem" (Vol. II, Eq. 3.16) which becomes,

$$\int u\Delta u \, d\tau + \int \text{grad } u \cdot \text{grad } u \, d\tau = \int u \frac{\partial u}{\partial n} \, d\sigma \tag{11}$$

if the two functions u and v occurring there are set equal to each other. We set $u = \Psi$ and extend the integration on the left over the entire exterior of our linear conductor, having made the potential Ψ unique with the aid of the branch cut S. The integral on the right is then to be carried out over the two sides 1 and 2 of S. We can omit the integration over the surface bounding the space considered at infinity in view of our knowledge regarding the behavior of Ψ at infinity.

The first term on the left of (11) vanishes since $\Delta\Psi = 0$; the second is identical with $2\,W/\mu$ since $\text{grad } \Psi = -\mathbf{H}$. Summing over sides 1 and 2 with due regard of the opposite signs of $\partial\Psi/\partial n$, the right side of (11) becomes

$$\int (\Psi_1 - \Psi_2) \frac{\partial \Psi}{\partial n} \, d\sigma = I \int H_n \, d\sigma, \tag{11a}$$

where the established rule regarding the correlation of the sign of the normal n and the direction of the current is to be observed. We thus obtain from (11)

$$W = \frac{\mu}{2} I \int H_n \, d\sigma = \frac{1}{2} I\Phi; \tag{12}$$

Here Φ is the flux of the magnetic induction through our conductor Λ.

A particularly simple definition of the coefficient of selfinduction L may be deduced from (12). For, if we compare (12) with the equation of definition (15.17) specialized for a single conductor,

$$W = \tfrac{1}{2}LI^2 \tag{13}$$

we find directly

$$\Phi = LI; \qquad L = \Phi/I. \tag{14}$$

We now pass from the single linear conductor considered so far to two linear conductors Λ_1 and Λ_2. We must then make space "singly connected" by means of two branch cuts S_1 and S_2, which are bounded by the currents I_1 and I_2. We shall call the normals correlated to I_1 and I_2 by the right screw rule n_1 and n_2. By superposition the magnetic fields of I_1 and I_2 form the total field $\mathbf{H} = \mathbf{H}_1 + \mathbf{H}_2$. In the integration over S_1 and S_2, to be carried out as in (11a), there appear the expressions

$$I_1 \int_{S_1} (\mathbf{H}_1 + \mathbf{H}_2)_{n1} \, d\sigma_1 \text{ and } I_2 \int_{S_2} (\mathbf{H}_1 + \mathbf{H}_2)_{n2} \, d\sigma_2.$$

Thus (12) is replaced by an energy expression of four terms, which we shall write, as in (15.17),

$$W = \tfrac{1}{2}(L_{11}I_1^2 + (L_{12} + L_{21})I_1I_2 + L_{22}I_2^2), \tag{15}$$

$$L_{11} = \frac{\mu}{I_1} \int_{S_1} H_{1n_1}\, d\sigma_1, \qquad L_{22} = \frac{\mu}{I_2} \int_{S_2} H_{2n_2}\, d\sigma_2, \tag{15a}$$

$$L_{12} = \frac{\mu}{I_2} \int_{S_1} H_{2n_1}\, d\sigma_1, \qquad L_{21} = \frac{\mu}{I_1} \int_{S_2} H_{1n_2}\, d\sigma_2. \tag{15b}$$

If **H** is expressed by Ψ, and Ψ by (8), it will be recognized that also with this definition

$$L_{12} = L_{21}.$$

We then obtain

$$4\pi L_{12} = \mu \int_{S_1} d\sigma_1 \int_{S_2} d\sigma_2 \frac{\partial^2}{\partial n_1\, \partial n_2} \frac{1}{r_{12}}, \tag{16}$$

where r_{12} is the distance between any one point on S_1 and any second point on S_2. Since the right side of (16) is symmetrical with respect to the indices 1 and 2 it represents also L_{21}.

By the representation in (15a, b) the magnetic fluxes Φ_1 and Φ_2 through S_1 and S_2 may be written

$$\Phi_1 = \mu \int_{S_1} (\mathbf{H}_1 + \mathbf{H}_2)_{n_1}\, d\sigma_1 = L_{11}\, I_1 + L_{12}\, I_2,$$

$$\Phi_2 = \mu \int_{S_2} (\mathbf{H}_1 + \mathbf{H}_2)_{n_2}\, d\sigma_1 = L_{21}\, I_1 + L_{22}\, I_2. \tag{17}$$

The magnetic flux is thus now expressed by two elements (for n linear conductors by n elements) in terms of the two (or n) currents.

C. Application to the Selfinductance of a Two-Wire Line

As in §15E we regard conductor and return conductor as *one* closed circuit and designate, just as there, the separation of the wires with a, the radius with b. Since we cannot proceed to the limit $b \to 0$ our method of the scalar potential, which is restricted to linear conductors, is strictly inapplicable. However, we may regard Eq. (14), quite apart from its origin in this method, as the definition of the selfinductance L, or, more exactly, of that part of L which has its origin in the exterior of the wires (see Fig. 22).

The portion of the branch cut S which is of importance to us has here been shaded. It extends in the xy-plane (the plane containing the two axes of the wires) from the periphery of one wire to that of the other and is to have the length 1 in the y-direction. The magnetic field **H** results, in general, from the *vectorial* superposition of the magnetic fields $\mathbf{H}_1$ and $\mathbf{H}_2$ of the two wires. On the plane area S, $\mathbf{H}_1$ and $\mathbf{H}_2$ have, however, the same direction, since $I_1 = -I_2$, and are perpendicular to S. With x and $a - x$,

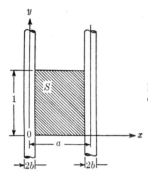

Fɪɢ. 22. Selfinductance of a two-wire line, computed from the magnetic flux through the branch cut S.

respectively, indicating the distance of the reference point on S from the conductor and the return conductor we obtain

$$H_n = \frac{I}{2\pi}\left(\frac{1}{x} + \frac{1}{a-x}\right);$$

hence we find from (14) for unity length of the two-wire line

$$L_a = \frac{\mu}{2\pi}\int_b^{a-b}\left(\frac{1}{x} + \frac{1}{a-x}\right)dx. \tag{18}$$

The subscript a of L indicates "external selfinductance". The evaluation of (18) yields:

$$L_a = \frac{\mu}{2\pi}\left(\log\frac{a-b}{b} - \log\frac{b}{a-b}\right) = \frac{\mu}{\pi}\log\frac{a-b}{b} \sim \frac{\mu}{\pi}\log\frac{a}{b}, \tag{19}$$

the last since $b \ll a$. If we compare this with (15.35), we see that the first part of the earlier formula corresponds to our "*external*" selfinductance and conclude from this that the second part will signify the "inner" selfinductance L_i.

We confirm this in the following manner: Employing the energetic definition of selfinductance in (15.17) we put

$$W_i = \tfrac{1}{2}L_i I^2. \tag{20}$$

W_i is the magnetic energy within a unit segment of our two-wire line. Within the individual wire the magnetic field is given, as in (15.33a), by

$$H = \frac{r}{b}\frac{I}{2\pi b},$$

if we neglect the magnetic field of the other single wire, which is weak in comparison. Hence the energy content of unit length of the individual wire becomes as in (15.33a)

$$\frac{\mu}{2}\int_0^b\int_0^{2\pi} H^2 r\, dr\, d\varphi = \mu\pi\int_0^b\left(\frac{r}{b}\frac{I}{2\pi b}\right)^2 r\, dr = \frac{\mu I^2}{4\pi b^4}\int_0^b r^3\, dr = \frac{\mu I^2}{16\pi}.$$

This is one half of the energy W_i in (20). It hence follows from (20) that

$$L_i = \frac{\mu}{4\pi}. \tag{20a}$$

Together with (19) this confirms our earlier result (15.35).

D. Application to the Electromagnetic Current Measurement of Wilhelm Weber

We proceed from Eq. (8) and assume to begin with that the surface S bounded by the current I is *plane*; the reference point for which Ψ is to be computed is supposed to be at a *great* distance from S. Then the direction of dn and the magnitude of $1/r$ become the same for all elements $d\sigma$ of S. The integration over $d\sigma$ may be carried out directly and yields

$$4\pi\Psi = IS\frac{\partial}{\partial n}\frac{1}{r}. \tag{21}$$

At the same time we recall Eq. (7.10b) for the potential of a single magnetic pole p. From it we obtain for the potential of a dipole of the very small separation l between poles and the direction n of the axis:

$$4\pi\Psi = M\frac{\partial}{\partial n}\frac{1}{r}, \qquad M = pl = \text{moment of the dipole.} \tag{22}$$

Comparison of (21) and (22) indicates: *The magnetic field of a current I bounding a surface S is, at large distance, equal to that of a dipole,* we might also say *a short bar magnet, which is placed perpendicular to the surface S and has the moment $M = IS$.*

A non-plane current path may be projected on three mutually perpendicular planes and the equivalent bar magnets may be arranged perpendicularly to the resulting plane current paths. Vectorial addition of their moments yields an obliquely oriented dipole, which at large distance again produces the same magnetic field as the original current path.

This *equivalence of current and magnetism* is the basis of the famous "electrodynamic determination of units" of Wilhelm Weber. Furthermore the electromagnetic system of units, which hails back to Weber, is based on it, i.e. on putting equal

$$I \cdot S = M. \tag{23}$$

The *electric* quantity I thus becomes a *magnetic* quantity. Or rather: For the electric quantity I a quantity M/S, which appears different in character, is substituted. This is possible only if a definite relation is established between the dimension of the magnetic pole, which as before we shall designate as P, and the dimension Q of charge. By (23) this relation is

$$Q \cdot \frac{\text{area}}{\text{time}} = P \cdot \text{length}$$

or

$$P = Q \cdot \frac{\text{length}}{\text{time}} = Q \cdot \text{velocity}.$$

We thus arrive at *Ampère's hypothesis* according to which magnetism is merely electricity in motion. However, we mentioned already on p. 47 that this hypothesis is today, after the discovery of the neutron as a basic element of all nuclear matter, no longer as binding as a hundred years ago; we also saw in §8B that Cohn's system of units is independent of this hypothesis and is recommended particularly by that fact.

   Our system of the four units MKSQ bears a peculiar relation to the electromagnetic cgs-system introduced by Weber. As we know, our unit Q = 1 coulomb = 1 ampere-second is defined as 1/10 of the electromagnetic unit of charge. The fundamental constants of vacuum, with due regard for the experimental fact $\varepsilon_0\mu_0 = 1/c^2$, were therefore found to be

$$\mu_0 = 4\pi \cdot 10^{-7} \frac{\text{joule } S^2}{Q^2 M} = 4\pi \cdot 10^{-7} \frac{\text{henry}}{M} \text{ , Eq. (15.18b)} \tag{24}$$

$$\varepsilon_0 = \frac{1}{\mu_0 c^2} = \frac{10^7}{4\pi c^2} \frac{Q^2 M}{\text{joule } S^2} \simeq \frac{10^{-9}}{36\pi} \frac{\text{farad}}{M} \text{ , Eqs. (7.17) and (10.3b)} \tag{25}$$

$$\sqrt{\frac{\mu_0}{\varepsilon_0}} = 4\pi c 10^{-7} \frac{\text{joule } S^2}{Q^2 M} \simeq 120\pi\Omega, \text{ Eqs. (7.19) and (4.5c).} \tag{26}$$

   However, we know that, apart from the electromagnetic (more briefly *magnetic*) system, also an electrostatic (more briefly *electric*) system is in use. Here the arbitrary, and only historically justifiable, convention (7.8) is made:

$$f = 4\pi\varepsilon_0 = 1 \tag{27}$$

and from this an *electrostatic unit* of charge e_{el} is defined. We ask how Q is to be expressed in this unit, i.e. what value Q/e_{el} may have, having fixed the value of Q measured in terms of the *electromagnetic unit* of charge e_{magn} by

$$\frac{Q}{e_{magn}} = \frac{1}{10}. \tag{28}$$

From Eqs. (25) and (27):

$$4\pi\varepsilon_0 = 1 = \frac{10^7}{c^2} \frac{(Q/e_{el})^2 \cdot 10^2 \text{ cm}}{10^7 \text{ erg sec}^2}.$$

From this and (28) it follows that

$$1 = \frac{1}{c^2}(e_{magn}/e_{el})^2, \qquad \frac{e_{magn}}{e_{el}} = c \simeq 3\cdot 10^{10} \frac{\text{cm}}{\text{sec}}. \qquad (29)$$

The unit of charge in the magnetic system is $3\cdot 10^{10}$ *times larger than the unit of charge in the electric system.* This corresponds to our metaphor of river and waterfall on p. 53. The numerical values of a given physical charge behave of course in inverse fashion. Thus the coulomb has, in the electric cgs-system, the numerical value

$$Q = \frac{1}{10}\cdot c = 3\cdot 10^9;$$

see also our data on the charge of the electron in §8D:

$$4.80\cdot 10^{-10} \text{ electric and } 1.60\cdot 10^{-20} \text{ magnetic units.}$$

The conversion rule for the numerical value e of an arbitrary charge in and MKSQ system to the corresponding numerical values e_{magn} and e_{el} is

$$10^9 \left(\frac{e^2}{4\pi\varepsilon_0}\right)_{MKSQ} = c^2 e_{magn}{}^2 = e_{el}{}^2. \qquad (30)$$

Since the quotient D/e on the one hand and the product Ee on the other are independent of our fourth unit, the conversion of D and E may readily be derived from Eq. (30); for the conversion of E (30) yields.

$$10 \, (4\pi\varepsilon_0 E^2)_{MKSQ} = \frac{1}{c^2} E_{magn}{}^2 = E_{el}{}^2. \qquad (31)$$

Rules (30) and (31) replace the rule given by H. A. Lorentz on p. 87 of Vol. V_2 of the *Mathematische Enzyklopädie*.

With this the unpleasant business of the electrical units may be regarded as definitely disposed of.

§17. *Detailed Treatment of the Field of a Straight Wire and a Coil*

We consider the apparently trivial case of an infinitely long straight wire carrying a stationary current with a return path in a coaxial hollow cylinder surrounding it. Let the radius of the wire be a, the inner radius of the hollow cylinder b, and the outer radius $c \to \infty$. For reasons of symmetry the magnetic field is known directly, and similarly the current density. As in §4, Eqs. (10) to (13), we have

$$r < a, \qquad H = \frac{r}{a}\frac{I}{2\pi a}, \qquad J = \frac{I}{\pi a^2},$$

$$a < r < b, \qquad H = \frac{I}{2\pi r}, \qquad J = 0, \tag{1}$$

$$b < r < \infty, \qquad H = \frac{I}{2\pi r}, \qquad J = 0.$$

The direction of H is everywhere azimuthal: $H = H_\varphi$.

In the interior the electric field is everywhere axial in direction and has by Ohm's law the value

$$E = E_z = \frac{J}{\sigma} = \frac{I}{\pi a^2 \sigma}, \qquad r \leqq a. \tag{2}$$

Similarly by Ohm's law we have within the hollow cylinder (conductivity σ_1)

$$E = \frac{J}{\sigma_1} = 0, \qquad b \leqq r < \infty. \tag{3}$$

In the region between wire and hollow cylinder the field is yet to be determined, from the differential equations

$$\mathbf{E} = -\text{grad } \Psi, \qquad \Delta\Psi = 0 \tag{4}$$

and the boundary conditions

$$\frac{\partial \Psi}{\partial z} = -E_z = \begin{cases} -\dfrac{I}{\pi a^2 \sigma} & \text{for} \qquad r = a, \\[2mm] 0 & \text{for} \qquad r = b. \end{cases} \tag{5}$$

Since these conditions are independent of z and φ we can write for the solution of $\Delta\Psi = 0$

$$\Psi = \Psi_1(r)z. \tag{6}$$

We can omit an additive term $\Psi_2(r)$ independent of z since we can satisfy all the conditions of the problem with formula (6). We then obtain from (5)

$$\Psi_1(a) = -\frac{I}{\pi a^2 \sigma}; \qquad \Psi_1(b) = 0. \tag{7}$$

The differential Eq. (4) demands

$$\frac{d}{dr} r \frac{d\Psi_1}{dr} = 0, \qquad \Psi_1 = A \log r + B,$$

which yields, with (7)

$$\Psi_1(r) = -\frac{I}{\pi a^2 \sigma} \log \frac{r}{b} \Big/ \log \frac{a}{b}. \tag{8}$$

Now the field $\mathbf{E}$ is known also in the intermediate space $a < r < b$. According to (6) and (8) it is represented by

$$E_z = -\Psi_1 = \frac{I}{\pi a^2 \sigma} \log \frac{r}{b} \Big/ \log \frac{a}{b}, \tag{9}$$

$$E_r = -\frac{\partial \Psi_1}{\partial r} z = \frac{Iz}{\pi a^2 \sigma r} \Big/ \log \frac{a}{b}. \tag{9a}$$

Thus the field is here by no means axial in direction, as in the interior of the wire; rather, its radial component is of the same order of magnitude as its axial component.

In passing from the interior to the exterior of the wire there occurs a jump in E_r and hence also in the corresponding component of the excitation $\mathbf{D}$, which indicates the existence of a *surface charge*:

$$\omega = D_r = \varepsilon E_r = \frac{\varepsilon Iz}{\pi a^3 \sigma} \Big/ \log \frac{a}{b}. \tag{10}$$

This surface charge decreases linearly along the wire, from positive to negative values, in formal language from $+\infty$ to $-\infty$. It depends only slightly, i.e. logarithmically, on the radius b of the outer return conductor. The zero point of the charge remains undetermined since the point $z = 0$ can be fixed arbitrarily. We may eventually identify it with the "center" of the wire, which, for infinite length, also remains indefinite.

We can obtain an idea as to the magnitude of this charge in the following manner: The dielectric constant in (10), which refers to the exterior of the wire, does not differ materially from that in the interior of the wire (though for metals it is rather hypothetical). Hence the quotient ε/σ does not differ materially from the relaxation time T_r defined in (4.9a), for the material of the wire; this is of the order of *micro*seconds. The product $\varepsilon I/\sigma$ occurring in (10), which represents a charge, is thus not of the order of an ampere-second $= Q$, but of the order of a microampere-second $= 10^{-6}$ Q. The peripheral charge of the wire and the radial field strength corresponding to it are hence very small. This is the reason why they generally remain unnoticed both in theory and in experiment although, as we shall see, they are essential for an understanding of the current transport.

Implicit in the existence of a radial electric field is the appearance of a potential difference between wire and return conductor. According to (9a) it is given by

$$V = \int_a^b E_r \, dr = -\frac{Iz}{\pi a^2 \sigma}. \tag{11}$$

We compare it with the charge on unit length of the wire, $e = 2\pi a \omega$, or the equal and opposite charge on unit length of the return conductor. We find

$$\frac{e}{V} = \frac{2\pi\varepsilon}{\log \dfrac{b}{a}}.$$ (11a)

This is the capacity per unit length of the cylindrical condenser formed by the wire and the return conductor (see Problem II.5). We speak here and in similar cases of a "distributed capacity."

Fig. 23 shows the shape of the *equipotential lines* Ψ = const. in the z, r-plane, given according to Eqs. (6) and (8) by

$$z \log \frac{b}{r} = C.$$ (12)

For $C = 0$, $z = 0$ and $r = b$, corresponding to the broken lines ABC and ABD of the figure. The equipotential lines for $C > 0$ accommodate themselves within the area bounded by them. The angles under which they meet the surface of the wire deviate increasingly from a right angle with increasing distance from A. The orthogonal trajectories to the equipotential lines represent the *lines of force*; the arrows on them indicate the direction from positive to negative surface charge ω. Two bounding curves passing through the "equilibrium point" B (see footnote 2 after Eq. (9.14b)) belong to this family of curves. According to (12) the equipotential lines satisfy the differential equation

$$\log \frac{b}{r} dz - z \frac{dr}{r} = 0;$$

on the other hand, the lines of force, which are orthogonal thereto, (replacement of dz/dr by $- dr/dz$) are given by

$$z\, dz + r \log \frac{b}{r} dr = 0.$$

In the neighborhood of B we find, with $\rho = b - r$,

$$z\, dz - \rho\, d\rho = 0.$$

Thus *two* lines of force, with tangent directions $z = \pm\rho$, pass through the point B forming a right angle with each other. At a greater distance from these bounding curves, above and below them, the lines of force pass more or less radially from the wire surface to the outer conductor.

The equipotential lines represent at the same time the paths of the *energy flux* $\mathbf{S}$; the arrows marked on them indicate the direction of $\mathbf{S}$. From the formula $\mathbf{S} = \mathbf{E} \times \mathbf{H}$, $\mathbf{S}$ is perpendicular to $\mathbf{H}$ and hence lies in the plane of the drawing, since $\mathbf{H}$ is everywhere perpendicular thereto; in addition $\mathbf{S}$ is perpendicular to $\mathbf{E}$ and hence has everywhere the direction of the family of curves $\Psi = \text{const.}$ An application of the right-screw rule for the three vectors $\mathbf{E}, \mathbf{H}, \mathbf{S}$ shows that the arrows are properly oriented.

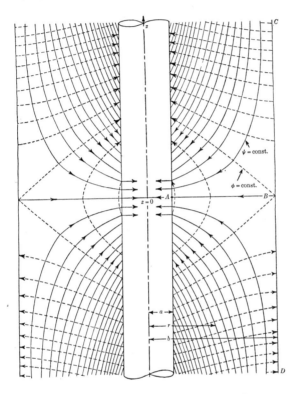

Fig. 23. Energy flux about a straight wire carrying stationary current with co-axial return conductor. Equipotential lines = stream lines of the energy drawn out full, electrical lines of force = excitation lines drawn dotted. They form tubes of constant charge starting and ending on the surface of the wire and the return conductor, respectively. Their number per unit length of the wire indicates the surface charge on the latter and its linear increase with distance from A, positive for $z < 0$, negative for $z > 0$.

Since $\mathbf{E} = \mathbf{J}/\sigma$ the lines of force within the wire (not drawn in the figure) are axial in direction; hence the vector $\mathbf{S}$ is directed inward, *perpendicular* to the surface of the wire; here also it lies in the equipotential surfaces $\Psi = \text{const.}$ The energy flux is dissipated in the interior of the wire since it becomes zero for $r = 0$, in view of $\mathbf{H} = 0$. This is indicated in the figure by the terminated arrows at A. Such arrows should be imagined along the entire surface $r = a$. The energy entering from the surface is converted into Joule heat in the interior of the wire.

According to (9) and (1) the magnitude of the energy flux for $r = a$ is

$$|\mathbf{S}| = E_z H_\varphi = \frac{I}{\pi a^2 \sigma} \frac{I}{2\pi a};$$

the energy supply to unit length of the wire from all sides hence becomes

$$2\pi a\,|\mathbf{S}| = I^2 R_1, \qquad R_1 = \frac{1}{q\sigma}, \qquad q = \pi a^2. \tag{13}$$

R_1 is the ohmic resistance of unit length of the wire; (13) thus yields, in fact, the Joule heat claimed by this unit length.

Accordingly we obtain the following total picture of the behavior of the energy: *Outside of the wire the energy flows from the electrodes $z = \pm\infty$ from all sides toward the surface of the wire. After entering it it flows radially toward the axis of the wire, being converted at the same time into heat. There is no energy flux parallel to the wire axis within the wire.*

This picture is materially different from the popular concept of the energy transfer in a wire carrying current. From the Maxwellian standpoint there is no doubt, however, about the inner consistency and unique validity of our picture. It indicates the fundamental change which Maxwell's theory has brought in the concepts conductor and nonconductor: *The conductors are nonconductors of energy. Electromagnetic energy is transported without loss only in nonconductors*; in conductors it is destroyed, or rather transformed. The notation "conductor" and "nonconductor" refers only to the behavior with respect to *charge*; it is misleading if applied to behavior with regard to *energy*.

We thus come to the conclusion that our simple example is after all not as trivial as it appeared.

The next-simplest case of the *circular* conductor is already beyond elementary treatment. Even in the limiting case of the linear conductor the geometrical formula for the solid angle involves an elliptic integral. In fact, if we use the polar coordinates r, φ, z for the reference point and ρ, $\alpha + \varphi - \pi$, 0 for the point of integration we find from Eq. (16.8), denoting the distance between point of integration and reference point by R,

$$\int \frac{\partial}{\partial n} \frac{1}{R}\, d\sigma = \int_0^a \rho\, d\rho \int_0^{2\pi} d\alpha\, \frac{\partial}{\partial z} (r^2 + \rho^2 + 2r\rho \cos \alpha + z^2)^{-\frac{1}{2}}$$

$$= 4 \frac{\partial}{\partial z} \int_0^a \frac{\rho\, d\rho}{\sqrt{(r+\rho)^2 + z^2}} \int_0^{\pi/2} \frac{d\beta}{\sqrt{1 - k^2 \sin^2 \beta}}$$

$$\beta = \alpha/2$$

$$k^2 = \frac{4r\rho}{(r+\rho)^2 + z^2}$$

The preceding integral with respect to β is a "complete elliptical integral of the first kind in the Legendre standard form," and k is the "modulus" thereof. We cannot of course here delve further into the treatment of this formula.

The exact treatment of a coil of wire of finite thickness with finite pitch would be even more complex. We hence pass directly to the limiting case of very small thickness and pitch, i.e. Ampère's solenoid, whose magnetic field was already discussed in Eq. (4.14), though only superficially. We now wish to compare it with the field of a permanent uniformly magnetized bar magnet of the same dimensions as the coil. We shall here assume the interior of the coil and the coil wire to be unmagnetic ($\mu = \mu_0$).

We shall show that the **H**-*field of the coil* corresponds to the **B**-*field of such a bar magnet*. As proof we write down the boundary conditions and differential equations for the two cases side by side:

$$
\begin{array}{cc}
\textit{Coil} & \textit{Bar Magnet} \\
(\mathbf{H}_a - \mathbf{H}_i)_s = N_1 I, & (\mathbf{B}_a - \mathbf{B}_i)_s = \mu_0(\mathbf{H}_a - \mathbf{H}_i - \mathbf{M})_s = -\mu_0 M, \\
(\mathbf{H}_a - \mathbf{H}_i)_n = 0, & (\mathbf{B}_a - \mathbf{B}_i)_n = 0, \\
\Delta\Psi = 0, & \text{div } \mathbf{B} = \mu_0 \text{ div } \mathbf{H} = -\mu_0 \Delta\Psi = 0.
\end{array}
$$

The first line relates to the *mantel surface*. Its left half states simply, in the terminology introduced at Eq. (4.4e), that the surface curl on the mantel surface of the coil is equal to $N_1 I$, where N_1 is the number of turns per unit length. The right half of the first line follows from our equation $\mathbf{B} = \mu_0(\mathbf{H}_i + \mathbf{M})$ for the interior of the bar magnet, which for the exterior becomes $\mathbf{B} = \mu_0\mathbf{H}_a$. It states that the surface curl of **B** on the mantel surface of the bar is equal to $-\mu_0\mathbf{M}$. A comparison of the two halves of the first line shows that, in the bar magnet, the quantity $-\mu_0 M/N_1$ corresponds to the current I in the coil.

We apply the second line in particular to the two *end surfaces*. Here both **B** and **H** are of course continuous for the solenoid, for the bar magnet only **B**. Since we have assumed $\mu = \mu_0$ the same equations apply also for the mantel surface.

The third line applies in the two cases both for the interior and for the exterior; here we must recall our assumption regarding the bar magnet that its magnetization was supposed to be uniform, since otherwise the term μ_0 div **M** would have to be added to the term μ_0 div **H** and this would spoil the comparison of the two cases.

We see therefore that our earlier Fig. 17, which represented the **B**-field of the bar magnet, reproduces simultaneously the **H**-field of the solenoid. Accordingly the earlier Eq. (4.14), which applies only for an infinitely long coil, is rounded out graphically by Fig. 17. This provides us now with a complete picture of the spreading of the lines of excitation at the ends of the coil and of their exit through the convex surface.

With respect to the representation of the coil field by the potential Ψ we wish to point out expressly its familiar multivalued character; the branch cut S, which makes it single valued, is a helical surface of infinitely small pitch which follows the turns of the wire. It follows that the equation

$$\oint \mathbf{H} \cdot d\mathbf{s} = 0$$

which is universally valid for the bar magnet loses its validity for the coil if the path of integration links one or several turns. Hence also the conclusion with regard to the demagnetizing character of the $\mathbf{H}$-field of the bar magnet is not applicable to the coil. For to reach this conclusion, we employed (see p. 85) a path which inside was along the axis of the bar and on the outside led back to the bar axis. This same path, for the coil, intersects the above-mentioned helical surface and therefore is not a permissible path of integration.

We can also make the following statement: The $\mathbf{H}$-field of the *bar magnet* is *lamellar* throughout, that of the *coil* is *not*; instead it has the *curl* I *concentrated* at the turns of wire. The $\mathbf{H}$-field of the coil is *solenoidal* throughout, since everywhere $\mathbf{B} = \mu_0 \mathbf{H}$, that of the bar magnet is *not*. For the uniformly magnetized bar the bar ends have a *surface distribution of divergence*. In view of the proportionality of $\mathbf{B}$ and $\mathbf{H}$ for the coil, Fig. 17 evidently also represents the $\mathbf{B}$-*field of the coil*. The $\mathbf{H}$-*field of the bar magnet*, Fig. 18, is materially different from Fig. 17.

The near-uniformity of the internal field evident in Fig. 17 suggests the computation of the magnetic energy W of the coil by the formula

$$W = \frac{\mu}{2} \mathbf{H}^2 V, \tag{14}$$

where $V = \pi a^2 l$ is the volume of the interior of the coil. In view of the relation $Hl = NI$ (Eq. (4.14)), where N is the number of turns along the full length of the coil, (14) takes the form

$$W = \frac{\pi a^2}{l} \mu N^2 \frac{I^2}{2}. \tag{14a}$$

A comparison with our energetic definition of selfinductance in (15.17) leads to the following value of the latter:

$$L = \frac{\pi a^2}{l} \mu N^2. \tag{15}$$

This formula applies, of course, as may also be seen from Fig. 17, only for a *very long* coil and hence can scarcely be used for forms encountered in practice. On the other hand, it retains its validity if the straight coil is bent into a *ring electromagnet*, because of the close approach to field uniformity within the latter. Here the factor μ in (15) also becomes significant if the ring is provided with a soft-iron core. Eq. (15) shows that this arrangement, first suggested by Ampère, has a much greater selfinductance and hence realizes a much greater concentration of energy than the electromagnet without iron *for equal coil current.*

This increase in energy concentration applies qualitatively also to the straight coil with soft-iron core. The quantitative computation of field and energy would, however, be much more complex, since a boundary-value problem, corresponding to the passage of the magnetic lines of force from iron into air, would be added to the summation problem which we have treated.

§18. Quasi-Stationary Currents

Most of the problems of electrical engineering and of laboratory physics lie within the domain of the *slowly variable fields*. It is true that there is no one-word answer to the question "slow compared with what?". For vibrations which are periodic in time or exhibit a damped periodicity the answer may be identified with the demand that the light path corresponding to the period τ of the vibration be large compared to the dimensions l of the apparatus:

$$c\tau \gg l. \tag{1}$$

It is then permissible to neglect the "retardation of the fields" to be introduced in §19. At the end of this paragraph we shall, however, treat successfully also very long lines, which do not obey this condition, on a quasi-stationary basis, by subdivision into differential segments.

In general terms the *quasi-stationary* approximation consists in calculating all fields as for *stationary* processes. In this manner it becomes possible to establish a *linear* relation between the expressions occurring in the integral form of the Maxwell equations. We refer to the magnetic flux Φ through a closed curve, the current I through a cross section, and the electromotive forces V_{ik} along segments of the path of integration, which add up to the loop e.m.f. V for this path:

$$\Phi = \int B_n \, d\sigma, \qquad I = \int J_n \, d\sigma, \qquad V_{12} = \int_1^2 E_s \, ds,$$

$$V = \sum V_{ik} = \oint E_s \, ds.$$

We know that, under stationary conditions, the following relations exist between them:

$$\Phi_1 = L_{11} I_1 + L_{12} I_2 + L_{13} I_3 + \cdots \text{ (magn. flux through circuit 1)}$$

$$I = \frac{V}{R} \text{ (Ohm's law)}, \quad I = \dot{V}K \text{ (condenser charge)}.$$

The inductances L, resistances R, and capacities K depend, apart from the material constants, only on the geometry of the field and are the result of an integration over the *space* coordinates. Hence only an integration with respect to the *time* remains to be carried out. The mathematical simplification achieved in this manner is considerable: While the exact treatment of rapidly variable fields demands the integration of the Maxwell *partial differential equations*, the integration of *ordinary differential equations* with constant coefficients suffices for slowly varying fields; for periodic processes these reduce even to *algebraic equations*.

This method was developed by Gustav Kirchhoff (see beginning of §15) and was applied to metal wire loops and networks composed of them. It is in general appropriate to choose the closed path of integration along the metal wires; an interruption of the metallic path by nonconducting gaps (condensers) does not interfere with the method, however. The case of *thin* wires is particularly convenient, since here the field within the wire is practically uniform.

The *first* Maxwell equation in integral form then yields for the chosen closed path of integration

$$-\dot{\Phi} = V = \text{loop e.m.f.} \tag{2}$$

The *second* Maxwell equation finds expression in the computation of the coefficients of induction insofar as it represents the production of magnetic fields by currents and in the computation of the resistances and capacities insofar as it represents the origin of these currents from the electric field.

For a simple current loop without branching the two sides of (2) may be expressed in terms of the current I which is the *same* for all cross sections:

$$-\dot{I} \sum L = I \sum R + \int I \, dt \sum \frac{1}{K} + E. \tag{3}$$

It is seen that here the selfinductance coefficients of all the magnetic fields can be combined in a single expression, which may be represented by an imaginary coil at an arbitrary point of the circuit. The same applies for the resistances and the capacitances. E is the presumably known e.m.f. between the terminals by which the current enters and leaves, i.e. the socalled *terminal e.m.f.* The introduction of this e.m.f. E conveniently avoids carrying the path of integration through apparatus whose action is not covered by the Maxwell theory (galvanic or thermoelements, photoelectric cells, electron tubes) or through "machines" which, though they function fully within the framework of the theory, would unduly complicate the problem.

If *several* loops are joined in a *network*, the latter may be regarded as made up of elements connecting the "junction points". Let the nth element carry a current I_n from one junction point to another. Since we do not as yet know its sign, we place a marker arrow on the element which is to indicate in what direction we shall reckon the current as positive. In view of the absence of current sources we have for the junction points

$$\sum I_n = 0 \tag{4}$$

and for *every* circuit made up of *arbitrary* elements Eq. (2) applies again in the form

$$\sum I_n R_n + \sum V_{nm} = -\dot{\Phi} \quad (= \text{e.m.f.}). \tag{5}$$

(4) and (5) are known as the "first" and "second" Kirchhoff equations. They date from the time preceding Maxwell's theory. Hence Kirchhoff places on the right side of (5) not $-\dot{\Phi}$, but the older concept of the *electromotive force* (e.m.f.) of all "current sources" which are inserted in the closed circuit. If we are dealing with currents which result from Faraday induction, e.g. in coils of machines, this e.m.f. becomes exactly identical with $-\dot{\Phi}$ and its concept is superfluous. It has, however, the advantage of covering also the effect of other current sources (batteries etc.) without requiring an examination of the physical processes taking place therein.

A. Energetic Interpretation of the Wave Equation

We consider in particular an unbranched circuit with selfinductance, capacitance, and ohmic resistance, which we imagine connected in series and, as in Eq. (3), concentrated at certain points of the circuit. Thus we insert a resistance box in place of the resistance which is distributed over the wire; a coil in place of the distributed selfinductance with which we became familiar in connection with the two-wire line in §15E; and we shall not consider distributed capacities in the circuit but assume instead the presence of an electric plate condenser.

The law of conservation of energy offers a particularly convenient approach to the treatment of such a system, just as to the treatment of material vibrations in mechanics. We write it in the form of Eq. (5.7a):

$$\dot{W}_m + \dot{W}_e + Q = \oint S_n \, d\sigma. \tag{6}$$

W_m is the magnetic energy concentrated in the coil, given according to Eq. (17.14a) by

$$W_m = \frac{L}{2} I^2. \tag{6a}$$

W_e is the electric energy within the plate condenser, according to Eq. (10.11) given by

$$W_e = \frac{1}{2K} e^2. \tag{6b}$$

Here $\pm e$ are the varying charges on the two condenser plates and

$$I = \frac{de}{dt}.$$ (6c)

Q is the Joule heat generated within the resistance box:

$$Q = RI^2.$$ (6d)

The energy flux $\mathbf{S}$ refers to the current source. By Eq. (15.40) we write for the energy supplied by it in unit time

$$\oint S_n \, d\sigma = EI \text{(volt-ampere = watt)}.$$ (6e)

E is the terminal e.m.f. of the current source mentioned above.

From (6a, b) follows

$$\dot{W}_m = LI\dot{I}, \qquad \dot{W}_e = \frac{1}{K} e\dot{e} = \frac{1}{K} eI.$$ (6f)

Substitution of (6d, e, f) in (6) leads to, after canceling of a factor I:

$$L\dot{I} + RI + \frac{1}{K} e = E,$$ (7)

and, after a second differentiation with respect to t, to the *wave equation*

$$L\ddot{I} + R\dot{I} + \frac{1}{K} I = \dot{E}.$$ (7a)

In the terminology of point mechanics there corresponds thus
$\dot{E}$ to the exciting force,
L to the inertia, more particularly the mass of the vibrating particle,
R to the damping, and
K to the coefficient of the restoring force.
As in mechanics, we distinguish between free and forced vibrations.

a. Free Vibrations

We set $E = 0$ and ask for the solution of the homogeneous equation

$$L\ddot{I} + R\dot{I} + \frac{1}{K} I = 0.$$ (8)

We can assume a trigonometric form for I, but know from point mechanics that it is definitely preferable to use instead the exponential form at the start and to pass to the real part of I only after the integration. We therefore set[1]

$$I = I_0 e^{i\omega_0 t}$$ (8a)

[1] Temporarily we adhere to custom in employing the usual positive sign of i in the exponent, although we generally prefer the negative sign. See e.g. §6, Eq. (11).

and obtain for the circular frequency $\omega_0 = 2\pi/\tau_0$ of the free vibration the quadratic equation

$$-L\omega_0^2 + iR\omega_0 + \frac{1}{K} = 0. \tag{8b}$$

For no damping we find

$$\omega_0^{\ 2} = \frac{1}{KL}, \qquad \tau = 2\pi\sqrt{KL}. \tag{9}$$

This is the Kirchhoff-Thomson formula. If it is necessary to take account of damping the solution of (8b) yields

$$\omega_0 = \frac{iR}{2L} \pm \sqrt{\frac{1}{KL} - \frac{R^2}{4L^2}}. \tag{9a}$$

The process is aperiodic or periodic, depending on whether

$$\frac{R}{2L} > \frac{1}{\sqrt{KL}} \quad \text{or} \quad \frac{R}{2L} < \frac{1}{\sqrt{KL}}. \tag{9d}$$

In the aperiodic case ω_0 is purely imaginary and the current (8a) decreases *monotonically*. In the periodic case usually found with condenser discharges the angular frequency is

$$\sqrt{\frac{1}{KL} - \frac{R^2}{4L^2}}. \tag{9c}$$

Since the ohmic damping occurs here only as a correction of the second order as compared with the first term on the right of (9a), (9c) can usually be replaced by (9)—in analogy with the mathematical pendulum, where the formula $\tau = 2\pi\sqrt{l/g}$ is not affected appreciably by air resistance etc; hence it is generally permissible to write in place of (8a):

$$I = I_0 e^{-Rt/(2L)} \cdot e^{\pm 2\pi it/\tau_0}, \tag{9d}$$

where τ_0 now represents the value (9) with no damping. The double sign in (9d) evidently becomes unimportant when finally passing to the real part, but permits taking account of the phase of the oscillation which, like the amplitude, may be prescribed arbitrarily, through superposition of the two solutions.

b. Forced Vibrations

Here we proceed preferably from Eq. (7) and set

$$E = E_0 e^{i\omega t};$$

ω is the angular frequency of the alternating current source. The circuit oscillates with the same rhythm as soon as its characteristic vibrations, which are determined by an arbitrary initial state, have decayed. We have therefore

$$I = I_0 e^{i\omega t}, \qquad \dot{I} = i\omega I, \qquad e = \frac{I}{i\omega} \qquad \text{(see (6c))}.$$

We thus obtain from (7)

$$\left(i\omega L + R + \frac{1}{i\omega K} \right) I = E,$$

for which we shall write more briefly

$$\mathbf{R}I = E. \tag{10}$$

(10) is *Ohm's law for alternating currents*; the real ohmic resistance R is here replaced by the complex impedance

$$\mathbf{R} = R + i \left(\omega L - \frac{1}{\omega K} \right). \tag{10a}$$

If we put

$$\mathbf{R} = | \mathbf{R} | e^{i\alpha}, \tag{10b}$$

we evidently have

$$| \mathbf{R} | = \sqrt{ R^2 + \left(\omega L - \frac{1}{\omega K} \right)^2 }, \qquad \tan \alpha = \frac{\omega L - 1/(\omega K)}{R} . \tag{10c}$$

We introduce the following designations:

$$R = \text{resistance}$$
$$\omega L - 1/(\omega K) = \text{reactance}$$
$$\omega L = \text{inductive reactance}$$
$$1/(\omega K) = \text{capacitative reactance}$$
$$| \mathbf{R} | = \text{impedance.}[1]$$

We give an interpretation of Eq. (10) in the complex Gaussian plane of Fig. 24. The two-dimensional "vector" I lags by the constant angle α behind the two-dimensional "vector" E. Of course only the real parts of E and I have physical meaning.

The engineer calls this representation a rotating vector diagram; The figure should in fact be thought of as rotating with the angular velocity ω as time progresses. The projections of the two-dimensional vectors E and I on the real axis give the instantaneous values of these quantities.

We shall establish furthermore how the energy flowing into the system

---

[1] In electrical engineering this term, incidentally, is used not only for $| \mathbf{R} |$, but also for our impedance operator $\mathbf{R}$ itself.

is used up. To this end we multiply Eq. (7) with I, where now I and E are to represent their real parts. We find

$$\frac{L}{2}\frac{d}{dt}I^2 + RI^2 + \frac{1}{2K}\frac{d}{dt}e^2 = IE. \tag{11}$$

If we average over a period $\tau = 2\pi/\omega$ of the vibration, we obtain:

$$\frac{1}{\tau}\int_0^\tau RI^2\,dt = \frac{1}{\tau}\int_0^\tau IE\,dt,$$

so that

$$R\overline{I^2} = \overline{IE}. \tag{11a}$$

The contributions derived from L and K vanish since they are given by differentials. The average power $\overline{IE}$ introduced into the system is thus dis-

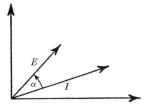

FIG. 24. Representation of the complex E and I in the Gaussian plane. Lag of I with respect to E by the angle α.

sipated entirely in the ohmic resistance R. The imaginary part of our impedance operator **R** has a purely wattless effect on the e.m.f., i.e. it does not consume energy on a time average. We can calculate directly

$$\overline{I^2} = \frac{1}{\tau}\int_0^\tau I_0^2\cos^2(\omega t - \alpha)\,dt = \frac{1}{2}I_0^2; \tag{11b}$$

$$I_{eff} = \frac{1}{\sqrt{2}}I_0 \text{ and correspondingly } E_{eff} = \frac{1}{\sqrt{2}}E_0; \tag{11c}$$

I_{eff} is called the "effective current", E_{eff} the "effective voltage". For the average power we find

$$\overline{EI} = \frac{1}{\tau}\int_0^\tau E_0 I_0\cos\omega t\cos(\omega t - \alpha)\,dt$$

$$= \frac{1}{\tau}E_0 I_0\left\{\int_0^\tau\cos^2\omega t\cos\alpha\,dt + \int_0^\tau\cos\omega t\sin\omega t\sin\alpha\,dt\right\} \tag{12}$$

$$= \tfrac{1}{2}E_0 I_0\cos\alpha = I_{eff}E_{eff}\cos\alpha.$$

Analog: Work = Path·Projection of the force on the path, where the projection must be carried out in the complex plane in the present example. This analogy applies however only for the time average, in which the second integral of the middle line of (12) vanishes. The latter signifies an oscillatior of the energy between storage (K, L) and the current source (E).

B. The Wheatstone Bridge

We distinguish between the four bridge arms a, b, c, d and the arm: e and f, containing the current source and the galvanometer, respectively The disappearance of the current in the galvanometer arm is attained by adjustment of the slide-wire contact S (or eventually the two sliding con tacts S and T). Thus a *null method*, viewed with such favor in physica

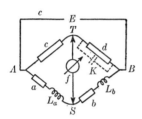

FIG. 25. Wheatstone bridge: a, b, c, d branches o bridge with resistance boxes and induction coils con nected in series and, eventually, capacitances con nected in parallel; f and e, branches containing galvan ometer and voltage source; S and T slide contacts.

measurements, is realized. The geometrical structure of the bridge is bes represented by a tetrahedron (Fig. 26), the six arms being transformed into the six sides without altering their relationship, as in analysis situs. The two arms e and f then become opposite sides, similarly the arms a, d and b c, while the arms a, b etc. become "adjoining sides".

FIG. 26. Space representation of Wheatstone bridge as tetrahedron; adjoining and opposite branches.

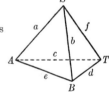

From the great range of applications of the Wheatstone bridge we selec two characteristic special cases; the trivial case of the comparison of two ohmic d.c. resistances will be a by-product:

a. Comparison of two selfinductances

b. Comparison of a selfinductance and a capacitance

a. Let the two selfinductances L_a and L_b be inserted in the adjoining arm: a and b, in series with the ohmic resistances a and b, as shown in Fig. 25 In order that there may be no current in the arm f, we adjust the sliding contacts S and T so that there is no difference of potential between them Starting from A we have, from (10) and (10a),

$$(a + i\omega L_a)I_1 = cI_2 \tag{13}$$

nd starting from B:

$$(b + i\omega L_b)I_1 = dI_2, \tag{13a}$$

o that

$$(a + i\omega L_a)\, d = (b + i\omega L_b)c. \tag{14}$$

f this equation is to be satisfied, the equality must exist individually for he real and for the imaginary parts. Thus:

$$ad = bc, \qquad L_a d = L_b c.$$

$$\frac{a}{b} = \frac{c}{d} = \frac{L_a}{L_b}. \tag{14a}$$

The first half of this double equation applies also for the equilibrium of an nductance-free bridge, irrespective of whether it is traversed by direct or by alternating current.

b. Let the selfinductance and the capacitance lie in two opposite arms of he bridge, e.g. a and d, in such fashion that the selfinductance L and the ohmic resistance a are connected in series, the capacitance K and the ohmic resistance d, in *parallel*. According to Kirchhoff (see p. 101) the potential drop across two parallel ohmic resistances R' and R'' with the urrents I' and I'' is given by

$$RI \quad \text{with} \quad I = I' + I'', \quad \frac{1}{R} = \frac{1}{R'} + \frac{1}{R''}.$$

The corresponding voltage drop for alternating current is evidently

$$\mathbf{R}I \quad \text{with} \quad I = I' + I'', \quad \frac{1}{\mathbf{R}} = \frac{1}{\mathbf{R}'} + \frac{1}{\mathbf{R}''}. \tag{15}$$

For our bridge arm d the total current at its end points is I_2 (see Fig. 25); we therefore set

$$I = I_2, \quad \mathbf{R}' = d, \quad \mathbf{R}'' = \frac{1}{i\omega K}, \quad \frac{1}{\mathbf{R}} = \frac{1}{d} + i\omega K$$

and obtain in place of Eq. (13a)

$$bI_1 = \frac{I_2}{\dfrac{1}{d} + i\omega K}. \tag{15a}$$

n combination with the unchanged Eq. (13) this leads to:

$$\frac{I_2}{I_1} = \frac{a}{c} + i\omega\,\frac{L}{c} = \frac{b}{d} + i\omega bK,$$

which yields, separating real and imaginary parts,

$$ad = bc = \frac{L}{K}. \tag{16}$$

In the notation of p. 138 the last term of this double equation is the *produc* of inductive and capacitative reactance. Eq. (16) shows that the deter mination of a capacity from bridge equilibrium rests, apart from ohmi€ resistances, on a known selfinductance (or vice versa).

In our examples we have restricted ourselves to cases in which the equi librium conditions, Eqs. (14a) and (16), do not depend on ω. In such case: the bridge equilibrium exists not only for the purely periodic alternating current here assumed, but for an arbitrary time variation, e.g. excitation by an interrupter. There are other cases in which the equilibrium condition: depend on ω; bridge equilibrium then exists only for sinusoidal alternating current.

C. Coupled Circuits

The law of conservation of energy sufficed to set up the oscillatory equa tion for one degree of freedom. For a system of two circuits (two degree: of freedom I_1 and I_2) the energy theorem no longer is adequate for setting up the differential equations, just as in mechanics. On the other hand Kirchhoff's law (5) yields directly

$$L_{11}\dot{I}_1 + L_{12}\dot{I}_2 + R_1 I_1 + \frac{e_1}{K_1} = E_1,$$

$$L_{22}\dot{I}_2 + L_{21}\dot{I}_1 + R_2 I_2 + \frac{e_2}{K_2} = E_2. \tag{17}$$

The generalization for more than two circuits is obvious.

From (17) we can calculate, on the one hand, the *free* vibrations of our coupled system, on the other, the forced vibrations produced by an e.m.f In the first case we set $E_1 = E_2 = 0$ and reduce the resulting homogeneous system of equations, with the assumption

$$I_1 = A_1 e^{i\omega_0 t}, \qquad I_2 = A_2 e^{i\omega_0 t}$$

to a biquadratic equation for ω_0 after elimination of the ratio A_2/A_1. We can be brief in discussing the conclusions derived therefrom since the same problem has been treated in detail in Vol. I, §20 and the results are repre sented there in Figs. 34 and 36. The characteristic *beat phenomena* of the coupled pendulums in the case of *resonance* occur, in terms of our present notation, when the periods of the uncoupled circuits are equal, which lead: to $K_1 L_{11} = K_2 L_{22}$, and when in addition the two "coupling coefficients" L_{12}/L_{11} and L_{21}/L_{22} are equal, which leads, in view of the universal equality of L_{12} and L_{21}, to $L_{11} = L_{22}$. These beat phenomena become particularly impressive if we assume, as for the coupled pendulums, that damping is light, in our case $R_1 \cong 0, R_2 \cong 0$.

Also for *forced* vibrations the treatment in Vol. I, §19 may serve as a model. If the free and forced frequencies are identical, $\omega = \omega_0$, damping becomes essential and the amplitude maximum and phase lag represented in Fig. 33 of Vol. I occur.

At the beginning of the §20 mentioned above it was pointed out that in the early stages of wireless telegraphy coupled mechanical oscillations commonly served as model for the coupled electrical oscillations which occurred in the open primary antenna circuit and the tuned secondary circuit introduced by Ferdinand Braun. It is true that for these rapid oscillations the quasistationary treatment is only a crude approximation; only the complete integration of the Maxwell equations in §19 can yield a satisfactory representation.

D. The Telegraph Equation

Quasistationary calculations may also be applied to short sections of a long two-wire line, for which condition (1) is not fulfilled. If the length of these sections is permitted to approach zero the total differential equations of the system become a *partial differential equation*. This was set up by W. Thomson even *before* Maxwell in the treatment of the propagation of telegraph signals in marine cables. Between two oppositely located and oppositely charged points of the double line there is a charging current with a change in voltage; Hence, in addition to the voltage $V(x)$, *also the current* (x) *varies continuously* along the length of the line. According to Kirchhoff's second law these two variables are related by

$$L \frac{\partial I}{\partial t} + RI + \frac{\partial V}{\partial x} = 0; \tag{18}$$

L and R relate to unit length of the double line. Furthermore, it follows from the absence of current sources, i.e. Kirchhoff's first law (4), that

$$\frac{\partial I}{\partial x} + K \frac{\partial V}{\partial t} + GV = 0. \tag{18a}$$

For the sake of completeness a conduction current GV through the eventually semiconducting dielectric has been added to the charging current $K \partial V / \partial t$. This term may also account for hysteresis losses in the dielectric. G is known as the "leakage" per unit length of the double line; K also refers to this unit length.

Elimination of V from (18) and (18a) leads to the partial differential equation

$$\left\{ LK \frac{\partial^2}{\partial t^2} + (RK + LG) \frac{\partial}{\partial t} + RG - \frac{\partial^2}{\partial x^2} \right\} I = 0, \tag{19}$$

which applies also for V. If, in particular, the dissipative coefficients R and G are set equal to zero, it assumes the simple form of the differential equation of the vibrating string:

$$\left\{LK \frac{\partial^2}{\partial t^2} - \frac{\partial^2}{\partial x^2}\right\} I = 0$$

and is integrated, for a phenomenon advancing in the positive x-direction by

$$I = af(x - ct), \qquad c = \sqrt{\frac{1}{LK}}. \tag{20}$$

According to (18) and (18a) the corresponding value of V is in this special case

$$V = Lcaf(x - ct) = \sqrt{\frac{L}{K}} I. \tag{21}$$

The ratio of voltage V and current I is thus the quantity $(L/K)^{\frac{1}{2}}$, which is independent of x and t and represents a resistance. It is known as the *wave resistance*.

The following practical conclusion may be drawn from (21): If a finite double line is terminated by an ohmic resistance of the magnitude $(L/K)^{\frac{1}{2}}$ no discontinuity occurs in the current and voltage variation at the end hence also *no reflection*.

Eq. (20) states that the current and hence also the voltage propagate themselves along the line *without distortion* and *without damping*. We may also ask for the condition for *undistorted damped* propagation along the line, i.e. that

$$I = e^{-\alpha x} f(x - ct) \tag{22}$$

represents a solution of the differential equation (19). By substituting (22) in (19) and setting the resulting factors of f'', f', and f equal to zero we find

$$c = \sqrt{\frac{1}{LK}}, \qquad \alpha = \frac{1}{2} \frac{RK + LG}{\sqrt{LK}} = \sqrt{RG}. \tag{23}$$

The velocity of propagation c is the same as in (20). The double equation for α leads to

$$RK + LG = 2\sqrt{RKLG}, \qquad \text{i.e.} \quad \sqrt{\frac{RK}{LG}} + \sqrt{\frac{LG}{RK}} = 2,$$

and hence

$$\sqrt{\frac{RK}{LG}} = 1 \qquad \text{or also} \qquad \frac{K}{G} = \frac{L}{R}. \tag{23a}$$

his signifies equal decay time for the pure displacement current $K\dot{V} +$
$V = 0$ (Eq. (18a) with $\partial I/\partial x = 0$) and of the pure conduction current
$\dot{I} + RI = 0$ (Eq. (18) with $\partial V/\partial x = 0$). By (23) and (23a) our damping
oefficient α then becomes equal to $R\sqrt{K/L}$.

In the general case the current variation changes with progress along
he line. It is then proper to analyze the process into component waves of
he form $\exp i(kx - \omega t)$ with complex, frequency-dependent k which are
eriodic in time and damped spatially. The total phenomenon is now no
onger *distortion-free*.

The ideal case of undamped *plane* waves is approached if the two wires
re imagined flattened into wide bands, the intermediate space being vac-
um and the band material a perfect conductor. Then, apart from the
narginal portions, the electric field in the intermediate space is uniform,
milarly the magnetic field. $\mathbf{E}$ is perpendicular to the bands in direction,
$\mathbf{I}$, parallel thereto. The current becomes $I = |\mathbf{H}|b$ (b = band width),
he voltage $V = |\mathbf{E}|d$ (d = separation of the bands). The charge per unit
ength is $e = \varepsilon_0|\mathbf{E}|b$ and the capacity $K = e/V = \varepsilon_0 b/d$. The magnetic
ux per unit length, i.e. through a rectangle with the sides 1 and d, becomes
$= \mu_0|\mathbf{H}|d$, so that the selfinductance $L = \Phi/I = \mu_0 d/b$. From K and
we compute, by (20) and (21),

$$\text{wave velocity } c = \frac{1}{\sqrt{KL}} = \frac{1}{\sqrt{\varepsilon_0\mu_0}},$$

$$\text{wave resistance } \frac{V}{I} = \sqrt{\frac{L}{K}} = \sqrt{\frac{\mu_0}{\varepsilon_0}\frac{d}{b}}.$$

erewith we have again come upon the quantity $(\mu_0/\varepsilon_0)^{\frac{1}{2}}$, which in §6 we
ad described as the wave resistance of vacuum for the propagation of a
lane wave. (To obtain agreement we must refer it now to a quadratic
ection of the wave surface, i.e. set $b = d$.)

We have inserted this sketchy note on the telegraph equation partly to
fer the concept of the wave resistance (more generally wave impedance
surge impedance) to its historical origin, partly to prepare the way for
e transition to rapidly variable fields in the next section.

§19. Rapidly Variable Fields. The Electrodynamic Potentials

Only in this paragraph do we make full use of the unabbreviated Max-
ell equations. We indicate a general method of integration, which how-
er is limited to the case of a uniform medium, e.g. vacuum. Hence
roughout space we put $\varepsilon = \varepsilon_0$, $\mu = \mu_0$ and in addition imagine the charge
nsity ρ and the current density $\mathbf{J}$ to be given in all of space and for all
mes $t < t_0$ (t_0 = instant of observation). In this formulation of the prob-
m we already take cognizance of the electron theory, which, however, we

shall take up only in the third part. We start from the Maxwell equation in the form (4.4) with the auxiliary conditions (4.4a, b, c). In view of our assumptions $\varepsilon = \varepsilon_0 = $ const, $\mu = \mu_0 = $ const we can write instead

$$\dot{\mathbf{B}} = -\text{curl } \mathbf{E}, \tag{1}$$

$$\frac{1}{c^2} \dot{\mathbf{E}} + \mu_0 \mathbf{J} = \text{curl } \mathbf{B}, \tag{2}$$

$$\text{div } \mathbf{E} = \frac{\rho}{\varepsilon_0}, \tag{3}$$

$$\text{div } \mathbf{B} = 0, \tag{4}$$

$$\text{div } \mathbf{J} + \frac{\partial \rho}{\partial t} = 0. \tag{5}$$

We satisfy Eq. (4) by our earlier formulation (15.3)

$$\mathbf{B} = \text{curl } \mathbf{A}. \tag{6}$$

Substituting this in (1) we obtain

$$\text{curl } (\mathbf{E} + \dot{\mathbf{A}}) = 0.$$

This has the necessary consequence that the vector after the curl sign is a gradient. Thus

$$\mathbf{E} = -\text{grad } \Psi - \dot{\mathbf{A}}. \tag{7}$$

We call Ψ *scalar potential*, $\mathbf{A}$ *vector potential*.

We substitute expressions (6) and (7) in (2) and obtain

$$-\frac{1}{c^2} (\ddot{\mathbf{A}} + \text{grad } \dot{\Psi}) + \mu_0 \mathbf{J} = \text{curl curl } \mathbf{A} = -\Delta\mathbf{A} + \text{grad div } \mathbf{A}. \tag{8}$$

We have here utilized the transformation of curl curl which has been repeatedly employed before (e.g. in Eq. (6.2)), but applies only for Cartesian components of the vector $\mathbf{A}$. We shall simplify Eq. (8) by splitting it up into two vector equations, namely into

$$\Delta\mathbf{A} - \frac{1}{c^2} \frac{\partial^2 \mathbf{A}}{\partial t^2} = -\mu_0 \mathbf{J} \tag{9}$$

and

$$\text{grad} \left(\text{div } \mathbf{A} + \frac{1}{c^2} \dot{\Psi} \right) = 0. \tag{9a}$$

If we refrain from the inappropriate addition of a function depending on only, i.e. of a kind of "integration constant", the second of these become

$$\text{div } \mathbf{A} + \frac{1}{c^2} \dot{\Psi} = 0. \tag{10}$$

Our initial equations (1), (2), and (4) are thus satisfied; there remains, apart from Eq. (5), Eq. (3). In view of (7) this takes the form

$$\Delta\Psi + \text{div } \dot{\mathbf{A}} = - \frac{\rho}{\varepsilon_0}$$

or, taking account of (10),

$$\Delta\Psi - \frac{1}{c^2} \frac{\partial^2\Psi}{\partial t^2} = - \frac{\rho}{\varepsilon_0}. \tag{11}$$

Our two potentials $\mathbf{A}$ and Ψ thus satisfy two differential equations of the same form. We call them "wave equations". As noted above, their right sides are given functions of x, y, z and of "past time," $t < t_0$. The desired solutions are related by condition (10).

We recognize that this condition is appropriate from the following: If we call its left side X and if we form

$$\text{div } (9) + \varepsilon_0\mu_0 \frac{\partial}{\partial t} (11),$$

we obtain

$$\Delta X - \frac{1}{c^2} \frac{\partial^2 X}{\partial t^2} = - \mu_0 \left(\text{div } \mathbf{J} + \frac{\partial\rho}{\partial t} \right). \tag{12}$$

The right side of this equation vanishes however because of our Eq. (5), which here at last is drawn into the consideration. Thus X also satisfies the *homogeneous* wave equation, which represents a wave process without external excitation, i.e. not a forced, but a free vibration. It can be foreseen from this that a suitable integration of the differential equations for $\mathbf{A}$ and Ψ, which excludes the appearance of free vibrations, not only leads to Eq. (12) for X being satisfied, but also to $X = 0$, i.e. the satisfying of Eq. (10). Nevertheless this equation is neither superfluous nor obvious, since the splitting up of Eq. (8), i.e. the transition from (8) to (9) and (11), rests expressly on condition (10).

A. The Retarded Potentials

With respect to the integration of our wave equations (9) and (11) we shall be brief, since the next section will indicate the rational procedure. We write down directly the result of the integration:

$$4\pi\varepsilon_0\Psi = \int \frac{[\rho] \, d\tau}{r}, \tag{13a}$$

$$4\pi \frac{1}{\mu_0} \mathbf{A} = \int \frac{[\mathbf{J}] \, d\tau}{r}. \tag{13b}$$

Ψ and **A** refer to the reference point x, y, z and the "reference time" t, for which we wish to calculate the values of Ψ and **A**. ξ, η, ζ is the point of integration and $d\tau$ is equivalent to $d\xi\,d\eta\,d\zeta$. The integration is carried out over all of infinite space and we have

$$r^2 = (x - \zeta)^2 + (y - \eta)^2 + (z - \zeta)^2.$$

$[\rho]$ and $[\mathbf{J}]$ are, however, *not* the values of charge and current density at the time of observation t, but at the *earlier* time

$$t' = t - r/c. \tag{13c}$$

r/c is the time required by the "light" to travel from the point of integration to the reference point. Hence the expressions (13) are called *retarded potentials*. They are calculated from charge and current density at a time which is *set back* by r/c relative to the time of observation.

The method of integration (13) is mathematically unique if, for physical reasons, the addition of *advanced potentials*, which correspond to the later time

$$t'' = t + r/c, \tag{13d}$$

is excluded. It should be noted, however, that such advanced potentials have tentatively been introduced by Dirac into the theory of the electron and play an important role in more recent investigations (see §37).

If we apply the same method of calculation (13) to our quantity X we find directly

$$X = 0$$

since the right side of (12) was equal to zero. This result is also mathematically unique with the exclusion of "advanced" solutions, which alone, in combination with the retarded solutions, could give rise to free vibrations. This may be taken as confirmation of our earlier statement that condition (10) is satisfied automatically in the integration of the differential equations for **A** and Ψ. We note finally that a full understanding of the structure of the above formalism including the significance of our retarded and advanced potentials can be obtained only on the basis of the theory of relativity. What up to now may have appeared arbitrary and asymmetric will there assume an astonishingly unique and symmetrical form.

B. *The Hertzian Dipole*

We will explain our method of integration (13) for a particular case, the case of the *Hertzian dipole*. This is obtained if we combine a moving charge $+e$ with a neighboring stationary charge $-e$ to form a moment $\mathbf{p}(t) = e\mathbf{l}$ varying with time, where $\mathbf{l}$ signifies the separation of the two charges.

We substitute $\mathbf{J} = \rho\mathbf{v}$ in (13b), denoting the space density of the moving charge by ρ and its velocity by $\mathbf{v}$, and obtain on carrying out the integration, where r and $\mathbf{v}$ may be regarded as constant in space,

$$\int \frac{[\mathbf{J}]\,d\tau}{r} = \frac{[\mathbf{v}]}{r}\int \rho\,d\tau = \frac{e[\mathbf{v}]}{r} = \frac{e}{r}\left[\frac{\partial \mathbf{l}}{\partial t}\right] = \frac{1}{r}\left[\frac{\partial \mathbf{p}}{\partial t}\right].$$

We hence obtain from (13b), if we take due account of the meaning of the bracket symbol as given by (13c)

$$4\pi\mathbf{A} = \frac{\mu_0}{r}\frac{\partial}{\partial t}\,\mathbf{p}\left(t - \frac{r}{c}\right). \tag{14}$$

It is historically customary and convenient to introduce, in place of the vector potential $\mathbf{A}$, the *Hertzian vector* $\mathbf{\Pi}$ by writing

$$\mathbf{A} = \mu_0\frac{\partial \mathbf{\Pi}}{\partial t}, \qquad 4\pi\mathbf{\Pi} = \frac{1}{r}\,\mathbf{p}\left(t - \frac{r}{c}\right). \tag{15}$$

With this notation we follow the great paper of Hertz, already discussed in §1, p. 5:[1] "The Forces of Electrical Oscillations, Treated by Maxwell's Theory."

In all of space except at the origin of the coordinate system $\mathbf{\Pi}$ satisfies, by (9), the differential equation

$$\Delta\mathbf{\Pi} - \frac{1}{c^2}\frac{\partial^2\mathbf{\Pi}}{\partial t^2} = 0, \tag{16}$$

which can also readily be verified from the explicit representation of $\mathbf{\Pi}$ given in (15). By (10) the corresponding value of Ψ becomes

$$\varepsilon_0\Psi = -\operatorname{div}\mathbf{\Pi}. \tag{16a}$$

From (6) and (7) we obtain then as the representation of the electromagnetic field

$$\mathbf{H} = \operatorname{curl}\dot{\mathbf{\Pi}}, \qquad \varepsilon_0\mathbf{E} = \operatorname{grad}\operatorname{div}\mathbf{\Pi} - \frac{1}{c^2}\frac{\partial^2\mathbf{\Pi}}{\partial t^2}. \tag{17}$$

As an example we assume that the path of the mobile charge e is rectilinear and make its direction, which is also that of the vector $\mathbf{\Pi}$, the axis of a spherical coordinate system r, ϑ, φ. We then have[2]

$$\Pi_r = \cos\vartheta\cdot\Pi, \qquad \Pi_\vartheta = -\sin\vartheta\cdot\Pi, \Pi_\varphi = 0,$$

[1] Ann. d. Physik *36*, p. 1, 1888; Gesammelte Werke, Vol. II, p. 147.
[2] The positive r-direction forms the angle ϑ with the direction of $\mathbf{\Pi}$, the positive ϑ-direction, the angle $\vartheta + \pi/2$; hence the factors $\cos\vartheta$ at Π_r and $\cos(\vartheta + \pi/2) = -\sin\vartheta$ at Π_ϑ.

where, according to (15), Π depends only on t and r, i.e. is independent of ϑ and φ. In these coordinates we obtain, by Problem I.3 of Vol. II,

$$\text{curl}_\varphi \,\Pi = \frac{-\sin \vartheta}{r}\left(\frac{\partial (r\Pi)}{\partial r} - \Pi\right) = -\sin \vartheta \frac{\partial \Pi}{\partial r}, \quad \text{curl}_r \,\Pi = \text{curl}_\vartheta \,\Pi = 0,$$

$$\text{div } \Pi = \frac{\cos \vartheta}{r^2}\frac{\partial}{\partial r}(r^2\Pi) - \frac{1}{r \sin \vartheta}\frac{\partial}{\partial \vartheta}(\sin^2\vartheta.\,\Pi) = \cos \vartheta \frac{\partial \Pi}{\partial r},$$

$$\text{grad}_r \text{ div } \Pi = \cos \vartheta \frac{\partial^2 \Pi}{\partial r^2}, \quad \text{grad}_\vartheta \text{ div } \Pi = \frac{-\sin \vartheta}{r}\frac{\partial \Pi}{\partial r}, \quad \text{grad}_\varphi \text{div } \Pi = 0.$$

Hence, by (17):

$$H_r = H_\vartheta = E_\varphi = 0 \tag{18}$$

and by (15)

$$4\pi H_\varphi = -\frac{\sin \vartheta}{r}\left(\frac{\partial}{\partial r}\dot{p} - \frac{1}{r}\dot{p}\right),$$

$$4\pi\varepsilon_0 E_r = \frac{\cos \vartheta}{r}\left(\frac{\partial^2}{\partial r^2}p - \frac{2}{r}\frac{\partial}{\partial r}p + \frac{2}{r^2}p - \frac{1}{c^2}\ddot{p}\right), \tag{19}$$

$$4\pi\varepsilon_0 E_\vartheta = -\frac{\sin \vartheta}{r}\left(\frac{1}{r}\frac{\partial}{\partial r}p - \frac{1}{r^2}p - \frac{1}{c^2}\ddot{p}\right).$$

We conclude from (18): The magnetic lines of force are circles about the direction of $\mathbf{p}$, while the electric lines of force lie in the meridional planes through this direction.

Because of the argument $t - r/c$ of p it is possible, in Eqs. (19), to transform the differentiation with respect to r into one with respect to t. We have

$$\frac{\partial \mathbf{p}}{\partial r} = -\frac{1}{c}\dot{\mathbf{p}}, \quad \frac{\partial^2 \mathbf{p}}{\partial r^2} = \frac{1}{c^2}\ddot{\mathbf{p}}. \tag{19a}$$

Then the term with $\dfrac{\partial^2 \mathbf{p}}{\partial r^2}$ cancels that with $\ddot{\mathbf{p}}$ in the equation for E_r in (19). At the same time we will limit ourselves to the *"distant zone"* (large distances from the origin, i.e. set $r \to \infty$. We will indicate the more precise meaning of this in a moment, in discussing the periodically oscillating dipole. Accordingly we neglect all terms in (19) which contain higher powers of $1/r$ than the first. We then obtain

$$4\pi H_\varphi = \frac{\sin \vartheta}{cr}\ddot{p}\left(t - \frac{r}{c}\right),$$

$$4\pi\varepsilon_0 E_r = 0, \tag{20}$$

$$4\pi\varepsilon_0 E_\vartheta = \frac{\sin \vartheta}{c^2 r}\ddot{p}\left(t - \frac{r}{c}\right).$$

The vectors **H** *and* **E** *are perpendicular to each other and to the radius vector*
r *from the origin.* Both **H** and **E** vanish on the axis $\vartheta = 0$ and $\vartheta = \pi$; the
H- and **E**-fields have their *maxima* in the equatorial plane $\vartheta = \pi/2$.

From (20) we calculate

$$\frac{E_\vartheta}{H_\varphi} = \frac{1}{\varepsilon_0\, c} = \sqrt{\frac{\mu_0}{\varepsilon_0}}. \tag{21}$$

This is the same ratio as that which was obtained from Eqs. (6.11) and
(6.13) for the ratio E_y/H_z. *The structure of the radiated electromagnetic field
is thus that of a plane light wave.* It is customary to say instead, in both
cases, that **E** and **H** are equal, which however is dimensionally meaningless.

The amount of energy radiated per unit area and per unit time becomes

$$\mathbf{S} = \mathbf{E} \times \mathbf{H} = E_\vartheta\, H_\varphi = \frac{1}{16\pi^2\, \varepsilon_0\, c^3}\, \frac{\sin^2 \vartheta}{r^2}\, \ddot{\mathbf{p}}^2. \tag{22}$$

The total energy radiated in unit time is obtained by integration over the
spherical surface of radius r:

$$S = \int S\, d\sigma = 2\pi r^2 \int S \sin \vartheta\, d\vartheta = \frac{\ddot{p}^2}{6\pi\varepsilon_0\, c^3}. \tag{23}$$

Since $\mathbf{p} = e\mathbf{l}$ ($\mathbf{l}$ = separation of the mobile and the stationary charge)
$\dot{p} = ev,\ \ddot{p} = e\dot{v}$, where, of course, in accord with the meaning of $\ddot{p}$, $\dot{v}$ denotes
the value of the acceleration at the earlier time $t - r/c$. We thus obtain
from (23)

$$S = \frac{e^2\, \dot{v}^2}{6\pi\varepsilon_0\, c^3}. \tag{24}$$

In atomic physics it is customary to write, in electric or magnetic cgs-
units,

$$S = \frac{2}{3}\, \frac{e^2\, \dot{v}^2}{c^3} \tag{24a}$$

or

$$S = \frac{2}{3}\, \frac{e^2\, \dot{v}^2}{c} \tag{24b}$$

which, according to (16.30), corresponds to (24). J. J. Larmor[1] first gave
this fundamental law of radiation in the form (24b).

Fig. 27 shows the radiation density S as function of ϑ. It is simply the

---

[1] Phil. Mag. 1897, p. 512. Larmor points out the relationship to Hertz's paper of
1888 in his book Aether and Matter, Cambridge, 1900, p. 225.

polar diagram of $\sin^2 \vartheta$. In the theory of wireless telegraphy (see Vol. VI, Chapter VI) it finds extensive application in the treatment of a linear antenna radiating freely into space. In fact such an antenna does *not* radiate energy in its own direction; the maximum of the radiation is directed *transversally*.

In place of a single dipole **p** we may, of course, also consider a discrete or continuous sequence of dipoles. In the latter case we write in place of (15):

$$4\pi\mathbf{\Pi} = \int_c \frac{d\mathbf{p}(t - r/c)}{r} . \tag{25}$$

Here the integration is to be extended over a given curve C and the difference in direction of the vectors $d\mathbf{p}$ must be considered.

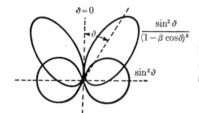

FIG. 27. Radiation of an electron accelerated longitudinally in the direction $\vartheta = 0$. Lower pair of curves: Hertz's formula (22), $v \ll c$. Upper pair of curves: corrected relativistically, v comparable with c.

A comparison of the preceding with Hertz's calculation in Cartesian coordinates, which is found in most textbooks, demonstrates the superiority of our vector representation of the problem or eventually, of our spherical polar coordinates, which fit the symmetry of the problem. It seems even more important that our presentation clearly indicates the dimensions of all field quantities, while the Gaussian system of units employed by Hertz obscures them.

C. Specialization for Periodic Processes

We obtain the simplest model for a *light source* by assuming that the electric moment **p** oscillates monochromatically with a certain circular frequency ω. For example we set

$$p(t) = A \cos \omega t = A \operatorname{Re} e^{-i\omega t},$$
$$p(t - r/c) = A \operatorname{Re} \exp \{-i\omega(t - r/c)\}. \tag{27}$$

If, as in (6.10a, b), we introduce the wave number $k = \omega/c$ and omit the sign indicating the real part, which is permissible for all field quantities except the quadratic ones **S**, S, we find

$$\frac{1}{r} p(t - r/c) = A \frac{e^{ikr}}{r} e^{-i\omega t}. \tag{27a}$$

We have thus arrived at the representation of the *spherical wave* in Vol. II, Eq. (13.18) if we also suppress the time factor in (27a). This may and will be done in the following. We then obtain from Eqs. (20) the following representation of the electromagnetic field:

$$E = E_\vartheta = -\frac{Ak^2}{4\pi\varepsilon_0} \sin \vartheta \, \frac{e^{ikr}}{r}, \qquad H = H_\varphi = -\frac{Ak\omega}{4\pi} \sin \vartheta \, \frac{e^{ikr}}{r}. \qquad (28)$$

This applies, as already noted above (20), for the "distant zone"; we are now however in a position to define this term exactly. For, if we let the wave-length

$$\lambda = \frac{2\pi}{\omega} c$$

correspond to the angular frequency ω, the distant zone includes all distances for which

$$r \gg \lambda, \qquad (29)$$

i.e. excludes only the immediate neighborhood of the light source. For aperiodic processes (29) is replaced by the inequalities

$$\frac{1}{c}\,|\,\ddot{p}\,| \gg \frac{|\,\dot{p}\,|}{r}, \qquad \frac{1}{c^2}\,|\,\ddot{p}\,| \gg \frac{|\,\dot{p}\,|}{cr} \gg \frac{|\,p\,|}{r^2}. \qquad (29a)$$

These statements justify precisely the approximations made in passing from (19) to (20).

As compared with a natural light source our model is specialized both with regard to its monochromatism and its intensity distribution. It radiates no energy in the directions $\vartheta = 0$ and $\vartheta = \pi$; for also now Fig. 27 and Eq. (22) apply to the radiation vector **S**. In (22) both the time factor and the phase factor e^{ikr} drop out in the time average. In fact, by (27),

$$\frac{\ddot{p}}{r} = -\frac{A\omega^2}{r} \cos \, (kr - \omega t)$$

and the time average of the square thereof is

$$A^2\omega^4/(2r^2) = A^2c^4k^4/(2r^2). \qquad (29b)$$

Since $k = 2\pi/\lambda$ this is *inversely proportional to the fourth power of the wave-length.*

If (29b) is substituted in (22) or (23) we obtain the famous law of Lord Rayleigh, explaining the *blue sky.* The sun rays falling on the particles of the air generate in them electric moments which vibrate in harmony and radiate light in turn. Their radiation is much stronger at the blue end of the spectrum than at the red end. Since $\lambda_{red} \cong 2\lambda_{blue}$ their ratio is about 2^4. The same law explains also the red color of the sun and moon when rising and setting. In its path through the atmosphere, which is here particularly long, the blue light is scattered much more strongly out of its direct path than the red light; primarily red sun- or moonlight reaches our eyes. We shall not inquire whether a certain selectivity of water vapor in the atmosphere plays an added role.

D. The Characteristic Vibrations of a Metallic Spherical Oscillator

The problem of *electromagnetic* characteristic vibrations became significant as a result of Hertz's experiments. A metallic body consisting of two oppositely charged halves (Hertzian oscillator) discharges with the formation of a spark and radiates exponentially damped vibrations toward infinity. How do we calculate their frequency and damping? In the case of the sphere, which we imagine as subdivided into two closely adjoining oppositely charged halves, the question is answered directly by formulas which are already familiar to us. It is true that we must not start here from Eq. (20) for the distant zone, but must employ the more general formulas (19), since the wave length of the characteristic vibration generated will understandably be of the order of magnitude of the radius of the sphere; the surface of the sphere thus belongs to the near zone. If the sphere is assumed to be perfectly conducting we have on the surface, i.e. for $r = a =$ radius of sphere, $E_\vartheta = 0$ for all ϑ. Hence, by (19), we have as boundary condition

$$\left(\frac{1}{r} \frac{\partial}{\partial r} - \frac{1}{r^2} - \frac{1}{c^2} \frac{\partial^2}{\partial t^2} \right) p = 0. \tag{30}$$

We let p take the same form as in (27), treating k not as a real number as up to now, but as an unknown complex number. The same applies then also for $\omega = ck$. To determine k (30) yields

$$\frac{ik}{a} - \frac{1}{a^2} + k^2 = 0. \tag{31}$$

The solution of this equation, which is quadratic in ka, is

$$ka = \frac{-i \pm \sqrt{3}}{2}. \tag{31a}$$

The imaginary part is negative, as we must demand, since we are dealing with a vibration which decreases with time. In the real part the positive sign is to be chosen in order that wave-length and frequency be positive. We thus obtain

$$\frac{2\pi a}{\lambda} = \frac{\sqrt{3}}{2}, \qquad \lambda = \frac{4\pi}{\sqrt{3}} a,$$

i.e. in fact λ of the order of magnitude of the radius of the sphere. Damping is very great; as follows from (31a), the amplitude decreases by a factor

$$e^{-\lambda/(2a)} = e^{-2\pi/\sqrt{3}}$$

in the course of a single vibration.

Herewith the nature of the *fundamental* vibration of our spherical oscillator is described. There is however also an infinite number of harmonics for which the sphere is not divided into two oppositely charged halves, but into 4, 6, $\cdots$ alternately charged zones. While the fundamental vibration corresponds to the Hertzian dipole, these harmonics cannot be derived from the Hertzian vector **II**. For them we must refer to Vol. VI and more particularly to Appendix II of Chapter V of that volume.

The case of the prolate spheroid, which comes closer to the Hertz oscillator than the spherical shape, was treated by Max Abraham,[1] after the problem of the spherical oscillator had been solved generally by J. J. Thomson as early as 1884.

E. Application to the Theory of X-Rays

The primary x-rays are produced by the incidence of cathode rays on the anticathode. Classically the initial velocity v of the incident electrons is reduced to a low value; the electrons experience a retardation $-v$. From Fig. 27 we expect that no radiation occurs in the direction of the cathode rays, insofar as this coincides with the direction of v. The proof of this is possible with extremely thin anticathodes (films a few microns in thickness), if the transmitted x-rays are observed; this has been shown by Kulenkampff and his students. For solid metal anticathodes the retardation takes place along a zigzag path; hence the variation with direction is smoothed out. We will show relativistically in §30, at Eq. (11), that the maximum of the radiation does not lie, as indicated by the pair of curves in Fig. 27, at $\vartheta = \pi/2$, but that it advances, instead, with increasing hardness of the cathode rays (increasing magnitude of v) more and more toward $\vartheta = 0$. The fact that the continuous or "brems"-spectrum discussed here has a short-wave limit is a consequence of the quantum theory, with which we shall not deal here. The same applies for the regular increase in hardness and intensity of the x-rays with the hardness of the cathode rays.

Here we shall only discuss the proof of the *transversal* nature of x-rays, which was given by Barkla in 1905, ten years after Röntgen's discovery. In planning his experiment Barkla assumed this transversality, drew the consequences of this assumption, and confirmed them by the experiment.

We consider, in Fig. 28, a broken line consisting of three mutually perpendicular segments, the "primary", "secondary", and "tertiary" segment. The *primary x-rays*, regarding whose polarization we shall make no assump-

---

[1] See Enzykl. d. mathem. Wiss., Vol. V$_2$, section 18, p. 498.

tion, travel along the first segment (for a very thin anticathode even these would be partly polarized). We imagine their electric field strength to be analyzed into its components along the directions 2 and 3 of the two other segments. They fall on a first scatterer Z_1, whose electrons they set into vibrations. Those parallel to 2 have no effect along the secondary segment, while those parallel to 3 produce on it *secondary x-rays*, which vibrate parallel to 3 and are *totally polarized*. They fall on a scatterer Z_2 and set its electrons into vibrations in direction 3. In this fashion *tertiary x-rays* are produced which, however, have the *intensity zero* along the tertiary segment. They have maximum intensity in the direction of the primary segment. This behavior of the tertiary x-rays proves both the transversal nature of the primary and the total polarization of the secondary x-rays.

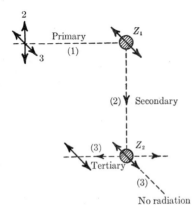

Fig. 28. Barkla's arrangement for demonstrating the transversal nature of x-rays. Z_1, Z_2 scatterers (spheres of paraffin).

The scatterers Z_1 and Z_2 were spheres of paraffine; for heavier materials the "characteristic radiation" might have falsified the result.

§*20. General Considerations on the Structure of Wave Fields of Cylindrical Symmetry. Applications to Alternating Current Impedance and Skin Effect*

In the following section we will concern ourselves almost exclusively with *surface waves* which are *guided* along bodies of *cylindrical shape*. Let the excitation be such that the process is periodic in time with the circular frequency ω. The calculation of the *propagation and damping* of the waves as they progress in the direction of the cylinder axis, which we shall choose as the direction of the x-coordinate axis, then becomes of primary interest. We leave the cross section of the cylindrical conductor (or also nonconductor) temporarily indeterminate. We express propagation and damping by a single *complex wave number* h, which differs from the real wave number $k = \omega/c$ in vacuum. We consider thus a wave type with the dependence on x and t

$$\exp \{i(hx - \omega t)\};$$

for the assumed cylindrical structure of the wave field, h has necessarily the same value outside of and inside of the guiding surfaces; the same applies of course to ω.

We define, in the plane perpendicular to the x-axis, an orthogonal coordinate system u, v; dx, du, dv, in this order, are to form a right-handed coordinate system. For the line element in space we write, in accord with (2.22) of Vol. II:

$$ds^2 = dx^2 + g_u^{\ 2}\, du^2 + g_v^{\ 2}\, dv^2. \tag{1}$$

g_u and g_v are here given functions of u and v. The cross section (which is constant, i.e. independent of x, for every conductor) may differ for different conductors. The coordinates u, v are to be fitted, in each particular case, to the shape of the cross section: polar coordinates for the single wire, bipolar coordinates for the two-wire line, Cartesian coordinates for semi-infinite space (limiting case of the single wire of infinitely great radius).

We set ourselves the problem of computing the transversal components E_u, E_v, H_u, H_v from the longitudinal components E_x, H_x. This is possible without making any special assumptions regarding the cross sections of the cylindrical conductors guiding the wave, without discussing the corresponding boundary conditions, and also without assuming that the wave equation is separable in the coordinates u, v. The general structure of the wave field so obtained applies not only for the exterior of the conductors, but also, with altered choice of the material constants, to their interior.

A. Longitudinal and Transversal Components

The longitudinal components E_x and H_x, which we shall designate by the single symbol X, satisfy, as Cartesian components, the simple wave equation

$$\left(\Delta - \varepsilon\mu \frac{\partial^2}{\partial t^2} - \sigma\mu \frac{\partial}{\partial t} \right) X = 0. \tag{2}$$

It has been written in this form for the interior of the conductors, but applies also to the exterior, where $\sigma = 0$, $\varepsilon = \varepsilon_0$, $\mu = \mu_0$. We put

$$\Delta = \frac{\partial^2}{\partial x^2} + \Delta_{uv},$$

where Δ_{uv} is the *two-dimensional* Laplace operator transformed to the curvilinear coordinates u, v. In view of the dependence of the phase factor on x and t we can write in place of (2)

$$(\Delta_{uv} + k^2 - h^2) X = 0, \qquad k^2 = \varepsilon\mu\omega^2 + i\mu\sigma\omega. \tag{3}$$

In the exterior of the conductors k is real $(= \sqrt{\varepsilon_0\mu_0}\,\omega = \omega/c)$, in the interior it is complex. However, by introducing the complex dielectric constant ε' from Eq. (6.18) we can employ the same formula

$$k = \sqrt{\varepsilon\mu}\,\omega \qquad (3a)$$

in both cases, if we set

$$\varepsilon = \varepsilon_0, \qquad \mu = \mu_0 \qquad \text{in the exterior,}$$

$$\varepsilon = \varepsilon', \qquad \mu = \mu \qquad \text{in the interior.} \qquad (3b)$$

$$\varepsilon' = \varepsilon + i\sigma$$

Actually to integrate Eq. (2) it would of course be necessary to pass to specific coordinates u, v, which are adapted to the shape of the conductor; we shall avoid this for the present, however. Nevertheless we may regard, for what follows, the longitudinal components E_x and H_x as known functions of space.

To calculate from them the transversal components we utilize the definition of the curl of an arbitrary vector $\mathbf{A}$ in any curvilinear coordinates p_1, p_2, p_3 contained in Eq. (2.26) of Vol. II:

$$\operatorname{curl}_1 \mathbf{A} = \frac{1}{g_2 g_3}\left(\frac{\partial(g_3 A_3)}{\partial p_2} - \frac{\partial(g_2 A_2)}{\partial p_3}\right) \qquad (4)$$

and set in accord with (1),

$$p_1 = x, \qquad p_2 = u, \qquad p_3 = v; \qquad g_1 = 1, \qquad g_2 = g_u, \qquad g_3 = g_v.$$

By cyclic interchange of the indices 1, 2, 3 and the coordinates x, u, v we obtain from (4) all the components of the curl occurring in the Maxwell equations.

We thus calculate from the first and second group of three Maxwell equations, taking account at the same time of the x, t dependence of the phase factor,

$$i\omega\mu H_v = \operatorname{curl}_v \mathbf{E} = ihE_u - \frac{1}{g_u}\frac{\partial E_x}{\partial u}$$

$$-i\omega\varepsilon E_u = \operatorname{curl}_u \mathbf{H} = -ihH_v + \frac{1}{g_v}\frac{\partial H_x}{\partial v}.$$

On the basis of our convention (3a, b) we substitute on the left $\omega = k/\sqrt{\varepsilon\mu}$ and obtain

$$i\left(k\sqrt{\frac{\mu}{\varepsilon}}H_v - hE_u\right) = -\frac{1}{g_u}\frac{\partial E_x}{\partial u}$$

$$i\left(h\sqrt{\frac{\mu}{\varepsilon}}H_v - kE_u\right) = \frac{1}{g_v}\sqrt{\frac{\mu}{\varepsilon}}\frac{\partial H_x}{\partial v}.$$

From these equations E_u and H_v are obtained by simple elimination:

$$i(k^2 - h^2)E_u = -\frac{h}{g_u}\frac{\partial E_x}{\partial u} - \frac{k}{g_v}\sqrt{\frac{\mu}{\varepsilon}}\frac{\partial H_x}{\partial v}$$

$$i(k^2 - h^2)\sqrt{\frac{\mu}{\varepsilon}}H_v = -\frac{k}{g_u}\frac{\partial E_x}{\partial u} - \frac{h}{g_v}\sqrt{\frac{\mu}{\varepsilon}}\frac{\partial H_x}{\partial v}.$$

(5)

Thus our objective regarding the two transversal components E_u and H_v has been attained, since on the right there occur only the longitudinal components which are assumed to be known.

The calculation for E_v and H_u is carried out similarly. Here the Maxwell equations

$$i\omega\mu H_u = -ihE_v + \frac{1}{g_v}\frac{\partial E_x}{\partial v}$$

$$-i\omega\varepsilon E_v = ihH_u - \frac{1}{g_u}\frac{\partial H_x}{\partial u}$$

are employed, and ω substituted in them once more in accord with the convention (3a, b). After eliminating one of the two unknowns H_u or E_v we obtain

$$i(k^2 - h^2)E_v = -\frac{h}{g_v}\frac{\partial E_x}{\partial v} + \frac{k}{g_u}\sqrt{\frac{\mu}{\varepsilon}}\frac{\partial H_x}{\partial u},$$

$$i(k^2 - h^2)\sqrt{\frac{\mu}{\varepsilon}}H_u = \frac{k}{g_v}\frac{\partial E_x}{\partial v} - \frac{h}{g_u}\sqrt{\frac{\mu}{\varepsilon}}\frac{\partial H_x}{\partial u}.$$

(6)

We will be able to make good use of these rather brief and abstract considerations in the following §21–25, where we shall replace our general coordinates u, v partly by polar, partly by bipolar coordinates. Thus, they show directly, e.g., that for polar coordinates r, φ and non-dependence of the field on φ, the pairs of equations (5) and (6), which in general are coupled, separate into one pair which contains only E_x, E_r, H_φ and another which contains only H_x, H_r, E_φ. This simplification corresponds to the symmetry of the single wire. In the present paragraph we shall deal with the still simpler case of rectangular coordinates $u = y$, $v = z$ for non-dependence of the field on z. This corresponds to the transition to the limit: radius of wire $\to \infty$ in Fig. 30. In all these special cases the preceding general relations take on a readily understood form and may, as we shall see, be verified directly.

Independently of the choice of u and v a further conclusion may be drawn which applies for all cylindrical perfect conductors: *The velocity of propagation on them is always equal to the velocity of light.* We note first that the electric lines of force must be perpendicular to the surface of the conductor and conclude hence $E_x = 0$. Since, furthermore, the energy flux can have no component in a direction toward the conductor, we must also have $H_x = 0$. To begin with this applies only to the surface of the conductors. We shall assume however that both longitudinal components vanish *everywhere* in the nonconductor without coming into conflict with the Maxwell equations. It then follows from (5) and (6) (excluding the trivial solution $E_u = E_v = H_u = H_v = 0$), that we must have $h = k$. The wave equation (3) which applies for every Cartesian component of the electric and magnetic field then passes over into the potential equation. We shall make use of this conclusion in a specific example (§25A).

B. The Wave Field of Semiinfinite Space and its Skin Effect

Let the metallic semiinfinite space be bounded by the plane $y = 0$. Let the positive y-axis point upwards into the empty space (eventually filled with air) $y > 0$ and the wave progress toward the right (positive x-axis). Of the two possible solutions (5) and (6) we choose the first, since it corresponds to the wire traversed by alternating current. We hence make the reasonable assumption

$$
\left.\begin{aligned}
E_x &= E(y) \\
E_y &= F(y) \\
H_z &= G(y)
\end{aligned}\right\} \exp\{i(hx - \omega t)\}; \qquad
\left.\begin{aligned}
E_z \\
H_x \\
H_y
\end{aligned}\right\} = 0 \qquad (7)
$$

The differential equation (3) for E_x then takes the form

$$
\frac{d^2 E}{dy^2} + (k^2 - h^2)E = 0, \qquad \text{with} \quad k^2 = \varepsilon\mu\omega^2 + i\mu\sigma\omega. \qquad (8)
$$

Its solution is

$$
E(y) = Ae^{i\sqrt{k^2 - h^2}\, y} + Be^{-i\sqrt{k^2 - h^2}\, y}. \qquad (8a)
$$

For $y < 0$ k^2 is complex, for $y > 0$ it is real. However, we shall, for the present, assume a very small $\sigma > 0$ even for $y > 0$ and pass over to the limit $\sigma \to 0$ at a later stage.[1]

We shall choose the sign of $\sqrt{k^2 - h^2}$ once and for all so that it has a *positive* imaginary part.

Since the state must remain finite for $y \to \pm\infty$ we must set in (8a)

$$
\text{for} \quad y > 0 : B = 0,
$$

$$
\text{for} \quad y < 0 : A = 0.
$$

[1] In this manner we circumvent some basic questions regarding the exterior of the wire which will be deferred until §22.

Furthermore, since E_x must be continuous at $y = 0$, B must be set equal to A in the two resulting expressions.

In the following we shall reserve the letter k for the real wave number for $y > 0$ in the limiting case $\sigma \to 0$. For the sake of distinction the value of k within the conductor will be denoted by k_L. Then (8a) assumes the final form:

$$E(y) = \begin{cases} A e^{i\sqrt{k^2-h^2}\,y} & y > 0, \\ A e^{-i\sqrt{k_L^2-h^2}\,y} & y < 0. \end{cases} \tag{9}$$

With $u = y$, $v = z$, $g_u = g_v = 1$, and $\dfrac{\partial}{\partial z} = 0$ we obtain from (5)

$$F(y) = \begin{cases} -\dfrac{hA}{\sqrt{k^2 - h^2}}\, e^{i\sqrt{k^2-h^2}\,y} & y > 0, \\[2ex] \dfrac{hA}{\sqrt{k_L^2 - h^2}}\, e^{-i\sqrt{k_L^2-h^2}\,y} & y < 0. \end{cases} \tag{9a}$$

$$\left.\begin{array}{c} \sqrt{\dfrac{\mu_0}{\varepsilon_0}} \\[2ex] \sqrt{\dfrac{\mu}{\varepsilon'}} \end{array}\right\} G(y) = \begin{cases} -\dfrac{kA}{\sqrt{k^2 - h^2}}\, e^{i\sqrt{k^2-h^2}\,y} & y > 0, \\[2ex] \dfrac{k_L A}{\sqrt{k_L^2 - h^2}}\, e^{-i\sqrt{k_L^2-h^2}\,y} & y < 0. \end{cases} \tag{9b}$$

The factors of G, on the left, denote "reciprocal wave impedances", the upper one the real reciprocal impedance of vacuum as on p. 36, the lower one the correspondingly defined complex quantity for our conductor. There we learned already that the dimensions of **H** and **E** must differ by a factor with this dimension.

We now take account of the continuity of H_z at $y = 0$. In view of (3a, b) this demands

$$\frac{\mu_0\sqrt{k^2 - h^2}}{k^2} = -\frac{\mu\sqrt{k_L^2 - h^2}}{k_L^2}. \tag{10}$$

From this we derive

$$\frac{1}{h^2} = \frac{\mu_0^2\,\dfrac{k_L^2}{k^2} - \mu^2\,\dfrac{k^2}{k_L^2}}{\mu_0^2\,k_L^2 - \mu^2\,k^2}. \tag{10a}$$

This value depends in a *symmetrical fashion* on the constants μ_0, k and μ, k_L of the two media air and metal. We find, in particular, for $\mu = \mu_0$

$$\frac{1}{h^2} = \frac{1}{k^2} + \frac{1}{k_L^2}. \tag{10b}$$

For a well conducting metal the conduction current (the σ-term in k_L) exceeds the displacement current (the ε-term). We may thus assume

$$k_L \cong \sqrt{i\mu\sigma\omega} = \sqrt{2i}\,\sqrt{\frac{\mu\sigma\omega}{2}} = (1 + i)\kappa, \qquad \kappa^2 = \frac{\mu\sigma\omega}{2}. \qquad (11)$$

We have then simultaneously

$$\kappa \gg k, \quad \text{i.e. by (10b)} \quad h \cong k. \qquad (11a)$$

In this case one obtains as phase velocity simply:

$$\frac{\omega}{h} \cong \frac{\omega}{k} = c,$$

as has already been pointed out at the end of section A.

From (9) and (9a) there follows with assumption (11a)

in the exterior of the conductor: $|E| \ll |F|$,

in the interior of the conductor: $|E| \gg |F|$.

The electric lines of force thus have essentially the direction of the y-axis outside of the conductor, that of the x-axis inside of the conductor.

We will now determine that value of the y-coordinate within the metal for which E_x and hence also J_x have decreased to a fraction $1/e$ of the value at the surface. We call this value of y, $-d$. According to our assumptions (11) and (11a):

$$\kappa\,d = 1, \qquad d = \frac{1}{\kappa} \ll \frac{1}{k}. \qquad (12)$$

Since $k = 2\pi/\lambda$ this is small compared to the wave-length λ corresponding to the frequency ω. *The current is confined to a thin skin at the periphery of the conductor, while the whole interior is practically free of current.* We speak of a *skin effect* which is known to play an important role in alternating current practice. The thickness of the skin is smaller in the degree that the frequency is higher (by (11) κ increases in proportion to $\sqrt{\omega}$). The following table gives some values of d for Cu ($\sigma = 57.5 \cdot 10^6 \Omega^{-1} \mathrm{M}^{-1}$, $\mu = \mu_0 = 4\pi \cdot 10^{-7} \Omega \mathrm{M}^{-1}\mathrm{S}$).

	Alternating Current Line	Telephone	Wireless Telegraphy	Hertzian Oscillations
$\dfrac{\omega}{2\pi} =$	60/sec	1000/sec	$3 \cdot 10^5$/sec	10^9/sec
$\lambda =$	$5 \cdot 10^3$ km	300 km	1 km	30 cm
$\kappa =$	116 m^{-1}	$4.7 \cdot 10^2$ m^{-1}	$8.15 \cdot 10^3$ m^{-1}	$4.7 \cdot 10^5$ m^{-1}
$d =$	8.6 mm	2.1 mm	0.13 mm	$2.1 \cdot 10^{-3}$ mm

We illustrate this by Fig. 29, which, however, does not indicate the circumstances of a semiinfinite space, but the practically more interesting ones for a wire of circular cross section. The straight line 00 for direct current or commercial alternating current (60/sec) passes over into the slightly concave curve 11 for telephone frequencies (1000/sec); curve 22 applies for high frequencies proper (e.g. 1 km wave-length) and shows a

FIG. 29. Variation of alternating-current amplitude in cross section of wire: 00, direct current; 11, telephone current; 22 high-frequency current.

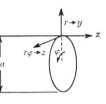

FIG. 30. Transition to the limit from the circular cross section of the wire (coordinates x, r, φ) to semi-infinite space (coordinates x, y, z).

pronounced skin effect. The three curves 00, 11, 22 have been drawn for the same total current I.

We also illustrate, by Fig. 30, the transition from the coordinates x, r, φ of the wire to the coordinates x, y, z of the semiinfinite space.

C. The Alternating-Current Impedance of a Semiinfinite Space

We cut out of the metallic semiinfinite space a rectangular parallelepiped which is infinitely long in the y-direction and whose upper end surface lies in the plane $y = 0$. Let the length of the side parallel to the z-direction be unity, that parallel to the x-direction be equal to the wave-length λ of the wave propagating itself in this direction. If, in the following, we neglect the imaginary part of h (the slight damping of the wave in the x-direction) we have $\lambda = 2\pi/h$. The total current flowing through the parallelepiped is:

$$I = \int_0^1 \int_0^{-\infty} J_x \, dz \, dy. \tag{13}$$

We define the resistance of the parallelepiped energetically with the aid of the heat generated within it, which is given by the Joule heat Q integrated over the parallelepiped and averaged over the time. By Poynting's theorem we have for the parallelepiped:

$$\dot{W}_e + \dot{W}_m + Q = -\int S_n \, d\sigma. \tag{13a}$$

Here the first two terms on the left drop out on taking the time average because of the periodicity of the process. The Poynting vector on the right

is only that across the xz-surface (shaded in Fig. 31); the contributions o
the xy-surfaces are zero since $H_x = H_y = 0$ and the contributions of th
yz-surfaces cancel because of the periodicity with respect to x. Hence

$$\int S_n\, d\sigma = -\int_0^1 dz \int_0^\lambda S_y\, dx = \int_0^\lambda E_x H_z\, dx.$$

If, for E_x and H_z, we use the values within the metal (which, as we know
agree with those in air at $y = 0$ and may be simplified because $|\,h\,| \ll |\,k_L\,|$)
we find from (9) and (9b), utilizing the representation (7)

$$\int S_n\, d\sigma = A^2 \int_0^\lambda \mathrm{Re}\{e^{-i\omega t+ihx}\} \cdot \mathrm{Re}\left\{\sqrt{\frac{\varepsilon'}{\mu}}\, e^{-i\omega t+ihx}\right\} dx.$$

If, as a sufficient approximation, we substitute here $i\sigma/\omega$ for ε' and writ
for $\sqrt{i}$ its value $(1 + i)/\sqrt{2}$, we can place the factor $(\sigma/(2\mu\omega))^{\frac{1}{2}}$ ahead o

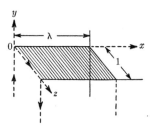

y

0

λ

x

1

z

Fig. 31. A block is cut out of the metallic semi
infinite space $y < 0$ by the two pairs of planes $x = 0$
$x = \lambda; z = 0, z = 1$. Computation of its resistance fo
a surface wave progressing in the x-direction.

the integral sign, while the factor $1 + i$ remains in the argument of Re. I
we make the further substitution

$$u = hx - \omega t, \qquad dx = \frac{du}{h} = \frac{\lambda\, du}{2\pi}, \qquad u_0 = -\omega t,$$

the preceding equation passes over into

$$\int S_n\, d\sigma = \frac{\lambda A^2}{2\pi} \sqrt{\frac{\sigma}{2\mu\omega}} \int_{u_0}^{u_0+2\pi} \cos u(\cos u - \sin u)\, du = \frac{\lambda A^2}{2} \sqrt{\frac{\sigma}{2\mu\omega}}. \tag{13b}$$

Since this value has become independent of t, it represents at the sam
time the time average of Q, which we require for the definition of th
resistance. Taking a time average of the energetic definition of R in (18.6d
and thus extending it for alternating current we write

$$R\bar{I}^2 = \bar{Q} = \frac{\lambda A^2}{2} \sqrt{\frac{\sigma}{2\mu\omega}}. \tag{14}$$

In order to compute the value of $\bar{I}^2$ which occurs here we make use o
the loop integral of **H** about the parallelepiped, e.g. in the plane $x = $

indicated in Fig. 31 by heavy arrows). In view of the direction of **H** only the edge $y = 0, 0 < z < 1$ of the parallelepiped yields a contribution. We thus obtain

$$I = \int_0^1 H_z \, dz = H_z = G(0) e^{-i\omega t}$$

$$= \sqrt{\frac{\varepsilon'}{\mu}} \, A e^{-i\omega t} = \sqrt{\frac{\sigma}{2\mu\omega}} \, A (1 + i) e^{-i\omega t}. \tag{14a}$$

Before taking the mean square of this we must pass to the real part:

$$I = \sqrt{\frac{\sigma}{2\mu\omega}} \, A (\cos \omega t + \sin \omega t).$$

We then obtain

$$\bar{I}^2 = \frac{\sigma}{2\mu\omega} A^2. \tag{14b}$$

Substitution in (14) yields

$$R = \lambda \sqrt{\frac{\mu\omega}{2\sigma}} = \frac{\lambda\kappa}{\sigma}, \qquad \text{referring to (11).} \tag{15}$$

The meaning of this formula becomes clear if we substitute the skin thickness d from Eq. (12). It then becomes

$$R = \frac{\lambda}{d\sigma}. \tag{15a}$$

is the "length" of our conductor segment measured in the direction of propagation of the waves. If we compare (15a) with the elementary formula for direct current

$$R_0 = \frac{l}{q\sigma}, \tag{15b}$$

we see that the cross section q becomes, in our alternating-current case, the rectangle

$$d \cdot 1 \ (d \text{ in the } y\text{-direction, 1 in the } z\text{-direction}). \tag{15c}$$

In place of the infinite cross section (the yz-surface of our parallelepiped) available to it the alternating current utilizes, in a sense, only the rectangle (15c); expressed differently: the alternating current, which drops off exponentially within the conductor, behaves, with respect to its resistance, just as a direct current which is distributed uniformly over the skin thickness d.

D. The Rayleigh Resistance of a Wire

We now pass from the resistance formula (15) for a conductor wit plane boundary to that for a *circularly cylindrical wire* of radius a. W assume here

$$a \gg d,$$

so that the skin effect may develop freely at the surface of the wire an its interior remains free of current. We must now, however, consider width $2\pi a$ of the parallelepiped, rather than the width 1, and imagine it current-carrying layer to be bent into the current-carrying surface laye

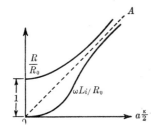

Fig. 32. Resistance and inner inductive rea tance for alternating current as function of fre quency. The abscissa is proportional to the squar root of the frequency. OA = bisector of the angl between the two axes. The inductive reactance curv approaches it asymptotically, whereas the resistanc curve runs parallel to it in the limit.

of the wire. Herewith we have changed the cross section $q = d \cdot 1$ define in (15c) to $q = d \cdot 2\pi a$ and the resistance R found in (15a) to R_a:

$$R_a = \frac{R}{2\pi a} = \frac{\lambda}{2\pi a \sigma\, d}. \tag{16}$$

If, furthermore, we introduce the direct current resistance of the wire

$$R_0 = \frac{\lambda}{\pi a^2\, \sigma} \tag{16a}$$

which, like R_a, we refer to the same length λ and the same conductivity as our R, we obtain simply:

$$\frac{R_a}{R_o} = \frac{1}{2}\frac{a}{d} = \frac{a}{2}\,\kappa. \tag{17}$$

This is *Rayleigh's resistance formula* for high-frequency alternatin current. Fig. 32 indicates its limits of validity. For small ω (stationary an quasistationary currents) $R_a = R_o$, in contradiction with (17): Our plotte curve at $\omega = 0$ is tangent to the horizontal at a distance 1 from the axi of abscissas (to a higher order of tangency). The approximate representa tion (17) applies only for sufficiently large ω. Because of our choice of th scale of abscissas the direction of the curve for increasing ω follows th asymptote OA, which is inclined by 45° to the axes. The intermediat

gion between small and large ω requires a more detailed analytical
eatment (see end of this section).

E. The Alternating Current Inductance

The method given so far could yield only the *resistance*. To obtain in
milar manner the *reactance* we would have to compute the *magnetic
nergy* W as a function of the current I:

$$W_m = \frac{L}{2} I^2$$

nd this not only for the conductor (inner selfinductance L_i, see p. 122),
ut also for the surrounding air space (outer selfinductance L_a). On the
ther hand our analysis of the external field, as carried out up to this
oint, would not be adequate for application to the wire, so that in the
ollowing we shall confine ourselves to the interior field and the inner
elfinductance L_i.

For this however, we have a more general and also simpler method
vailable, namely that of the impedance operator of Eq. (18.10):

$$\mathbf{R}I = E, \qquad \mathbf{R} = R - i\omega L_i .^{1}$$

Ve put E equal to the field strength E_x (voltage per unit length of our
onductor) at its surface $y = 0$, so that, suppressing the phase factor, we
ave by Eq. (9) $E = A$, where now both R and L have to be referred to
nit length. We obtain I from (14a), where we again suppress the time
actor:

$$I = \sqrt{\frac{\sigma}{2\mu\omega}} A(1 + i) = \frac{\sigma}{2\kappa} A(1 + i).$$

Ve find as the ratio of the two

$$\frac{E}{I} = (1 - i) \frac{\kappa}{\sigma},$$

o that, by (18),

$$\mathbf{R} = \frac{E}{I} = (1 - i) \frac{\kappa}{\sigma}, \qquad R = \omega L_i = \frac{\kappa}{\sigma}. \tag{19}$$

Vith respect to R this agrees with (15) if λ is replaced by our present unit
f length, and shows at the same time that the inner *inductive reactance*

---

[1] We have changed the sign of the imaginary unit as compared with §18 in order
o be able to employ the preceding formulas for I and E directly. In them the time
actor was written in the form $\exp(-i\omega t)$, while in §18 $\exp(+i\omega t)$ occurred in the
orresponding formulas.

ωL_i is equal to the *resistance* R. This applies generally for a conductc bounded by a plane, but applies also to the circularly cylindrical wire i *Rayleigh's limiting case of sufficiently high frequency*.

F. Further Treatment of the Alternating Current Field of a Circularly Cylindrical Wire

In order to close the subject of this section it is necessary to utiliz some formulas which will not be derived systematically until §22. We ar concerned first of all with the longitudinal alternating current field in wire of radius a:

$$E_x = C J_0(kr). \tag{2C}$$

J_0 denotes the Bessel function of order 0 which is continuous at $r = 0$ (Just as previously, the phase factor should be imagined to be added.) I the following k is to signify the *complex* wave number in the interior of th wire, not, as before, the wave number in air. The coefficient C in (20) i determined from the current density $J = \sigma E_x$ in the wire and the tota current I by the later Eq. (22.3d):

$$C = - \frac{kI}{2\pi a\sigma} \frac{1}{J_0'(ka)}. \tag{20a}$$

From (20) and (20a) we obtain for the current density J

$$\frac{J}{J_0} = - \frac{ka\, J_0(kr)}{2\, J_0'(ka)}; \tag{21}$$

$J_0 = I/(\pi a^2)$ is the direct-current value of J.

Eq. (21) may serve to check our Fig. 29. For low frequencies the argu ments kr and ka are small in absolute value. Then the expansion (22.3c and its derivative may be employed:

$$J_0(\rho) = 1 - \left(\frac{\rho}{2}\right)^2 + \frac{1}{4}\left(\frac{\rho}{2}\right)^4 - \frac{1}{36}\left(\frac{\rho}{2}\right)^6 \pm \cdots$$

$$J_0'(\rho) = -\frac{\rho}{2}\left(1 - \frac{1}{2}\left(\frac{\rho}{2}\right)^2 + \frac{1}{12}\left(\frac{\rho}{2}\right)^4 \pm \cdots\right) \tag{21a}$$

substituted in (21) this leads, with due regard of (11), to

$$\frac{J}{J_0} = \frac{1 - \left(\frac{kr}{2}\right)^2 + \frac{1}{4}\left(\frac{kr}{2}\right)^4 - \cdots}{1 - \frac{1}{2}\left(\frac{ka}{2}\right)^2 + \frac{1}{12}\left(\frac{ka}{2}\right)^4 - \cdots} = \frac{1 - \frac{i}{2}(\kappa r)^2 - \frac{1}{16}(\kappa r)^4}{1 - \frac{i}{4}(\kappa a)^2 - \frac{1}{48}(\kappa a)^4}. \tag{21b}$$

[1] We employ here and in the following (unlike Vol. VI) the symbol J, in accorc with the practice followed in Vol. II of these lectures and in American physics anc engineering literature generally.

'rom this follows

$$\left|\frac{J}{J_0}\right|^2 = \frac{1 + \dfrac{1}{8}\,(\kappa r)^4}{1 + \dfrac{1}{48}\,(\kappa a)^4}. \tag{21c}$$

s compared to the direct-current straight line 00 in Fig. 29 there occurs
nus a dip, which on the axis of the wire ($r = 0$) has the depth $(\kappa a)^4/48$,
nd a rise at the periphery of the wire ($r = a$) to a height which is five
imes as great.

At high frequencies we obtain asymptotically, according to (22.7) and
22.6a):

$$J_0(kr) = (\pi kr/2)^{-\frac{1}{2}} \cos\,(kr - \pi/4) \tag{22}$$

$$\frac{J}{J_0} = \frac{ka}{2}\left(\frac{a}{r}\right)^{\frac{1}{2}} \frac{\cos\,(kr - \pi/4)}{\sin\,(ka - \pi/4)} \qquad \left|\frac{J}{J_0}\right| = \frac{\kappa a}{\sqrt{2}}\left(\frac{a}{r}\right)^{\frac{1}{2}} e^{-\kappa(a-r)} \tag{22a}$$

n view of the sharp decrease for $r < a$ the whole interior of the wire is
ractically free of current; the magnitude at the edge is $\kappa a/\sqrt{2}$ times as
reat as in the direct-current case.

The same formulas also yield a closed expression for the operator **R** in
£q. (18). By (20) and (20a) we have

$$\mathbf{R} = \frac{E_x}{I} = -\frac{k}{2\pi a\sigma}\frac{J_0(ka)}{J_0'(ka)} = -\frac{ka}{2}\frac{J_0(ka)}{J_0'(ka)}\,R_0, \tag{23}$$

vhere R_0 is once more the direct-current resistance per unit length, i.e.
$/(\pi a^2\sigma)$. Hence we have for *low* frequencies (by (21a), expanding in powers
f κa):

$$\frac{R}{R_0} - \frac{i\omega L}{R_0} = 1 - \frac{i}{4}\,(\kappa a)^2 + \frac{1}{48}\,(\kappa a)^4.$$

The separation of the real and imaginary parts yields:

$$\frac{R}{R_0} = 1 + \frac{1}{48}\,(\kappa a)^4, \qquad \frac{\omega L}{R_0} = \frac{1}{4}\,(\kappa a)^2. \tag{23a}$$

)n the other hand, we obtain, from (22) and (22a), for very *high* frequencies
imply:

$$\frac{R}{R_0} - \frac{i\omega L}{R_0} = -i\frac{ka}{2} = (1 - i)\frac{\kappa a}{2}$$

$$\frac{R}{R_0} = \frac{\omega L}{R_0} = \frac{\kappa a}{2}. \tag{23b}$$

This agrees with our earlier results (17) and (19), obtained for the con ductor with the plane boundary.

Our approximations (23a, b) permit us to check also Fig. 32 and t interpolate for the intermediate region between low and high frequencies At low frequencies, by (23a), the resistance curve approaches the straigh line $R = R_0$ as a parabola of the fourth order and at high frequencies i approaches the straight line OA in Fig. 32 from above. The curve for th inner inductive reactance starts at low frequencies as a parabola of th second order[1] and approaches[2] at high frequencies the same straight lin from below.

§21. The Coil Carrying Alternating Current

Inasmuch as in §17 we had to defer the treatment of the direct-curren field of the circular wire as mathematically too complicated, the rigorou treatment of the alternating-current field of a long coil appears to be ou of the question. We hence make the same approximation as on p. 25 i.e. replace the coil, which we assume to consist of a single closely-woun layer, by an infinitely long hollow cylinder of uniform metal. Let its axi (x-axis) be vertical and let it be traversed by horizontal circular current whose intensity distribution we shall determine. Let the inner radius o the hollow cylinder be a, the outer radius $a + d$.

A. The Field of the Coil

As in the direct-current case we assume that the magnetic excitation i zero outside of the coil, uniform[3] within it and parallel to the coil axis, s that we may write

$$H = 0 \text{ for } r \geq a + d, \qquad H = H_x = H_a e^{-i\omega t} \text{ for } r \leqq a. \qquad (1$$

We must then assume the **H**-field to be parallel to the cylinder axis als in the metallic conductor, i.e.

$$H = H_x = H(r)e^{-i\omega t} \text{ for } a < r < a + d.$$

[1] This statement applies to the product ωL plotted in the figure; L itself has fo $\omega = 0$, in accord with the meaning of $\kappa^2 = \mu\sigma\omega/2$, the non-vanishing value

$$L = \frac{R_0}{4} \frac{\mu\sigma a^2}{2} = \frac{\mu}{8\pi},$$

in agreement with §16C.

[2] This approach is a true tangency for $\omega \to \infty$; on the other hand, the curve fo the resistance remains even for $\omega \to \infty$ a finite amount $R_0/4$ above the straight lin OA, as would be shown by a more precise formulation of the approximation (23b).

[3] This customary and practically unavoidable assumption for infinite length o the coil is, strictly speaking, not permissible in Maxwell's theory. It contradicts th equation, applying for the nonconducting interior space, $\dot{D} = $ curl **H** and is, in view of curl **H** $= 0$ equivalent to the neglect of the displacement current $\dot{D}$.

H_x must satisfy the general wave equation (20.2), which yields, for $H(r)$, the differential equation

$$\Delta H(r) + k^2 H(r) = 0, \qquad k^2 = \varepsilon\mu\omega^2 + i\mu\sigma\omega. \tag{2}$$

Transformed to polar coordinates x, r, φ it is integrated in terms of Bessel functions. We do not require here, however, the particular solution $J_0(kr)$ as in (20.20), but the general solution containing two constants, which we write preferably in the form

$$C_1 H_0^1(kr) + C_2 H_0^2(kr). \tag{2a}$$

H^1, H^2 are the two Hankel cylinder functions, about which we shall give some information in the next section. In particular we shall familiarize ourselves there with their asymptotic behavior for large values of the argument $\rho = kr \to \infty$:

$$H_0^1(\rho) \to \sqrt{\frac{2}{\pi\rho}}\, e^{i(\rho-\pi/4)}, \qquad H_0^2(\rho) \to \sqrt{\frac{2}{\pi\rho}}\, e^{-i(\rho-\pi/4)}. \tag{2b}$$

Since for high-frequency alternating current invariably $|k|\, a \gg 1$, we can limit ourselves in the integration of (2) to these asymptotic values and can write

$$H(r) = \sqrt{\frac{2}{\pi k r}}\, (C_1 e^{i(kr-\pi/4)} + C_2 e^{-i(kr-\pi/4)}). \tag{3}$$

From Eqs. (1) we have the boundary conditions

$$H(a+d) = 0 \qquad \text{and} \qquad H(a) = H_a.$$

They are satisfied if, by special choice of C_1, C_2, expression (3) is transformed into

$$H(r) = \sqrt{\frac{a}{r}}\, H_a\, \frac{\sin\,[k(a+d-r)]}{\sin\,(kd)}. \tag{4}$$

Having found in this manner H_x as function of r (we may also say as a function of the polar coordinates x, r, φ) we can now utilize §20 A. It is true that now we are not dealing, as there, with a wave advancing along the x-axis, but with ordinary (stationary) alternating current, for which the wave number h given there vanishes. In fact our present Eq. (2) passes over into the earlier (20.3) for $X = H_x$, $h = 0$. Furthermore we must not employ, as in the preceding paragraph, Eq. (20.5) (electric type), but Eq. (20.6) (magnetic type). This yields the transversal components H_r, E_φ expressed in terms of the longitudinal component H_x. We are particularly interested in E_φ. With $g_u = 1$, $g_v = r$, $H = 0$, (20.6) yields:

$$E_\varphi = -\frac{i}{k}\sqrt{\frac{\mu}{\varepsilon'}}\, \frac{dH(r)}{dr}\, e^{-i\omega t} = \frac{J}{\sigma}. \tag{5}$$

If we carry out the differentiation with respect to r only in the factor $\sin\{k(a + d - r)\}$ (the square-root factor is "slowly variable") and if we denote the value of J on the inner surface of the coil by J_a, we obtain

$$J = J_a \sqrt{\frac{a}{r}} \frac{\cos\{k(a + d - r)\}}{\cos(kd)}, \qquad J_a = i\sigma\sqrt{\frac{\mu}{\varepsilon'}} H_a \frac{\cos(kd)}{\sin(kd)} e^{-i\omega t}. \qquad (6)$$

Passing to the discussion of the field, we distinguish two cases:

$$a. \ |k| \, d \gg 1 \qquad \text{and} \qquad b. \ |k| \, d \ll 1;$$

we of course continue to adhere to the original assumption $|k| \, a \gg 1$.

a. High-Frequency Alternating Current with not too Small Coil Thickness. Since we take the imaginary part of k to be positive we have then

$$|e^{-ikd}| \gg |e^{ikd}|$$

and, more particularly near the inner coil boundary also:

$$|e^{-ik(a+d-r)}| \gg |e^{ik(a+d-r)}|.$$

Hence (4) yields, for $r \cong a$,

$$H(r) \cong H_a e^{ik(r-a)}, \qquad (7)$$

which signifies a steep exponential falling off at the inner boundary of the coil. By (6) the current density shows a similar steep exponential decline:

$$J \cong J_a e^{ik(r-a)}, \qquad (7a)$$

$$J_a = i\sigma \sqrt{\frac{\mu}{\varepsilon'}} H_a e^{-i\omega t}. \qquad (7b)$$

We thus have a *pronounced skin effect* at the inner surface of the coil.

b. Small Coil Thickness and Relatively Low-Frequency Alternating Current. We may then expand Eqs. (4) and (6) in powers of kd and, particularly in the neighborhood of $r = a$, also in powers of $k(a + d - r)$. We indicate only the first term of these expansions:

$$H(r) \cong H_a \frac{a + d - r}{d}, \qquad J \cong J_a. \qquad (8)$$

Fig. 33 illustrates this graphically: At the left is shown the behavior of $H(r)$, at the right that of J. The curves 0 correspond to the limiting case b, the curves 2 to case a, and the curves 1 to an intermediate case. All three pairs of curves refer to the same H within the coil and hence also to the same total current I in the coil.

B. Resistance and Inner Inductive Reactance of the Coil

We wish to compute these quantities per unit length of the coil and must first know the total current passing through this unit length. It is obtained by integration of (5) with respect to r:

$$\int_a^{a+b} J \, dr = C\{H(a+d) - H(a)\}e^{-i\omega t} = -CH_a e^{-i\omega t};$$

in view of the meaning of k and ε' we have

$$-C = \frac{i\sigma}{k}\sqrt{\frac{\mu}{\varepsilon'}} \cong 1. \tag{9}$$

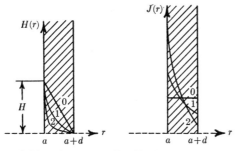

FIG. 33. Magnetic field and current distribution in a single-layer coil: At the left, magnetic field $H(r) = H_z$, at the right current density $J(r) = J_\varphi$. The curves 0 and 2 correspond to the limiting cases of direct current and high-frequency alternating current, curves 1 to intermediate frequencies with the same total current as in the two limiting cases.

To refer this total current to the current I flowing in a *single wire* of the coil we set it equal to NI, where N is the number of turns per unit length:

$$I = \frac{1}{N} H_a e^{-i\omega t}. \tag{9a}$$

With this value of I we write the equation

$$(R - i\omega L)I = E,$$

where we must set the voltage E equal to the field strength on the inner surface of the coil, i.e. equal to J_a/σ. We thus obtain from (6) and (9)

$$R - i\omega L = \frac{Nk}{\sigma}\frac{\cos(kd)}{\sin(kd)} = i\frac{Nk}{\sigma}\frac{e^{ikd} + e^{-ikd}}{e^{ikd} - e^{-ikd}}. \tag{10}$$

For a discussion of this expression we write, as in (20.11), $k = (1 + i)\kappa$. The denominator of (10) then becomes

$$e^{(-1+i)\kappa d} - e^{(1-i)\kappa d}.$$

We multiply numerator and denominator with the complex conjugate of this quantity, whereupon they can be expressed in terms of the trigonometric and hyperbolic functions of $2\kappa d$. We obtain

$$R - i\omega L = (1 - i) \frac{N\kappa}{\sigma} \frac{\sinh (2\kappa d) + i \sin (2\kappa d)}{\cosh (2\kappa d) - \cos (2\kappa d)} \tag{11}$$

so that, separating real and imaginary parts,

$$R = \frac{N\kappa}{\sigma} \frac{\sinh (2\kappa d) + \sin (2\kappa d)}{\cosh (2\kappa d) - \cos (2\kappa d)}, \tag{11a}$$

$$\omega L = \frac{N\kappa}{\sigma} \frac{\sinh (2\kappa d) - \sin (2\kappa d)}{\cosh (2\kappa d) - \cos (2\kappa d)}. \tag{11b}$$

For direct current ($\kappa d \to 0$) (11a) yields

$$R_0 = \frac{N}{\sigma d}, \tag{12}$$

corresponding to the direct-current resistance per unit length of a wire of rectangular cross section with the width d of the coil (not to be confused with the layer thickness d of p. 162) and the height $1/N$. We have therefore

$$\frac{R}{R_0} = \kappa d \frac{\sinh (2\kappa d) + \sin (2\kappa d)}{\cosh (2\kappa d) - \cos (2\kappa d)}, \tag{12a}$$

$$\frac{\omega L}{R_0} = \kappa d \frac{\sinh (2\kappa d) - \sin (2\kappa d)}{\cosh (2\kappa d) - \cos (2\kappa d)}. \tag{12b}$$

More particularly we consider the two limiting cases a and b of p. 172:
a. $\kappa d \gg 1$. Then $\sinh(2\kappa d) = \cosh(2\kappa d) \to \infty$, so that

$$\frac{R}{R_0} = \frac{\omega L}{R_0} = \kappa d, \tag{13a}$$

b. $\kappa d \ll 1$. By expansion in powers of $2\kappa d$, and retaining only the first nonvanishing term we find from (12a, b)

$$\frac{R}{R_0} = 1 + \frac{4}{45} (\kappa d)^4, \qquad \frac{\omega L}{R_0} = \frac{2}{3} (\kappa d)^2. \tag{13b}$$

These results (13a, b) are represented qualitatively once more by Fig. 32. Also now the direct-current straight line $R/R_0 = 1$ is approximated at low frequencies to the fourth order and the axis of abscissas is parabolically tangent to the curve $\omega L/R_0$ at the origin. At high frequencies both curves again approach the 45° line asymptotically. Only the scale of abscissas, now given by κd, differs from that given before since the skin effect occurs here unilaterally, on the inner surface of the coil.

However much more important than the inner selfinductance, to which we have limited ourselves, is of course the *external selfinductance of the coil*. In the discussion of the coil traversed by direct current (§17, Eq. (15)) we were exclusively interested in the latter. With the limitation mentioned in footnote 3, p. 170 (neglect of the displacement current) we can transfer the value found there directly to the alternating-current case.

With our idealization of the problem (closed ring currents in horizontal planes) we could circumvent the determination of the electric field within the coil and leave the boundary conditions for this field out of consideration. Actually the currents are, in view of the finite pitch of the coil, not exactly horizontal and there exists, from turn to turn, an external electric field of complex character which is predominantly axial in direction. Our elementary treatment evidently does not suffice to determine this field.

C. The Multilayer Coil

We imagine the several layers, n in number, to be placed one over the other without intervening space, idealized as hollow cylinders of thickness d/n, and traversed in series by alternating current I of uniform magnitude. Let the insulation of the successive layers be perfect, though by assumption infinitely thin; the same is to apply for the insulation of successive turns in any single layer.

In view of the magnitude of the loop integral of $\mathbf{H}$ about a single layer, i.e. over a rectangle of height 1 and width d/n, the magnetic field decreases by NI for each layer (N = number of turns per unit length along the axis of the coil). Hence we find at the boundaries of successive layers, from the inside outward:

$$H_0 = H, \qquad H_1 = H - NI, \qquad H_2 = H - 2NI, \cdots, H_n =$$
$$H - nNI = 0. \tag{14}$$

The factor $\exp(-i\omega t)$ must be thought of as added here, and in the following. In the νth layer the magnetic field satisfies the differential equation and the boundary conditions:

$$\Delta H + k^2 H = 0, \qquad H = \begin{cases} H_{\nu-1} & r = a + (\nu - 1)\dfrac{d}{n}, \\[2mm] H_\nu & r = a + \nu \dfrac{d}{n}. \end{cases} \tag{15}$$

The differential equation is again integrated by the superposition of the two Hankel functions, for which we can substitute their asymptotic values from Eq. (2b) since kr is very large; furthermore we can treat the denominator $\sqrt{kr}$ as slowly variable in comparison with the exponential functions and include it along with the remaining constants ($\exp\{-\mathrm{i}\pi/4\}$ and $\sqrt{2/\pi}$) in the amplitudes C_1, C_2. Eq. (3) is then simplified to

$$H = C_1 e^{ikr} + C_2 e^{-ikr}.$$

The following expression, which at the same time satisfies the boundary conditions (15) and hence represents the magnetic field in the νth layer, is of this form:

$$H = H_{\nu-1} \frac{\cos\left\{k\left(r - a - (\nu - \tfrac{1}{2})\dfrac{d}{n}\right)\right\}}{\cos\left(k\dfrac{d}{2n}\right)}$$

$$- NI \frac{\sin\left\{k\left(r - a - (\nu - 1)\dfrac{d}{n}\right)\right\}}{\sin\left(k\dfrac{d}{n}\right)} \tag{16}$$

In particular we obtain for the first layer, $a < r < a + d/n$, $\nu = 1$

$$H = H_a \frac{\cos\left\{k\left(r - a - \dfrac{d}{2n}\right)\right\}}{\cos\left(k\dfrac{d}{2n}\right)} - NI \frac{\sin\{k(r - a)\}}{\sin\left(k\dfrac{d}{n}\right)}. \tag{16a}$$

We need consider only this formula if now we wish to determine the electric field E_φ at the inner surface of the coil and from this the impedance operator $\mathbf{R} = R - i\omega L$ of the multilayer coil.

To begin with we obtain for the current density J_φ, for which we shall consider right away the value for $r = a$, from (5) (see also (9)):

$$J_a = C\left(\frac{dH}{dr}\right)_a = -Hk \frac{\sin\left(k\dfrac{d}{2n}\right)}{\cos\left(k\dfrac{d}{2n}\right)} + NIk \frac{1}{\sin\left(k\dfrac{d}{n}\right)}.$$

We here set $H = nNI$ (see last of Eqs. (14)) and obtain after simple trigonometric transformation,

$$J_a = NIk \frac{n\cos\left(k\dfrac{d}{n}\right) - (n - 1)}{\sin\left(k\dfrac{d}{n}\right)}.$$

Hence we obtain

$$E = E_\varphi = \frac{1}{\sigma} J_a \quad \text{and} \quad \mathbf{R} = \frac{E}{I} = \frac{Nk}{\sigma} \frac{n \cos\left(k\dfrac{d}{n}\right) - (n-1)}{\sin\left(k\dfrac{d}{n}\right)},$$

which, for $n = 1$, is identical with Eq. (10).

We divide $\mathbf{R}$ by the direct-current value R_0 of $\mathbf{R}$, obtained by passing to the limit $\omega \to 0$, i.e. $k \to 0$:

$$R_0 = \frac{Nn}{\sigma d},$$

and find

$$\frac{\mathbf{R}}{R_0} = kd \frac{\cos\left(k\dfrac{d}{n}\right) - \left(1 - \dfrac{1}{n}\right)}{\sin\left(k\dfrac{d}{n}\right)}. \tag{17}$$

The separation of (17) into real and complex parts is rather complicated. If as before we set $k = (1 + i)\kappa$, it yields

$$\left.\begin{array}{l} R/R_0 \\ \omega L/R_0 \end{array}\right\} = \kappa d \frac{\begin{array}{c} \sinh\left(2\kappa\dfrac{d}{n}\right) \pm \sin\left(2\kappa\dfrac{d}{n}\right) - 2\left(1 - \dfrac{1}{n}\right) \\ \cdot \left(\sinh\left(\kappa\dfrac{d}{n}\right)\cdot\cos\left(\kappa\dfrac{d}{n}\right) \pm \cosh\left(\kappa\dfrac{d}{n}\right)\cdot\sin\left(\kappa\dfrac{d}{n}\right)\right) \end{array}}{\cosh\left(2\kappa\dfrac{d}{n}\right) - \cos\left(2\kappa\dfrac{d}{n}\right)}$$

Thus, thanks to our extensive (possibly excessive) idealization of the problem, we have obtained a quite simple final formula. The frequency band for which our formula is valid has an upper limit determined by the characteristic frequency of the coil; as we approach the latter our notion of equal current in all turns obviously becomes invalid. It should also be emphasized that our formula presumes the regular superposition of the layers and does not cover the spiral interweaving of the turns (Dolezalek, litz wire) which is preferred for practical reasons (suppression of the skin effect).

§22. The Problem of Waves on Wires

As is well known, the experiments of Heinrich Hertz dealt with "surface waves" progressing along wires as well as with "space waves" propagated freely through the air. Hertz expected their velocity also to be equal to c, but could confirm this result neither experimentally nor theoretically. The reason for his experimental failure was the influence of the walls of the laboratory; the reason for his theoretical failure, an excessive idealization of the problem. He treated the wire as infinitely thin and hence could not set up electromagnetic boundary conditions. This was first accomplished in a paper by the author[1] which yielded for the velocity of propagation a value nearly equal to—more precisely, slightly less than—c. It was here essential that a phase velocity exceeding c could be excluded by a condition at infinity. The experimental difficulties were overcome by E. Lecher (see §25) by using a *two-wire line*. In the present section we shall confine ourselves to Hertz's problem of the *single wire*.

A. The Field within and outside of the Wire

While in the preceding section the Maxwell equations were utilized only in part, inasmuch as not only was the displacement current within the conductor neglected, but to some extent also that in free space, we must now adhere strictly to these equations. The problem is symmetrical about the axis of the wire. We make it the x-axis of a cylindrical coordinate system x, r, φ. Then for all components $\partial/\partial\varphi = 0$ and only the components

$$E_x , E_r , H_\varphi \tag{1}$$

differ from zero. As in §20 we set them equal to products of the common factor

$$e^{-i\omega t + ihx} \tag{1a}$$

with a function of r only. Let the time variation be purely periodic, i.e. ω be real, and the phase propagation take place along the positive x-axis; h must then have a positive real part.

We deal first with the Cartesian longitudinal component E_x. It satisfies the wave equation (20.3), in which we put $u = r, v = \varphi$. If, in place of r, we introduce the dimensionless variable

$$\rho = \sqrt{k^2 - h^2}\, r \tag{1b}$$

and set

$$E_x = F(\rho)e^{-i\omega t + ihx}, \tag{2}$$

we obtain for F the differential equation

$$\frac{1}{\rho}\frac{d}{d\rho}\left(\rho\,\frac{dF}{d\rho}\right) + F = 0 \tag{3}$$

[1] Ann. d. Physik, Vol. 67, pp. 233–290, 1899.

or, with the differentiation carried out,

$$\frac{d^2F}{d\rho^2} + \frac{1}{\rho}\frac{dF}{d\rho} + F = 0. \tag{3a}$$

This, as well as the more general equation

$$\frac{d^2F}{d\rho^2} + \frac{1}{\rho}\frac{dF}{d\rho} + \left(1 - \frac{n^2}{\rho^2}\right)F = 0 \tag{3b}$$

is known as Bessel's differential equation. We have dealt with it already in Vol. II, §27. The solution which is continuous for $\rho = 0$ was represented there, in (27.7), by the series

$$J_n(\rho) = \frac{1}{n!}\left(\frac{\rho}{2}\right)^n - \frac{1}{1!\,(n+1)!}\left(\frac{\rho}{2}\right)^{n+2} + \frac{1}{2!(n+2)!}\left(\frac{\rho}{2}\right)^{n+4} - \cdots \tag{3c}$$

For all that follows n may be assumed to be an integer. For $n = 0$ we obtain the representation (21a) for $J_0(\rho)$ employed in §20, from which we see directly

$$J_1(\rho) = -\frac{d}{d\rho}J_0(\rho), \qquad \int_0^\rho \rho J_0(\rho)\,d\rho = \rho J_1(\rho). \tag{3d}$$

Within the wire, $r < a$ (a = radius of wire), where E_x must nowhere become infinite, our function F is thus determined but for a constant by its differential equation (3):

$$F = C J_0(\rho) \qquad \text{for} \qquad 0 < r < a. \tag{4}$$

Outside of the wire (in air) $\sigma = 0$, $\varepsilon = \varepsilon_0$, $\mu = \mu_0$, so that $k = \omega/c$ is real. For the sake of differentiation we shall denote the complex value of k, which applies within the wire, by k_L as in §20. Since the condition of continuity at $r = 0$ plays no role outside of the wire, Eq. (3) has to be integrated generally. This is done by the two "Hankel functions of order 0", already mentioned on p. 171:

$$H_0^1(\rho) \qquad \text{and} \qquad H_0^2(\rho).$$

We shall deal in detail with these and the general Hankel functions of order n in Vol. VI, §19. It must here suffice to enumerate some of their principal properties:

a. The functions H_0^1 and H_0^2 become *logarithmically infinite* for $\rho = 0$, since we have, for small ρ:

$$H_0^{1,2}(\rho) = 1 \pm \frac{2i}{\pi}\log\frac{\gamma\rho}{2} + \cdots = \pm\frac{2i}{\pi}\log\frac{\gamma\rho}{\pm 2i}\cdots \tag{5}$$

γ is related $_x$ to the Euler-Mascheroni constant

$$\lim_{n\to\infty}\left(1 + \frac{1}{2} + \frac{1}{3} + \cdots + \frac{1}{n} - \log n\right) = 0.5772 \cdots;$$

For

$$\log \gamma = 0.5772 \cdots, \gamma = 1.781 \cdots \tag{5a}$$

(log here denotes, as always, the natural logarithm.)

b. The Hankel functions are branched in the complex ρ-plane, just as the logarithm. To make them single-valued we must provide a branch cut in the ρ-plane, e.g. along the negative imaginary axis. If we set $\rho = |\rho|e^{i\vartheta}$, we thus limit the angle ϑ to the values $-\pi/2 < \vartheta < 3\pi/2$. In this sense we speak of the principal branch of the Hankel functions, just as we speak of the principal branch of the logarithm.

c. The Hankel functions $H^1_n(\rho)$ and $H^2_n(\rho)$ of order n are defined as solutions of Eq. (3b) in such fashion that also their representation contains a logarithmic term for $\rho \to 0$, i.e. that the branching mentioned in b applies also to them. However, the determining factor for their singularity is not this logarithmic term, but the term which becomes most strongly infinite:

$$H^1_n(\rho) = \frac{(n-1)!}{i\pi}\left(\frac{2}{\rho}\right)^n, \qquad H^2_n(\rho) = \frac{(n-1)!}{-i\pi}\left(\frac{2}{\rho}\right)^n. \tag{6}$$

The logarithmic singularity and the branching disappear for the sum of the two functions H_n. The regular solution of the differential equation (3b) is obtained in the form

$$J_n(\rho) = \tfrac{1}{2}(H^1_n(\rho) + H^2_n(\rho)). \tag{6a}$$

d. For $\rho \to \infty$ we have as asymptotic representation of the two principal branches:

$$H^1_n(\rho) = \sqrt{\frac{2}{\pi\rho}}\,e^{i(\rho-(n+\frac{1}{2})\pi/2)}, \qquad H^2_n(\rho) = \sqrt{\frac{2}{\pi\rho}}\,e^{-i(\rho-(n+\frac{1}{2})\pi/2)}. \tag{7}$$

Thus H^1_n vanishes for large ρ in the positively-imaginary ρ-halfplane, H^2_n, in the negatively imaginary ρ-halfplane. The two together vanish for large ρ only on the real axis. J_n becomes, by (6a), infinite everywhere at infinity except on the real axis. In view of (6a) and (7) we have at infinity in the positively imaginary ρ-halfplane

$$\frac{J_n(\rho)}{J'_n(\rho)} = +i. \tag{7a}$$

After these insertions, which unfortunately were necessary for what follows, we return to our actual problem. We make the convention that the sign of the square root in (1b) is always to be chosen so that its imaginary part is *positive*. This covers the case that the root is itself *complex*. We must however also take into account the possibility that the root is *real* and that $h < k$.

In the first case the only possible formula for the exterior of the wire is

$$F(\rho) = AH_0^1(\rho), \qquad a < r < \infty, \qquad A = \text{const}, \qquad (8)$$

since, by (7), $H_0^2(\rho)$ becomes infinitely great for $r \to \infty$.

In the second case

$$F(\rho) = AH_0^1(\rho) + BH_0^2(\rho), \qquad a < r < \infty. \qquad (9)$$

is a possible formula since now, for real ρ, *both* H vanish, by (7), as $\rho^{-\frac{1}{2}}$. A and B are for the present arbitrarily disposable constants.

What is the meaning of this case? Since according to (2) the phase velocity of the wave is equal to ω/h and since ω/k is equal to c, it implies

phase velocity > velocity of light.

We supplement our representation of E_x by that of the transversal components E_r and H_φ, which is obtained most readily with the aid of the general rule (20.5) (where however the terms with H_x are of course omitted). If for the present we designate the functions of ρ appearing in E_r and H_φ with $G(\rho)$ and $E(\rho)$ this yields

$$G(\rho) = \frac{ih}{k^2 - h^2} \frac{\partial F(\rho)}{\partial r} = \frac{ih}{\sqrt{k^2 - h^2}} F'(\rho), \qquad (10)$$

$$\sqrt{\frac{\mu}{\varepsilon}} E(\rho) = \frac{ik}{k^2 - h^2} \frac{\partial F(\rho)}{\partial r} = \frac{ik}{\sqrt{k^2 - h^2}} F'(\rho). \qquad (11)$$

B. The Boundary Condition at Infinity

As for the surface wave in §20 we are also now dealing with a process which draws its energy from the end of the wire at $x = -\infty$. We hence shall demand that the total energy flux through a cylindrical surface $r = R$ coaxial with the wire vanishes:

$$S = 2\pi \{rS_r\}_{r=R} = 0. \qquad (12)$$

If as an abbreviation we designate the phase $hx - \omega t$ by Φ we obtain

$$S_r = E_x H_\varphi = \text{Re}\{F(\rho)e^{i\Phi}\} \ \text{Re}\left\{ \sqrt{\frac{\varepsilon_0}{\mu_0}} \frac{-ik}{\sqrt{k^2 - h^2}} \frac{dF(\rho)}{d\rho} e^{i\Phi}\right\}.$$

We consider the *second case*, in which h, $\sqrt{k^2 - h^2}$, and ρ were real. Eq. (9) then applies for $F(\rho)$ and we obtain, indicating all immaterial constant factors by . . . and utilizing the asymptotic expressions (7) with $n = 0$:

$$S = \cdots \frac{r}{\rho} \{A^2 \cos^2(\Phi + \rho - \pi/4) - B^2 \cos^2(\Phi - \rho + \pi/4)\}. \qquad (12a)$$

It is significant here that the factor r/ρ remains finite for arbitrarily large $r = R$ and that the phase Φ may take on any real values. Our requirement (12) can then be satisfied only by

$$A = B = 0.$$

This signifies: *In the second case wire waves cannot occur; they would have to be fed by an artificial arrangement of energy sources at infinity*, which contradicts the physical meaning of the process.

Conditions are different in the first case, where ρ has a *positive* imaginary part. According to Eqs. (8) and (7) the field outside of the wire decreases here exponentially as $r \to \infty$; the energy flux S vanishes in similar manner. Only this case is relevant for us. By Eqs. (2), (4), (8), (10), and (11) we calculate the corresponding field within and outside of the wire, and distinguish the complex k_L within from the real k outside, as well as the complex ε'/μ within from the real ε_0/μ_0 outside as in (20.9a, b). Furthermore it is convenient to redefine the constants C and A so that all components are multiplied with $\dfrac{\sqrt{k_L^2 - h^2}}{ih}$ inside and with $\dfrac{\sqrt{k^2 - h^2}}{ih}$ outside of the wire. We thus obtain the following tabulation:

$0 < r < a,\quad \rho = \sqrt{k_L^2 - h^2}\, r$	$a < r < \infty,\quad \rho = \sqrt{k^2 - h^2}\, r$
$E_x = \dfrac{\sqrt{k_L^2 - h^2}}{ih}\, CJ_0(\rho)$	$E_x = \dfrac{\sqrt{k^2 - h^2}}{ih}\, AH_0(\rho)$
$E_r = CJ_0'(\rho)$	$E_r = AH_0'(\rho)$
$\sqrt{\dfrac{\mu}{\varepsilon'}}\, H_\varphi = \dfrac{k_L}{h}\, CJ_0'(\rho)$	$\sqrt{\dfrac{\mu_0}{\varepsilon_0}}\, H_\varphi = \dfrac{k}{h}\, AH_0'(\rho)$

$$(13)$$

H_0 is identical with H_0^1; the prime at J_0 and H_0 indicates differentiation with respect to the argument ρ.

C. The Boundary Condition at the Surface of the Wire

E_x and H_φ must be continuous for $r = a$. Hence we must require

$$\rho_L CJ_0(\rho_L) = \rho A H_0(\rho), \qquad \sqrt{\frac{\varepsilon'}{\mu}}\, k_L CJ_0'(\rho_L) = \sqrt{\frac{\varepsilon_0}{\mu_0}}\, kAH_0'(\rho) \tag{14}$$

$$\rho_L = \sqrt{k_L^2 - h^2}\, a, \qquad \rho = \sqrt{k^2 - h^2}\, a.$$

By eliminating the amplitude factors A and C we obtain the *transcendental equation*

$$\frac{\rho H_0(\rho)}{H_0'(\rho)} = \sqrt{\frac{\varepsilon_0 \mu}{\varepsilon' \mu_0}}\, \frac{k}{k_L}\, \frac{\rho_L J_0(\rho_L)}{J_0'(\rho_L)}. \tag{14a}$$

We regard this as the determining equation for the as yet unknown wave number h. However, Eq. (14a) can be greatly simplified by taking account of the fact that if the material of the wire is a good conductor ρ_L is a large complex number with positive imaginary part so that Eq. (7a) is applicable. With its aid the right side of (14a) may be transformed into

$$i\sqrt{\frac{\varepsilon_0\mu}{\varepsilon'\mu_0}}\frac{k}{k_L}\,\rho_L \cong i\sqrt{\frac{\varepsilon_0\mu}{\varepsilon'\mu_0}}\,ka. \tag{14b}$$

Since $|\varepsilon'| \gg \varepsilon_0$ its absolute value is small compared with 1. Furthermore the left side may also be simplified. Since it must be small we may use Eq. (5). We then obtain for the left side of (14a)

$$\rho^2 \log\frac{\gamma\rho}{2i} = -\frac{2}{\gamma^2}\,u\log u \quad\text{with}\quad u = \left(\frac{\gamma\rho}{2i}\right)^2. \tag{14c}$$

Comparison with (14b) then yields as the final form of our transcendental equation

$$u\log u = v \quad\text{with}\quad v = -\frac{i\gamma^2}{2}\sqrt{\frac{\varepsilon_0\mu}{\varepsilon'\mu_0}}\,ka. \tag{15}$$

For its solution it is possible to employ a peculiar method reminiscent of the continued fraction. This rests on the fact that $\log u$ varies slowly in comparison with u. Hence if an nth approximation u_n has been found, an $n + 1^{\text{st}}$ approximation may be obtained from

$$u_{n+1}\log u_n = v. \tag{15a}$$

We may begin, for example with $u_0 = v$ and put, in accord with (15a),

$$u_1 = \frac{v}{\log v}; \tag{15b}$$

the exact initial value is of little importance since it is corrected step by step in the subsequent approximations. Furthermore, by (15a):

$$u_2 = \frac{v}{\log u_1} = \frac{v}{\log\dfrac{v}{\log v}}, \qquad u_3 = \frac{v}{\log\dfrac{v}{\log\dfrac{v}{\log v}}} \quad\text{etc.} \tag{15c}$$

Consider, for example, a copper wire with radius $a = 1$ mm and the frequency given farthest to the right on the table on p. 162, which corresponds to a wave-length of 30 cm and a value $ka = 2.1\cdot10^{-2}$. For the corresponding value of κ taken from the same table, (15) yields

$$v = -(1 + i)\cdot7.2\cdot10^{-7}.$$

We begin with

$$u_1 = (1 + i)\cdot3.6\cdot10^{-8}, \tag{16}$$

where log v, occurring in (15b), has been approximated by -20. We then find from (15a)

$$u_2 = (4.1 + 4.5i) \cdot 10^{-8}, \qquad u_3 = (4.2 + 4.6i) \cdot 10^{-8}; \qquad (16a)$$

We thus have already arrived at the limit of convergence of our "continued fraction."

From this value of u we find by (14c)

$$\rho^2 = -\frac{4}{\gamma^2} u = -(5.3 + 5.8i) \cdot 10^{-8}; \qquad (16b)$$

and by (1b)

$$h^2 = k^2 + \frac{1}{a^2} (5.3 + 5.8i) \cdot 10^{-8}, \qquad h = k\{1 + (6.0 + 6.6i) \cdot 10^{-5}\}. \qquad (17)$$

We form

$$\frac{\omega}{h} = \frac{\omega}{k} \{1 - (6.0 + 6.6i) \cdot 10^{-5}\} \qquad (17a)$$

and conclude from this, since $\omega/k = c$, that the *phase propagation lags behind c by only* $6 \cdot 10^{-5}$ *c*. On the other hand we see from our field factor $\exp(ihx)$ that the amplitude is reduced by a factor $1/e$ only after traversal of a distance x given by

$$k \cdot 6.6 \cdot 10^{-5} x = 1, \qquad x = 720 \text{ m.} \qquad (17b)$$

This corresponds completely to what is expected: *Nearly undamped propagation with c as phase velocity.*

There are however also conditions for which this expectation proves to be erroneous. Consider, for example, a Wollaston wire of platinum with a radius $a = 2 \cdot 10^{-4}$ cm; the conductivity of platinum is 8 times less than that of copper. Let the wave-length in air be 1 meter. Then

$$\kappa = 9.2 \cdot 10^2 \text{ cm}^{-1}, \qquad \sqrt{\frac{\varepsilon_0 \mu}{\varepsilon' \mu_0}} = (1 - i) \cdot 0.34 \cdot 10^{-4}, \qquad \rho_L = (1 + i) \cdot 0.2.$$

The argument of J_0 in (14) is then no longer large, so that we must use Eq. (3c) in place of (7a). It yields

$$\frac{J_0(\rho_L)}{J_0'(\rho_L)} = -(1 - i) \cdot 5.0$$

and for the right side of (15), with the meaning of u unchanged, the value $v = -i \cdot 7.0 \cdot 10^{-9}$. Our transcendental equation thus becomes

$$u \log u = v, \qquad v = -i \cdot 7.0 \cdot 10^{-9}.$$

If, once again, we put $u_1 = -v/20$, we obtain by (15c),

$$u_2 = (-0.29 + 3.5i) \cdot 10^{-10} \cong u_3$$

and

$$\rho^2 = -\frac{4}{\gamma^2} u = (0.36 - 4.4i) \cdot 10^{-10},$$

a value of the same order of magnitude as that found before in (16b). However the further calculation becomes quite different because of the smallness of $a = 2 \cdot 10^{-4}$ cm. In place of (17) we find in cm^{-2}

$$h^2 = k^2 - 0.0009 + 0.011\ i.$$

It is now no longer adequate to retain a first term in a binomial expansion, since here $k^2 = (2\pi/\lambda)^2 = 0.0039$. Instead we obtain

$$h = 0.085 + 0.065i.$$

From this follows for the length of wire along which the wave amplitude has been reduced by a factor $1/e$:

$$\frac{1}{0.065} = 15 \text{ cm}$$

and for the *phase velocity* $\omega/0.085$. Division by the velocity of light ω/k yields

$$\frac{k}{0.085} = 0.74$$

as the ratio of the velocity of wave propagation and the velocity of light. *The former lags behind the latter by 26 per cent.*

The reason for this abnormal behavior evidently lies in the extreme thinness of the wire, which increases the alternating current impedance and prevents the development of a normal skin effect. The interior of the wire is then no longer free of current; the current distribution no longer has the character of curve 22 of Fig. 29, but that of curve 11. The field is then no longer "immunized" against Joule heat loss. Damping and propagation become anomalous.

§23. General Solution of the Wire-Wave Problem

In the preceding section we have derived that particular solution which is related to Hertz's original problem of wire waves. The question arises as to whether there is a more general solution. This question was proposed to D. Hondros as subject for his Munich thesis.[1] At the time it seemed of

---

[1] Ann. d. Phys. *30*, p. 905, 1909.

purely theoretical interest, remote from any practical application. It has been found since that it possesses close analogies with the theory of cavity conductors (§24), at present a favored field of communications engineering. Furthermore, it may be utilized advantageously for the theory of the Lecher two-wire line which first converted Hertz's single wire into a functioning system. Also the "wire waves in nonconductors" which fit into Hondros's formulation of the problem have found practical application.

A. Primary Wave and Electrical Secondary Waves

We call the solution in §22 the *principal wave*; for it $|\rho|$ was small and hence $|\rho_L|$ large. In the converse case that $|\rho|$ is large we speak of a *secondary wave*. Then, by (22.7), the left side of (22.14a) becomes equal to $-i\rho$, i.e. also large in absolute value. The denominator on the right side of (22.14a) must then become very small. ρ_L must hence approximate one of the infinitely many and frequently tabulated roots of

$$J'_0(\rho) = 0, \qquad \rho = w'_1, \qquad w'_2, \cdots, \tag{1}$$

generally w'_ν, where we shall note in particular

$$w'_1 = 3.83. \tag{1a}$$

By Eq. (22.3d) these w'_ν are identical with the roots of $J_1(\rho) = 0$. (We want to reserve the symbol w_ν for the roots of $J_0(\rho) = 0$.) In view of the meaning of ρ_L given by Eq. (22.14) we have then approximately

$$h_\nu^2 \cong k_L^2 - \left(\frac{w'_\nu}{a}\right)^2. \tag{2}$$

For a well-conducting wire h_ν is thus approximately equal to k_L and we can write, for all moderate values of ν (see (20.11)):

$$h_\nu \cong (1 + i)\kappa. \tag{2a}$$

From the formula for phase and damping $\exp(-i\omega t + ihx)$, which is to be interpreted as before, we find: *All secondary waves are exceedingly strongly damped in their progress along the wire*; their amplitude decreases by the factor $1/e$ in the short distance $1/\kappa$. *Their phase velocity ω/κ is small compared to the velocity of light $c = \omega/k$*, the ratio of the two being equal to k/κ.

The secondary waves also behave oppositely to the principal wave with respect to the character of the field. By (2a) we have *outside of the wire*

$$\rho = \sqrt{k^2 - h_\nu^2}\, r \cong (-1 + i)\kappa r$$

and by (22.7)

$$H_0^1(\rho) \sim e^{-(1+i)\kappa r} \qquad \text{for} \qquad a < r < \infty.$$

By (22.8) this signifies a *skin effect outside of the wire*. In its interior, on the other hand, the argument of $J_0(\rho)$ is real since

$$\rho = \sqrt{k_L^2 - h^2}\, r = w'_\nu \frac{r}{a} \ (w'_\nu \text{ real})$$

and hence $J_0(\rho)$ is of the order of magnitude 1 for $0 \leqq r \leqq a$.

The entire interior is filled by current. Considerable Joule heat is generated here, which explains the rapid damping of the wave in its progress along the wire and makes any observation of the secondary waves *illusory*.

B. Magnetic Waves

While we derived the electrical principal and secondary waves from the general Eq. (20.5), we obtain the magnetic waves from (20.6). We here set $E_x = 0$, H_x equal but for a constant to the Bessel function J_0 inside, and equal to the first Hankel function H_0 outside of the wire. If we take the corresponding transversal components from (20.6) we obtain, with $u = r$, $v = \varphi$, $g_u = 1$, $g_v = r$:

$$0 < r < a, \qquad \rho = \sqrt{k_L^2 - h^2}\, r \qquad\qquad a < r < \infty, \qquad \rho = \sqrt{k^2 - h^2}\, r$$

$$\sqrt{\frac{\mu}{\varepsilon'}}\, H_x = \frac{\sqrt{k_L^2 - h^2}}{ih}\, DJ_0(\rho) \qquad\qquad \sqrt{\frac{\mu_0}{\varepsilon_0}}\, H_x = \frac{\sqrt{k^2 - h^2}}{ih}\, BH_0(\rho)$$

$$\sqrt{\frac{\mu}{\varepsilon'}}\, H_r = DJ'_0(\rho) \qquad\qquad\qquad \sqrt{\frac{\mu_0}{\varepsilon_0}}\, H_r = BH'_0(\rho) \qquad\qquad (3)$$

$$-E_\varphi = \frac{k_L}{h}\, DJ'_0(\rho) \qquad\qquad\qquad -E_\varphi = \frac{k}{h}\, BH'_0(\rho).$$

From the requirement of continuity of H_x and E_φ we now obtain the boundary conditions

$$\rho_L DJ_0(\rho_L) = \sqrt{\frac{\varepsilon_0 \mu}{\varepsilon' \mu_0}}\, \rho BH_0(\rho), \qquad k_L DJ'_0(\rho_L) = kBH'_0(\rho). \qquad (4)$$

Here we have put $\rho = \sqrt{k^2 - h^2}\, a$, $\rho_L = \sqrt{k_L^2 - h^2}\, a$, as in (22.14). The reader may prove to his own satisfaction that the second of these conditions assures at the same time the continuity of B_r, taking account of the relation $k^2 = \varepsilon_0 \mu_0 \omega^2$, which we shall also use in the following.

Elimination of B and D in (4) leads to the transcendental equation

$$\rho \frac{H_0(\rho)}{H'_0(\rho)} = \sqrt{\frac{\varepsilon' \mu_0}{\varepsilon_0 \mu}}\, \frac{k}{k_L}\, \frac{\rho_L J_0(\rho_L)}{J'_0(\rho_L)}, \qquad (5)$$

whose right side differs materially, even in order of magnitude, from that of (22.14a). We ask whether (5) has a solution of the type of the *principal wave* $h \cong k$, i.e. $\rho \ll 1$. Then the left side would, by (22.5), be of the order of $\rho^2 \log \rho$, i.e. in absolute value $\ll 1$, while the right side, by (22.7a) and since $\rho_L \cong k_L a$, would be approximately equal to $\{\varepsilon' \mu_0/(\varepsilon_0 \mu)\}^{\frac{1}{2}} ka$, i.e. in absolute value $\gg 1$ since $|\,\varepsilon'\,| \gg \varepsilon_0$. This applies even for soft iron where μ is much larger than μ_0. The assumption $h \cong k$ thus leads to a contradiction. *There is no magnetic principal wave*; the magnetic waves all have the character of *secondary waves*:

$$|\,\rho\,| \gg 1, \qquad J_0'(\rho_L) \cong 0, \qquad h_\nu^2 \cong k_L^2 - \left(\frac{w_\nu'}{a}\right)^2,$$

as in (1). The earlier comments on the *electric* secondary waves may be transferred without change to the *magnetic secondary waves*. Their field also shows a skin effect outside of the wire and is rapidly damped in the interior by Joule heat.

C. Asymmetric Waves of the Electromagnetic Type

We now consider processes without rotational symmetry about the axis of the wire and employ for the representation of E_x the more general solutions of the Bessel differential equation

$$J_n(\rho)\,\cos\,(n\varphi) \qquad \text{or} \qquad H_n(\rho)\,\cos\,(n\varphi) \tag{6}$$

in place of the functions $J_0(\rho)$ and $H_0(\rho)$. We readily convince ourselves then that the former three-component solutions E_x, E_r, H_φ and H_x, H_r, E_φ are no longer sufficient, but that all six components of $\mathbf{E}$ and $\mathbf{H}$ must occur in the solution. We must now combine with the formula (6) for E_x the formula

$$\left.\begin{matrix} J_n(\rho) \\[4pt] H_n(\rho) \end{matrix}\right\}\,\sin\,(n\varphi) \tag{6a}$$

for H_x, so as to give all terms in (20.5) the common factor $\cos\,(n\varphi)$, all terms in (20.6) the common factor $\sin\,(n\varphi)$. We thus obtain from (20.5, 6) for the interior of the wire, with $\rho = \sqrt{k_L^2 - h^2}\,r$

$$E_x = \frac{\sqrt{k_L^2 - h^2}}{ih}\,CJ_n(\rho)\,\cos\,(n\varphi)$$

$$E_r = \left\{CJ_n'(\rho) + \frac{k_L\,n}{h\rho}\,DJ_n(\rho)\right\}\cos\,(n\varphi)$$

$$-E_\varphi = \left\{\frac{n}{\rho}\,CJ_n(\rho) + \frac{k_L}{h}\,DJ_n'(\rho)\right\}\sin\,(n\varphi)$$

$$\sqrt{\frac{\mu}{\varepsilon'}}\,H_x = \frac{\sqrt{k_L^2 - h^2}}{ih}\,DJ_n\,(\rho)\sin\,(n\varphi)$$

$$\sqrt{\frac{\mu}{\varepsilon'}}\,H_r = \left\{\frac{n}{\rho}\,\frac{k_L}{h}\,CJ_n(\rho) + DJ_n'(\rho)\right\}\sin\,(n\varphi)$$

$$\sqrt{\frac{\mu}{\varepsilon'}}\,H_\varphi = \left\{\frac{k_L}{h}\,CJ_n'(\rho) + \frac{n}{\rho}\,DJ_n(\rho)\right\}\cos\,(n\varphi).$$

$$\tag{7}$$

The phase factor $\exp(-i\omega t + ihx)$ is again to be thought as included. The constant coefficients in E_x and H_x have been so chosen that (7) passes over into (22.13) for $D = 0$ and $n = 0$ and (after the permissible interchange of cos and sin) into (3) for $C = 0$ and $n = 0$.

Proceeding likewise for the *exterior* of the wire we obtain, with $\rho = \sqrt{k^2 - h^2}\, r$

$$E_x = \frac{\sqrt{k^2 - h^2}}{ih} A H_n(\rho) \cos(n\varphi)$$

$$E_r = \left\{ A H'_n(\rho) + \frac{kn}{h\rho} B H_n(\rho) \right\} \cos(n\varphi)$$

$$-E_\varphi = \left\{ \frac{n}{\rho} A H_n(\rho) + \frac{k}{h} B H'_n(\rho) \right\} \sin(n\varphi)$$

$$\sqrt{\frac{\mu_0}{\varepsilon_0}}\, H_x = \frac{\sqrt{k^2 - h^2}}{ih} B H_n(\rho) \sin(n\varphi) \tag{8}$$

$$\sqrt{\frac{\mu_0}{\varepsilon_0}}\, H_r = \left\{ \frac{n}{\rho} \frac{k}{h} A H_n(\rho) + B H'_n(\rho) \right\} \sin(n\varphi)$$

$$\sqrt{\frac{\mu_0}{\varepsilon_0}}\, H_\varphi = \left\{ \frac{k}{h} A H'_n(\rho) + \frac{n}{\rho} B H_n(\rho) \right\} \cos(n\varphi).$$

We now turn to the boundary conditions between interior and exterior at $r = a$. With $\rho = \sqrt{k^2 - h^2}\, a$ and $\rho_L = \sqrt{k_L^2 - h^2}\, a$ we obtain from the continuity of E_x and H_x

$$\rho_L C J_n(\rho_L) = \rho A H_n(\rho), \tag{9}$$

$$\rho_L D J_n(\rho_L) = q\rho B H_n(\rho), \qquad q = \sqrt{\frac{\varepsilon_0 \mu}{\varepsilon' \mu_0}}. \tag{9a}$$

and from the continuity of E_φ and H_φ the two conditions

$$C \frac{n}{\rho_L} J_n(\rho_L) + D \frac{k_L}{h} J'_n(\rho_L) = A \frac{n}{\rho} H_n(\rho) + B \frac{k}{h} H'_n(\rho), \tag{10}$$

$$C \frac{k_L}{h} J'_n(\rho_L) + D \frac{n}{\rho_L} J_n(\rho_L) = A \frac{qk}{h} H'_n(\rho) + B \frac{qn}{\rho} H_n(\rho). \tag{10a}$$

The constants A, B, C, and D are to be eliminated from these four equations (9), (9a), (10), and (10a), most simply in the form of a four-row determinant. We divide their columns immediately by H_n and J_n and find:

$$\begin{vmatrix} \rho & 0 & \rho_L & 0 \\[2mm] 0 & q\rho & 0 & \rho_L \\[2mm] \dfrac{n}{\rho} & \dfrac{k}{h}\dfrac{H'_n}{H_n} & \dfrac{n}{\rho_L} & \dfrac{k_L}{h}\dfrac{J'_n}{J_n} \\[2mm] q\dfrac{k}{h}\dfrac{H'_n}{H_n} & \dfrac{qn}{\rho} & \dfrac{k_L}{h}\dfrac{J'_n}{J_n} & \dfrac{n}{\rho_L} \end{vmatrix} = 0 \tag{11}$$

This transcendental equation is to be regarded as the *equation determining the wave number h*, which occurs not only explicitly, but also implicitly in ρ, ρ_L, H'/H, and J'/J. The electric and magnetic components of the wave are *coupled* by it. An *uncoupling* occurs only in the symmetrical case $n = 0$, for which (11) may be separated:

$$\left\{ \rho k_L \frac{J'_0}{J_0} - \rho_L qk \frac{H'_0}{H_0} \right\} \left\{ \rho_L k \frac{H'_0}{H_0} - \rho qk_L \frac{J'_0}{J_0} \right\} = 0. \qquad (11a)$$

When set individually equal to zero the two parentheses correspond exactly with the transcendental equations for the symmetric magnetic and the symmetric electrical case, i.e. with the earlier Eqs. (5) and (22.14a).

We utilize (11) below only to answer the question whether in the unsymmetrical case a state of the character of the *principal wave* is possible. We thus assume

$$h \cong k, \qquad \rho \cong 0, \qquad |\rho_L| \gg 1, \qquad \frac{J'_n}{J_n} \to -i, \qquad \frac{H'_n}{H_n} \to -\frac{n}{\rho} \, ;$$

the last two statements follow from Eqs. (22.6) and (22.7a). We then can neglect, in the first two rows of (11), not only $q\rho$, but also ρ in comparison with ρ_L. Then the determinant (11) breaks up into the product of the two subdeterminants

$$\begin{vmatrix} \rho_L & 0 \\ 0 & \rho_L \end{vmatrix} = \rho_L^2 \quad \text{and} \quad q\left(\frac{n}{\rho}\right)^2 \begin{vmatrix} 1 & -\dfrac{k}{h} \\ -\dfrac{k}{h} & 1 \end{vmatrix} = -\frac{qn^2}{a^2 h^2}. \qquad (12)$$

The product of the two yields

$$(h^2 - k_L^2) \, \frac{qn^2}{h^2}$$

Set equal to zero this leads, for $n = 0$, to $h = \pm k_L$, which contradicts our requirement $h \cong k$. *There is hence no asymmetric principal wave.* We must refer to Hondros's thesis for the rather complicated solution of the transcendental equation for the secondary waves.

D. Wire Waves on a Nonconductor

The dissipation of the secondary waves on the metallic wire by Joule heat raises the question as to whether secondary waves on a *dielectric* wire might be observable. According to Hondros and Debye this question is to be answered in the affirmative.[1]

---

[1] D. Hondros and P. Debye, Ann. d. Phys. *32*, p. 465, 1910.

We shall consider a "water wire" (which may be imagined surrounded by an infinitely thin-walled glass cylinder). In view of the absence of absorption h is real, so that $\sqrt{k^2 - h^2}$ is either real or purely imaginary. The first possibility ($h < k$, propagation with a velocity exceeding that of light) is excluded by the prohibition of radiation, in accord with §22B. Hence $\sqrt{k^2 - h^2}$ and our former $\rho = \sqrt{k^2 - h^2}\, a$ become purely imaginary. On the other hand, $\rho_L = \sqrt{k_L^2 - h^2}\, a$ is real. For we have now, since $\sigma = 0$

$$k_L^2 = \varepsilon\mu\omega^2 = \frac{\varepsilon\mu}{\varepsilon_0\mu_0} \cdot \varepsilon_0\mu_0\omega^2 = n^2 k^2,$$

where n is now to denote the *refractive index*, in accord with Maxwell's law in Eq. (6.7) (not, as up to now, the order of the Bessel functions!). For water we have in the high-frequency range (decimeter waves) $n \cong 9$.

We introduce the two *real* quantities

$$\xi = \sqrt{h^2 - k^2}\, a, \qquad \eta = \sqrt{n^2 k^2 - h^2}\, a, \tag{13}$$

which will serve as rectangular coordinates for a graphical representation.

All our earlier formulas, in particular those for symmetrical waves, remain valid for our present case of real ξ, η insofar as they do not contain approximations. Eq. (22.14a) now takes the form

$$i\xi \frac{H_0(i\xi)}{H_0'(i\xi)} = \frac{\eta}{n^2} \frac{J_0(\eta)}{J_0'(\eta)}. \tag{14}$$

For $\xi \to 0$ and $\xi \to \infty$ its left side varies, according to (22.5) and (22.7) as

$$\frac{i\xi \log \dfrac{\gamma\xi}{2}}{1/i\xi} = \xi^2 \log \frac{2}{\gamma\xi} \quad \text{and as} \quad \frac{i\xi}{i} = \xi, \quad \text{respectively.}$$

It thus becomes equal to zero and infinity along with ξ. Hence the right side of (14) must also vanish for $\xi = 0$; this is not the case when $\eta = 0$ (since $J_0'(0) = 0$), but only when

$$J_0(\eta) = 0. \tag{15}$$

On the other hand, the right side of (14) becomes infinite for

$$J_0'(\eta) = -J_1(\eta) = 0. \tag{15a}$$

We shall represent the variation given by (14) graphically in the $\xi\eta$-plane. On the ordinate axis we mark the roots of (15) and (15a), which alternate with each other. As on p. 186 we call the sequence of points

$$w_1, w_2, w_3, \cdots \qquad \text{and} \qquad w_1', w_2', w_3', \cdots$$

and note in particular

$$w_1 = 2.40. \tag{15b}$$

We draw lines parallel to the axis of abscissas through the points $\eta = w'_i$, $\xi = 0$. The points $\eta = w_1$, $\xi = 0$ are initial points of curve branches of the desired representation, which must approach the horizontal straight lines $\eta = w'_i$ asymptotically for $\xi = \infty$.

There exists however, according to (13), also the relation

$$\xi^2 + \eta^2 = (n^2 - 1)k^2 a^2 = (n^2 - 1)\left(\frac{2\pi a}{\lambda}\right)^2 \tag{16}$$

between ξ and η. This means that the desired solutions of (14) must also lie on circles about the origin of the $\xi\eta$-plane with radius

$$r_\lambda = 2\pi\sqrt{n^2 - 1}\,\frac{a}{\lambda}, \tag{16a}$$

where λ is the "wave-length in air" corresponding to our state of vibration. Hence we must let the curve branches intersect with the circles (16).

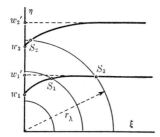

FIG. 34. Wire waves on a nonconductor. Plane of the real coordinates $\xi = \sqrt{h^2 - k^2}\,a$, $\eta = \sqrt{n^2 k^2 - h^2}$ a. Construction of the roots of Eq. (14) with the aid of the roots w_1, w_2, $\cdots$ of Eq. (15), $J_0(\eta) = 0$, and the roots w'_1, w'_2, $\cdots$ of Eq. (15a), $J_1(\eta) = 0$.

Depending on the magnitude of λ there are zero, one, two, or even more intersections. Our figure shows directly:

For $r_\lambda < w_1$ there is no intersection.

For $w_1 < r_\lambda < w_2$ there is one intersection.

For $w_2 < r_\lambda < w_3$ there are two intersections etc.

According to (15b) the first root of $J_0(\eta) = 0$ is $w_1 = 2.40$. The *maximum* permissible value of λ which corresponds to this least value of r_λ for which a wire wave is just still possible is, according to (16a):

$$\lambda_{\max} = \frac{2\pi}{2.40}\sqrt{n^2 - 1}\,a$$

For our "water wire" with radius $a = 1$ cm this is

$$\lambda_{\max} = \frac{2\pi\sqrt{80}}{2.40} = 23.4 \text{ cm.}$$

Longer wave-lengths than this cannot be propagated along it. Thus we find ourselves in the range of the "decimeter waves" which is of such great interest at present.

As λ is reduced there is *one* possibility of propagation, represented by the first intersection S_1 in the figure; as λ is reduced still further, corresponding to $r_\lambda > 5.52$ being increased, there are two possibilities, given by the two intersections S_2, S_3 in the figure. The first, at small ξ, yields $h \cong k$ (velocity along the wire nearly equal to c), the second at larger ξ (h materially larger than k, velocity of propagation appreciably less than c) corresponds to an η which is nearly equal to the first root of $J_1(\eta) = 0$, which according to Eq. (23.1a) is $w_1' = 3.83$. In the first of these two cases the electric lines of force are nearly *perpendicular* to the surface of the wire and the decrease of the field outward is slow (the asymptotic decrease $e^{-\xi r}$ is attained only for large r). In the second case, where ξ is quite large, we have a *skin effect* outwards, as for our *auxiliary waves* for the metallic wire. The first case has the character of the *principal wave* on the outside, with the difference that the interior of the dielectric is filled by current (η real and of moderate magnitude). Similarly, for still smaller λ, the large number of vibration states then possible are arranged between the limiting cases of principal and auxiliary waves. At the same time the phase velocity of the wave varies between the velocity c in vacuum and that "in water."

The above results predicted by Hondros and Debye were verified most successfully by G. Southworth in the Bell Laboratories. Also in Germany such dielectric wire waves have been profitably applied in communications.

§24. On the Theory of Wave Guides

In the preceding section we have seen that electromagnetic fields may be held together and guided by the surface of a non-conducting rod and that they protect themselves against outward radiation by a skin effect. This protection will be complete if we embed the non-conductor in a metallic tube, whereupon the condition of a sufficiently high dielectric constant may be omitted and the dielectric within the tube may also be air. We thus arrive at the configuration of the *wave guides*, which have become important in high frequency practice.

We consider in particular the cylindrical wave guide, since its treatment may be deduced directly from the preceding formulas. Let a be the radius of the metallic envelope, which for the present will be assumed to be a *perfect* conductor, and h the wave number of the propagation. It is real since the wave is damped neither by Joule heat nor by radiation. There are electric and magnetic waves of symmetric type and also of asymmetric type.

We write for the *symmetric electric waves*, as in (22.13), omitting the amplitude coefficient C and the phase factor $\exp(-i\omega t + ihx)$:

$$E_x = \frac{\sqrt{k^2 - h^2}}{ih} J_0(\rho), \qquad E_r = J_0'(\rho), \qquad \sqrt{\frac{\mu_0}{\varepsilon_0}} H_\varphi = \frac{k}{h} J_0'(\rho). \quad (1)$$

For known E_x these formulas also follow directly from the relationship of transversal and longitudinal components in (20.5). The boundary conditions reduce to the single equation $E_x = 0$ for $r = a$, since for H_φ the required condition of continuity is satisfied by a surface current induced in the envelope. Thus, with $\rho = \sqrt{k^2 - h^2}\, a$ we have

$$J_0(\rho) = 0, \qquad \rho = w_1, w_2, \cdots w, \cdots \text{ with } w_1 = 2.40. \tag{2}$$

From the definition of ρ it follows that

$$h_\nu^2 = k^2 - \left(\frac{w_\nu}{a}\right)^2, \qquad h_\nu < k, \qquad \frac{\omega}{h_\nu} > \frac{\omega}{k} = c. \tag{2a}$$

The phase velocity ω/h_ν along the tube thus *exceeds the velocity of light.* As for our dielectric wire there is a *lower* limit for the wave number k, i.e. an *upper* limit for the corresponding "primary[1] wave-length" $\lambda = 2\pi/k$. It corresponds to $h = h_1 = 0$ (phase velocity infinite) and yields by (2a)

$$k_{\min} = \frac{w_1}{a}, \qquad \lambda_{\max} = \frac{2\pi}{k_{\min}} = \frac{2\pi}{w_1}\, a = \frac{2\pi}{2.40}\, a. \tag{3}$$

Since a is of the order of magnitude of centimeters all the following considerations apply to the *centimeter-wave* region. The number of possible states (or "modes") of type (1) depends on the frequency ω or, what is the same, on the primary wave number $k = \omega/c$. According to (2a) this number is equal to the number of roots w_ν which are less than ka.

The *magnetic symmetrical waves* are represented, according to (23.3),

$$\sqrt{\frac{\mu_0}{\varepsilon_0}}\, H_x = \frac{\sqrt{k^2 - h^2}}{ih}\, J_0(\rho), \qquad \sqrt{\frac{\mu_0}{\varepsilon_0}}\, H_r = J_0'(\rho), \qquad -E_\varphi = \frac{k}{h}\, J_0'(\rho), \tag{4}$$

corresponding to the general scheme of Eq. (20.6). The single boundary condition which must here be fulfilled is $E_\varphi = 0$ for $\rho = \sqrt{k^2 - h^2}\, a$. It demands

$$J_0'(\rho) = 0, \qquad \rho = w_1', w_2', \cdots w_\nu', \cdots \text{ with } w_1' = 3.83 \tag{5}$$

and yields, as in (2a), values of h_ν which are $< k$. The upper limit for the primary wave-length lies somewhat lower than for the electric type. It is

$$\lambda_{\max} = \frac{2\pi}{w_1'}\, a = \frac{2\pi}{3.83}\, a. \tag{5a}$$

To pass over to the asymmetric types we start from Eqs. (23.7). In view of the reduced number of boundary conditions we may now however set

---

[1] By the "primary" wave-length we understand that of the *exciting* oscillation, which of course has the same frequency ω as the wave guide oscillation excited by it. This primary wave-length is actually simply a measure of the frequency ω which is familiar to the engineer and convenient in dimension. The wave-length in the wave guide can be determined uniquely only in the axial direction and is $\lambda = 2\pi/h$, whereas the primary wave-length is $2\pi/k = 2\pi c/\omega$. For $\lambda_{\text{prim}} = \lambda_{\max}$, $\lambda_{\text{ax}} = \infty$ since $h = 0$.

one of the two amplitudes C and D equal to zero, the other equal to 1. This simplifies the formulas considerably and leads to an *asymmetric electric* ($D = 0$) and an *asymmetric magnetic* case ($C = 0$).

For the asymmetric electric type we obtain:

$$E_z = \frac{\sqrt{k^2 - h^2}}{ih} J_n(\rho) \cos(n\varphi), \qquad H_z = 0$$

$$E_r = J'_n(\rho) \cos(n\varphi), \qquad\qquad \sqrt{\frac{\mu_0}{\varepsilon_0}}\, H_r = \frac{k}{h}\frac{n}{\rho} J_n(\rho) \sin(n\varphi) \qquad (6)$$

$$E_\varphi = -\frac{n}{\rho} J_n(\rho) \sin(n\varphi), \qquad \sqrt{\frac{\mu_0}{\varepsilon_0}}\, H_\varphi = \frac{k}{h} J'_n(\rho) \cos(n\varphi)$$

and for the asymmetric magnetic case, if, for convenience, $n\varphi$ is exchanged for $n\varphi + \pi/2$,

$$E_z = 0, \qquad\qquad \sqrt{\frac{\mu_0}{\varepsilon_0}}\, H_z = \frac{\sqrt{k^2 - h^2}}{ih} J_n(\rho) \cos(n\varphi)$$

$$E_r = -\frac{k}{h}\frac{n}{\rho} J_n(\rho) \sin(n\varphi), \qquad \sqrt{\frac{\mu_0}{\varepsilon_0}}\, H_r = J'_n(\rho) \cos(n\varphi) \qquad (7)$$

$$E_\varphi = -\frac{k}{h} J'_n(\rho) \cos(n\varphi), \qquad \sqrt{\frac{\mu_0}{\varepsilon_0}}\, H_\varphi = -\frac{n}{\rho} J_n(\rho) \sin(n\varphi)$$

For $n = 0$ (6) and (7) pass over into (1) and (4).

The boundary conditions for $r = a$, $\rho = \sqrt{k^2 - h^2}\, a$ require

for (6): $E_z = E_\varphi = 0$,   i.e. $J_n(\rho) = 0$,

for (7) $E_\varphi = 0$,   i.e. $J'_n(\rho) = 0$.

As in (2) and (5) we call the roots of these two equations again w_ν and w'_ν and distinguish them when necessary from the former by the addition of the argument n, writing thus $w_\nu(n)$ in place of $w_\nu(0)$, $w'_\nu(n)$ in place of $w'_\nu(0)$.

The following table indicates the relative position of the smallest roots in the doubly-indexed twofold system w, w':

$J'_1 = 0$	$J_0 = 0$	$J'_0 = J_1 = 0$
$w'_1(1) = 1.84$	$w_1(0) = 2.40$	$w'_1(0) = 3.83 = w_1(1),$
$w'_2(1) = 5.33$	$w_2(0) = 5.52$	$w'_2(0) = 7.02 = w_2(1).$

It shows, contrary to expectation, that the *magnetic asymmetric wave* with $n = 1$, and not the *electric symmetric wave* with $n = 0$, possesses the smallest root. Also for the second root $\nu = 2$, in the second row of the table, the sequence of these two waves is the same as for the root $\nu = 1$. The third column shows finally that the electric symmetric wave is succeeded by the *magnetic symmetric wave* with $n = 0$, which yields the same root as the *electric asymmetric wave* for $n = 1$ since $J'_0 = -J_1$.

In view of this the wave-length $\lambda_{\max}$ given by (3) is not the absolute upper limit for all waves that can be propagated in the guide, but rather the wave-length

$$\lambda_{\max} = \frac{2\pi}{1.84} a. \tag{8}$$

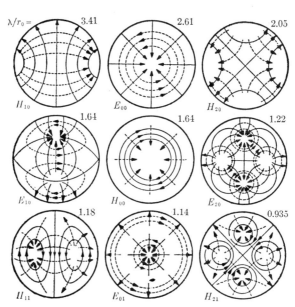

Fig. 35. Transversal fields of cylindrical-guide waves, ordered according to the limiting wave-length $\lambda_{\max}/a$. Full lines, *electric*; broken lines, *magnetic* lines of force, both represented in the cross section of their antinodes. Open beginnings or ends of the lines of force indicate their being bent out of or into the direction of the axis. Customary engineering notation: TM magnetic, TE electric type; first subscript, azimuthal number of nodal lines; second subscript, radial number of nodal lines within the guide.

Thus, if the frequency is increased continuously (continuous reduction of the primary wave-length) and if the exciting elements are suitably disposed in space, the electric symmetric wave is not the first to appear, but instead the magnetic asymmetric wave with $n = 1$. The electric symmetric wave follows and, after it, the magnetic symmetric wave simultaneously with the first asymmetric electric wave $n = 1$, as illustrated in the series of pictures in Fig. 35.

So far we have discussed only exactly circularly cylindrical tubes. Every deviation from circular symmetry occurring along the tube is equivalent to a disturbance in the symmetry of the excitation and hence occasions the appearance of new secondary waves of different symmetry, which leads to difficulties in the practical application of the desired modes of oscillation.

Also a tube of elliptical cross section may be treated directly by the general method of §20A, since the wave equation is separable in the elliptical coordinates u, v (see Vol. II, Problem IV 3). It is only necessary to put E_x, H_x equal to a *Mathieu function* (function of the elliptical cylinder) $F(iu)$ or equal to the product of two such functions $F(iu) \cdot F(v)$; the transversal components E_u, H_v; H_u, E_v may then be written down immediately with the aid of Eqs. (20.5) and (20.6).

We shall still glance briefly at tubes with *rectangular cross section*. Since there can be no question of a symmetrical wave in view of the shape of the rectangle (sides b and c in the y- and z-directions) we give right away the general formulas (n and m arbitrary integers) corresponding to Eqs. (6) and (7); the amplitudes of E_x in (9) and H_x in (10) have, just as in (6) and (7), been chosen in a manner convenient for what follows. As before, the phase factor $\exp i(hx - \omega t)$ must be imagined as included.

$$E_x = \pi^2 \left(\frac{n^2}{b^2} + \frac{m^2}{c^2}\right) \sin\left(n\pi \frac{y}{b}\right) \sin\left(m\pi \frac{z}{c}\right), \qquad H_x = 0,$$

$$E_y = ih \frac{n\pi}{b} \cos\left(n\pi \frac{y}{b}\right) \sin\left(m\pi \frac{z}{c}\right), \qquad \sqrt{\frac{\mu_0}{\varepsilon_0}} H_y = -\frac{k}{h} E_z, \quad (9)$$

$$E_z = ih \frac{m\pi}{c} \sin\left(n\pi \frac{y}{b}\right) \cos\left(m\pi \frac{z}{c}\right), \qquad \sqrt{\frac{\mu_0}{\varepsilon_0}} H_z = +\frac{k}{h} E_y.$$

$$\sqrt{\frac{\mu_0}{\varepsilon_0}} H_x = \pi^2 \left\{\frac{n^2}{b^2} + \frac{m^2}{c^2}\right\} \cos\left(n\pi \frac{y}{b}\right) \cos\left(m\pi \frac{z}{c}\right), \qquad E_x = 0,$$

$$\sqrt{\frac{\mu_0}{\varepsilon_0}} H_y = -ih \frac{n\pi}{b} \sin\left(n\pi \frac{y}{b}\right) \cos\left(m\pi \frac{z}{c}\right), \qquad E_y = \frac{k}{h} \sqrt{\frac{\mu_0}{\varepsilon_0}} H_z, \quad (10)$$

$$\sqrt{\frac{\mu_0}{\varepsilon_0}} H_z = -ih \frac{m\pi}{c} \cos\left(n\pi \frac{y}{b}\right) \sin\left(m\pi \frac{z}{c}\right), \qquad E_z = -\frac{k}{h} \sqrt{\frac{\mu_0}{\varepsilon_0}} H_y.$$

The wave number h is determined in both cases by the differential equation $\Delta X + k^2 X = 0$, which must be satisfied for every one of the Cartesian components of **E** and **H**. Substitution of either (9) or (10) readily leads to

$$h^2 + \pi^2 \left(\frac{n^2}{b^2} + \frac{m^2}{c^2}\right) = k^2 = \left(\frac{2\pi}{\lambda}\right)^2, \qquad \lambda = \text{primary wave-length.}$$

The maximum value of λ for given n and m, below which the tube is capable of oscillations, occurs for $h = 0$ and is

$$\lambda_{\max} = \frac{2}{\sqrt{\frac{n^2}{b^2} + \frac{m^2}{c^2}}}.$$

If, as we may assume, $b > c$ the absolute maximum is attained for $n = 1$, $m = 0$ and is

$$\lambda_{max} = 2b.$$

As Leon Brillouin has noted, these and similar oscillations in guides can be constructed elegantly and instructively by the superposition of ordinary plane space waves which interfere at the tube walls.

The same idea leads also directly from the progressive waves derived above to characteristic standing waves in e.g. a rectangular parallelepiped or a circular cylinder of finite length. Essentially, the wave number h must simply be replaced by an integer multiple of π/a where a is the length of the third side of the parallelepiped or the length of the cylinder, respectively. We will discuss this in greater detail in Problems II 7 and II 8 and treat, in Problem II 9, the radially symmetric characteristic vibrations of the sphere as well. The characteristic vibrations of the rectangular parallelepiped, in particular, find a useful application in microwave practice for the determination of the frequency of the primary excitation by the resonance principle.

More difficult questions arise in the practical application of wave guides, where, instead of perfectly conducting walls, the finite conductivity of real metals and the heat loss in them must be considered; the latter has, up to the present, prevented the propagation of waves in guides over great distances. Also the shaping of the ends of the guides into conical or horn-shaped openings raises questions upon which we shall not enter here.[1]

§25. The Lecher Two-Wire Line

Mathematically this is the generalization for *high-frequency alternating* currents of the *quasistationary* two-wire line treated in §18. Its advantage, compared with the *single wire* traversed by alternating current, rests in the fact that the field outside of the wires decreases more rapidly than for the single wire, so that the disturbances by the surroundings discussed on p. 178 are avoided.

The *phase* of the alternating current in the two wires advances in the *same direction*, say in the positive x-direction; the direction of the current itself, on the other hand, as in §15E, is *opposite* in the two wires. We might say: For the same x positive charge flows in one wire through a given cross section, negative charge in the other. Also the charge accumulated on the surface has at any moment, for equal x, the opposite sign in the two wires.

---

[1] We refer to the comprehensive textbook of S. A. Schelkunoff, "Electromagnetic Waves," Van Nostrand, New York, 1943, which was published as a Bell Monograph and is widely employed in the United States, as well as to the lectures of L. de Broglie, "Problèmes de propagation guidée des ondes électromagnétiques," Paris, Gauthier-Villars, 1941.

We call this a *push-pull excitation*. However, the mode in which charge of equal sign flows in both wires (and is accumulated at their surface) may also be realized. We then speak of *parallel excitation*. The conditions of excitation determine which of the two states occurs. Any asymmetry of excitation results in the appearance of both wave types. However, we only call the push-pull arrangement a "Lecher system." For parallel excitation the situation is quite similar to that for the single wire (§22) and is fraught with the same experimental drawbacks.

G. Mie[1] succeeded in giving a complete theoretical treatment of the Lecher problem as early as 1900. The following representation,[2] which is both simplified and rounded out to some extent, deviates from that of Mie more in form than in substance. Like Mie, we introduce a system of *bipolar coordinates*, to which the circumferences of the two cross sections, assumed circular, belong. These coordinates would be the ideal mathematical medium if the wave equation were separable in them. Unfortunately this is not the case (see Vol. II, Problem IV.1). We hence must employ methods of approximation which rest on the replacement of the wave equation by the potential equation in the yz-plane. However, this approximation is valid only for sufficiently good conductivity of the material of the wire and in the exterior of the wires. Inside we must calculate with ordinary cylindrical polar coordinates. The comparison of the two formulas at the surface of the wires leads to a clear-cut equation for the determination of h, the wave number, which in the push-pull case becomes even simpler than for the single wire, being algebraic in place of transcendental. In the parallel case it is practically identical with that for the single wire.

G. Gentile Jr. has proposed a procedure which differs from ours and from Mie's.[3] In accord with the general methods of perturbation theory he superposes on the symmetric wave propagated along the first wire the totality of asymmetric Hondros waves from §23, each multiplied by a disposable coefficient. He seeks to fit these coefficients to the boundary conditions on the first and second wires, in which process he has to utilize the generalized addition theorems of the Bessel and Hankel functions. This leads him to an infinite system of simultaneous linear equations for the coefficients. However he and his collaborator T. Magri failed to obtain an approximate solution of it. On the other hand, our procedure leads to a direct and explicit determination of the infinite number of coefficients which must be introduced.

[1] Ann. d. Phys. *2*, 201, 1900.
[2] It rests on a detailed study of the problem by Mr. J. Jaumann; he also has made available to me the elegant treatment of the limiting case $\sigma \rightarrow \infty$ given in the succeeding section A, which we owe his late father, the wellknown physicist G. Jaumann of Brünn.
[3] Nuovo Cimento, Vol. I, pp. 161 and 190, 1943.

A. The Limiting Case of Infinite Conductivity

For $\sigma \to \infty$ the waves propagate themselves with the velocity of light c, so that $h = k$, as has already been pointed out at the end of §20A and follows directly for the single wire from Eq. (22.15). Then the three-dimensional wave equation for each Cartesian field component becomes the two-dimensional potential equation, in accord with Eq. (20.3). This may be solved by the method of conformal mapping for arbitrary cross-section peripheries (which need not be circular, nor even the same for the two wires). The method can also be applied when the excitation is not purely periodic and monochromatic, i.e., when our phase factor $\exp\{i(hx - \omega t)\}$ is replaced by an arbitrary function $f(x - ct)$.

It is true that the longitudinal components E_x, H_x, which were considered first in Eq. (20.3), vanish in the limit $h \to k$, since for infinite conductivity the electric lines of force are perpendicular to the surface of the two conductors, and the magnetic lines of force also lie in the planes $x =$ const. Hence we have for the longitudinal components, in a first approximation

$$E_x = 0, \qquad H_x = 0. \tag{2}$$

On the other hand, the Cartesian transversal components E_y, E_z, H_y, H_z may be determined almost directly from the fact that as solutions of the two-dimensional potential equation they form an electrostatic and a corresponding magnetostatic field. They are most simply combined in the vector formula[1]

$$\mathbf{E} + i\sqrt{\frac{\mu_0}{\varepsilon_0}}\, \mathbf{H} = \operatorname{grad} w, \qquad w = u + iv. \tag{2a}$$

$w = w(\zeta)$ is a function of the complex variable $\zeta = y + iz$, which may be constructed by conformal mapping; the transversal $\mathbf{E}$- and $\mathbf{H}$-components are obtained as gradients of the real and imaginary parts, u and v, of this complex function.

The conformal mapping for our two identical circular cross sections is known to us from §19 of Vol. II. Fig. 26 given there is reproduced in the following figure with the notation to be employed here. Both systems of lines of force are circles. The electrical lines of force $v =$ const. proceed from the fixed points Q_1, Q_2 of the family of circles; the magnetic lines $u =$ const. have their centers on the real axis of the ζ-plane, on which Q_1 and Q_2 also lie. Let the center of the system M be the origin $\zeta = 0$. Our function w is given by Eq. (19.10) of Vol. II which in the present notation (u, v, ζ, ζ_0 in place of ρ, φ, z, c) and with a convenient choice of the constant A takes the form

$$w = \log \frac{\zeta - \zeta_0}{\zeta + \zeta_0}. \tag{3}$$

[1] As before, the factor $(\mu_0/\varepsilon_0)^{1/2}$ must be applied to H for dimensional reasons.

$\pm \zeta_0$ are the (real) values of ζ corresponding to Q_2 and Q_1, respectively. u and v, as real and imaginary parts of w, have the same meaning as the parameters ρ and φ of the *bipolar coordinate system* defined in Eq.(19.10b) of Vol. II. Of the magnetic lines of force, those have been drawn heavy in Fig. 36 which are supposed to correspond to the cross sections of the wires (radius a). Their centers O_1, O_2 do not coincide with the points Q_1, Q_2. We call the latter, as sources of the electric lines of force, *source points* (three-dimensionally they are *source lines* parallel to the axes of the wires).

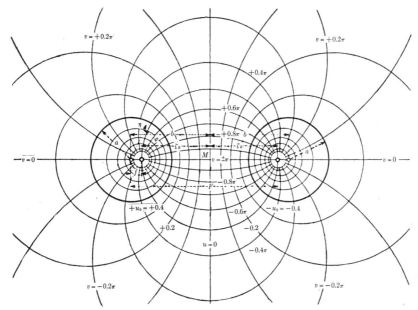

FIG. 36. The families of circles of the bipolar coordinates $u = \text{const}$, $v = \text{const}$ with the fixed points (source points) Q_1, Q_2. The peripheries of the two wire cross sections $u = \pm u_0$ are indicated by heavier lines; O_1, O_2 are their centers, a their radius. The center M of the figure is the origin of the complex variable $\zeta = x + iy$.

In the sense of Eq. (9.8) Q_1 and Q_2 are their *mutual electrical images* with reference to the *two* circular cross sections, i.e. are transformed into each other by the "transformation of reciprocal radii." With the designations

$$O_1Q_1 = O_2Q_2 = f, \qquad O_1Q_2 = O_2Q_1 = F, \qquad O_1M = O_2M = b$$

we have hence

$$fF = a^2, \qquad f + F = 2b, \qquad F - f = 2\zeta_0. \tag{3a}$$

Solution of a simple quadratic equation then leads to

$$F = b + \sqrt{b^2 - a^2}, \qquad f = b - \sqrt{b^2 - a^2}, \qquad \zeta_0 = \sqrt{b^2 - a^2}. \tag{3b}$$

B. The Exterior of the Wires

According to Problem IV.1 of Vol. II the line element of the bipolar co-ordinates may be written

$$ds^2 = g^2(du^2 + dv^2), \qquad \frac{1}{g} = \frac{\cosh u - \cos v}{(F - f)/2}. \qquad (4)$$

As compared with the general orthogonal line element in Eq. (20.1) we thus have here the special (isometric) case $g_u = g_v = g$.

The electric lines of force, represented in Fig. 36 by the family of circles $v = $ const, have the direction of increasing u, the magnetic lines of force, represented by the orthogonal family of circles $u = $ const, have the direction of increasing v. For the particular line elements ds_u, ds_v of the two systems of lines of force we obtain from (4)

$$\frac{ds_u}{du} = \frac{ds_v}{dv} = g. \qquad (4a)$$

We write Eq. (2a) separately for the u- and v-directions:

$$E_u + i\sqrt{\frac{\mu_0}{\varepsilon_0}} H_u = \frac{\partial w}{\partial u}\frac{du}{ds_u} = \frac{1}{g},$$

$$E_v + i\sqrt{\frac{\mu_0}{\varepsilon_0}} H_v = \frac{\partial w}{\partial v}\frac{dv}{ds_v} = \frac{i}{g}.$$

Separation of real and imaginary parts leads to

$$E_u = \sqrt{\frac{\mu_0}{\varepsilon_0}} H_v = \frac{1}{g} = \frac{\cosh u - \cos v}{(F - f)/2}, \qquad E_v = \sqrt{\frac{\mu_0}{\varepsilon_0}} H_u = 0. \qquad (5)$$

The last is obvious since the electric lines of force have the u-direction, the magnetic, the v-direction. Furthermore, in spite of the vanishing of the longitudinal components E_x and H_x noted in Eq. (2), we wish to obtain a somewhat closer approximation for them as solutions of the two-dimensional potential equation. This is, in terms of u and v,

$$\Delta_{uv} = \frac{\partial^2}{\partial u^2} + \frac{\partial^2}{\partial v^2} = 0.$$

It is integrated by particular solutions of the form

$$\left.\begin{array}{l}\sinh (nu)\\ \cosh (nu)\end{array}\right\} \cos (nv), \qquad \left.\begin{array}{l}\cosh (nu)\\ \sinh (nu)\end{array}\right\} \sin (nv). \qquad (5a)$$

Here n is a positive integer. For $n = 0$ these functions are replaced by the linear function

$$au. \qquad (5b)$$

The addition of a term bv is not permissible since E_x and H_x must be unique functions of space, whereas the coordinate v changes by $\pm 2\pi$ after revolving once about one of the wires, i.e. is multivalent. Furthermore, the addition of a constant c to (5b) is excluded since E_x and H_x must vanish at infinity ($u = v = 0$). In view of the symmetry of our problem we write E_u as an odd function of u and as an even function of v, H_v, vice versa, as an even function of u and an odd function of v:

$$E_x = E_0 u + E_1 \sinh u \cos v + E_2 \sinh (2u) \cos (2v) + \cdots$$

$$\sqrt{\frac{\mu_0}{\varepsilon_0}}\, H_x = \qquad H_1 \cosh u \sin v + H_2 \cosh (2u) \sin (2v) + \cdots \tag{6}$$

In justification we consider two symmetrically placed points u, v and $-u$, v to the right and to the left in the figure. In the push-pull case, which alone interests us to begin with, the currents flow in opposite directions in the two wires; the same applies to the x-components of the displacement currents outside of the wires. Hence E_x is, in our two points, equal and opposite. On the other hand H_x has the same sign in the two points in view of their position to the left and to the right of the two wires. Consider now two points u, v and u, $-v$ in the figure, above and below the straight line $v = \{{}^0_\pi\}$. In them E_x has the same sign and H_x opposite signs. The formulation (6) is hence justified.

We already know of the coefficients E_0, E_1, $\cdots$, H_1, $\cdots$, from (2), that they vanish in the first order for $h \to k$. To obtain more detailed information we best turn back to the general relations (20.5) and (20.6) which before served for the calculation of the transversal components from the longitudinal ones, and which we shall now employ to determine the longitudinal components to the first order from the transversal components known to the zero order of approximation. Since $g_u = g_v = g$, substitution of (5) and (6) in (20.5) and (20.6) leads to

$$\begin{aligned} i(k^2 - h^2) &= -hE_0 - (hE_1 + kH_1) \cosh u \cos v \cdots \\ i(k^2 - h^2) &= -kE_0 - (kE_1 + hH_1) \cosh u \cos v \cdots \end{aligned} \qquad \text{from (20.5)}$$

$$\begin{aligned} 0 &= \qquad (hE_1 + kH_1) \sinh u \sin v \cdots \\ 0 &= \qquad -(kE_1 + hH_1) \sinh u \sin v \cdots \end{aligned} \qquad \text{from (20.6)}$$

We conclude therefore:

$$E_0 \cong \frac{i\,(h^2 - k^2)}{h} \cong \frac{i(h^2 - k^2)}{k} \cong 2i(h - k), \tag{7}$$

$$H_1 \cong -\frac{h}{k} E_1 \cong -\frac{k}{h} E_1 \cong -E_1. \tag{8}$$

The sign $\cong$ signifies here "equal but for higher terms in $h - k$". It is readily seen that the same applies for H_2, H_3, $\cdots$ as for H_1. We have thus determined E_0 and H_1, H_2, H_3, $\cdots H_n$. The E_1, E_2, $\cdots E_n$ remain indefinite from this point on and are disposable for what follows.

To conclude the consideration of the exterior of the wires we write down the expression for H_v on the periphery of the first wire as we approach the latter from the outside: Since this periphery is a magnetic line of force, we have here $u = \text{const}$, say $= +u_0$. We utilize the abbreviation

$$p = e^{-u_0} \tag{9}$$

and obtain from (5)

$$\sqrt{\frac{\mu_0}{\varepsilon_0}}\, H_v = \frac{1}{F - f}\left(p + \frac{1}{p} - 2\cos v\right) \quad \text{for} \quad u = +u_0. \tag{9a}$$

C. The Interior of the Wires

Since, as in the preceding sections, we have within the wire

$$k_L = \sqrt{\varepsilon\mu\omega^2 + i\mu\sigma\omega} = \sqrt{\varepsilon'\mu}\,\omega$$

in place of k and since $|\,k_L\,| \gg h$, we do not attain our goal with solutions of the potential equation in bipolar coordinates, but must employ actual solutions of the wave equation in ordinary cylindrical coordinates. Hence we introduce at the center e.g. of the first wire a new complex variable $\eta = re^{i\varphi}$ with the origin O_1 and we must deal with the mutual transformation of our two systems, polar and bipolar coordinates, particularly at the periphery of the wires. This is furnished by Eq. (3) if there we express ζ in terms of the new variable η. Referring to Fig. 36 and Eqs. (3a, b) we set

$$\zeta + b = \eta, \quad \zeta \mp \zeta_0 = \eta - b \mp \sqrt{b^2 - a^2} = \begin{cases} \eta - F \\ \eta - f \end{cases}$$

and obtain from (3)

$$e^w = e^{u+iv} = \frac{\zeta - \zeta_0}{\zeta + \zeta_0} = \frac{\eta - F}{\eta - f}\,; \tag{10}$$

inversion leads to

$$\eta = re^{i\varphi} = \frac{fe^{u+iv} - F}{e^{u+iv} - 1}. \tag{10a}$$

The coordinates r, φ are thus expressed in terms of the coordinates u, v and vice versa.

Thus by forming the absolute value of (10) and squaring it we find for the periphery of the first wire, where we should have $u = +u_0$ and $r = a$,

$$e^{-2u_0} = p^2 = \frac{ae^{i\varphi} - f}{ae^{i\varphi} - F} \cdot \frac{ae^{-i\varphi} - f}{ae^{-i\varphi} - F} = \frac{a^2 + f^2 - 2af\cos\varphi}{a^2 + F^2 - 2aF\cos\varphi}.$$

Since p is independent of φ this equation is satisfied only if, after multiplying through with the denominator, the factors of $\cos \varphi$ on the two sides are equal, i.e.

$$-2ap^2F = -2af, \qquad p = \sqrt{\frac{f}{F}}.$$

or, in greater detail, in view of the reciprocity relation in Eq. (3a),

$$p = \sqrt{\frac{f}{F}} = \frac{a}{F} = \frac{f}{a}. \tag{11}$$

From the same Eq. (10) we find for $r = a$ since $p < 1$:

$$e^{iv} = p\frac{e^{i\varphi} - F/a}{e^{i\varphi} - f/a} = \frac{pe^{i\varphi} - 1}{e^{i\varphi} - p} = (p - e^{-i\varphi})(1 + pe^{-i\varphi} + p^2e^{-2i\varphi} + \cdots)$$

$$= p + (p^2 - 1)e^{-i\varphi} + (p^3 - p)e^{-2i\varphi} + \cdots \tag{11a}$$

and hence

$$\cos v = p + (p^2 - 1)\cos \varphi + p(p^2 - 1)\cos (2\varphi)$$
$$+ p^2(p^2 - 1)\cos (3\varphi) + \cdots \tag{12}$$

On the other hand we find from (10a) for $r = a$, making use of (11):

$$e^{i\varphi} = \frac{p - e^{-iv}}{1 - pe^{-iv}}, \qquad e^{in\varphi} = (p - e^{-iv})^n(1 - pe^{-iv})^{-n}. \tag{13}$$

The real part of this equation is, for $n = 1, 2, 3,$

$$\cos \varphi = p - (1 - p^2)\cos v - (p - p^3)\cos (2v)$$
$$- (p^2 - p^4)\cos (3v) + \cdots$$
$$\cos (2\varphi) = p^2 - 2(p - p^3)\cos v + (1 - 4p^2 + 3p^4)\cos (2v) + \cdots \tag{14}$$
$$\cos (3\varphi) = p^3 - 3(p^2 - p^4)\cos v + (3p - 9p^3 + 6p^5)\cos (2v) + \cdots$$

As the general expression for the field inside of the first wire we use the superposition of the system of partial waves in (23.7), where however we need write down only the expressions for E_x and H_φ :

$$E_x = \frac{\sqrt{k_L^2 - h^2}}{ih} \sum_{n=0}^{\infty} C_n J_n(\rho) \cos (n\varphi),$$

$$\sqrt{\frac{\mu}{\varepsilon'}} H_\varphi = \sum_{n=0}^{\infty} \left\{ \frac{k_L}{h} C_n J'_n(\rho) + \frac{n}{\rho} D_n J_n(\rho) \right\} \cos (n\varphi). \tag{15}$$

By superposing here all possible asymmetric Hondros waves we implicitly and in the most general fashion take account of the unilateral effect of the "second wire" on the interior of the "first".

D. The Boundary Condition $H_v = H_\varphi$

Previously the amplitudes C, D could be chosen arbitrarily for the individual partial wave. Here we do not consider the individual partial wave, but the superposition of all of them, and must fix the amplitudes C_n, D_n of each partial wave by the requirement of continuity in passing over to the exterior field.

To this end we compare the expression (9a) for H_v, after having replaced in it cos v with the series (12), with the representation of H_φ in (15). Taking account also of (11) we obtain as factor of cos $(n\varphi)$ in (9a)

$$\sqrt{\frac{\varepsilon_0}{\mu_0}} \frac{2p^{n-1}}{F-f}(1-p^2) = \sqrt{\frac{\varepsilon_0}{\mu_0}} \frac{2p^n}{a}. \tag{16}$$

We write for the corresponding factor in (15):

$$\sqrt{\frac{\varepsilon'}{\mu}} C_n \frac{k_L}{h} J'_n(\rho)\left\{1 + \frac{h}{k_L} \frac{n}{\sqrt{k_L^2 - h^2}a} \frac{D_n}{C_n} \frac{J_n}{J'_n}\right\}. \tag{16a}$$

Here J_n/J'_n has the order of magnitude unity, as was noted at (22.7a). Of the multiplying factors the first two are very small quantities; the same applies for D_n/C_n, as we shall confirm later on. A comparison of (16) and (16a) thus yields directly

$$C_n = -\sqrt{\frac{\varepsilon_0}{\varepsilon'} \frac{\mu}{\mu_0}} \frac{h}{k_L} \frac{2p^n}{aJ'_n(\rho)}. \tag{17}$$

The case $n = 0$ requires special consideration because of the constant term appearing in the expression (12) for cos v. We here obtain in place of (16) and (16a)

$$\sqrt{\frac{\varepsilon_0}{\mu_0}} \frac{1}{F-f}\left(-p + \frac{1}{p}\right) = \sqrt{\frac{\varepsilon_0}{\mu_0}} \frac{1}{a} \text{ and } -\sqrt{\frac{\varepsilon'}{\mu}} \frac{k_L}{h} C_0 J'_0(\rho), \text{ respectively}$$

and hence

$$C_0 = -\sqrt{\frac{\varepsilon_0}{\varepsilon'} \frac{\mu}{\mu_0}} \frac{h}{k_L} \frac{1}{aJ'_0(\rho)}. \tag{17a}$$

With these expressions for C_n and C_0 the continuity condition $H_v = H_\varphi$ is satisfied.

E. The Boundary Condition for E_x and the Law of Phase Propagation

Substitution of (17) and (17a) in (15) leads to the expression for E_x

$$-\sqrt{\frac{\varepsilon_0}{\varepsilon'} \frac{\mu}{\mu_0}} \frac{\sqrt{k_L^2 - h^2}}{ik_L a}\left(\frac{J_0(\rho)}{J'_0(\rho)} + 2\sum_{n=1}^{\infty} \frac{J_n(\rho)}{J'_n(\rho)} p^n \cos(n\varphi)\right), \tag{18}$$

which, in view of $| k_L | \gg | h |$ and $J_n/J'_n \cong i$, may be simplified to

$$-\sqrt{\frac{\varepsilon_0}{\varepsilon'}\frac{\mu}{\mu_0}}\frac{1}{a}\left(1 + 2\sum_{n=1}^{\infty} p^n \cos(n\varphi)\right). \qquad (18a)$$

Here we imagine the Fourier series (14) substituted for $\cos(n\varphi)$ and compare then (18a) with the representation (6) for the exterior of the wires, where we put $u = +u_0$:

$$E_0 u_0 + E_1 \sinh u_0 \cos v + E_2 \sinh(2u_0) \cos(2v) + \cdots \qquad (18b)$$

Equating of the factors of $\cos v$, $\cos(2v)$, $\cos(3v)$ on the two sides yields

$E_1 \sinh u_0$

$$= \frac{2p}{a}\sqrt{\frac{\varepsilon_0}{\varepsilon'}\frac{\mu}{\mu_0}}\{1 - p^2 + 2p^2(1 - p^2) + 3p^4(1 - p^2) + \cdots\},$$

$$\tag{19}$$

$$E_2 \sinh(2u_0) = \frac{2p^2}{a}\sqrt{\frac{\varepsilon_0}{\varepsilon'}\frac{\mu}{\mu_0}}\{1 - p^2 - (1 - 4p^2 + 3p^4) + \cdots\},$$

$$E_3 \sinh(3u_0) = \frac{2p^3}{a}\sqrt{\frac{\varepsilon_0}{\varepsilon'}\frac{\mu}{\mu_0}}\{1 - p^2 + \cdots\}.$$

Thus the coefficients E_n of the exterior field, left indefinite up to this point, are determined for $n > 0$; their values may be further simplified by utilizing

$$\sinh u_0 = \frac{1}{2p}(1 - p^2), \qquad \sinh(2u_0) = \frac{1}{2p^2}(1 - p^4), \cdots$$

What, however, is the status of the coefficient E_0 , which was fixed already by Eq. (7) and hence is not disposable for satisfying the boundary condition? Fortunately it contains the as yet undetermined quantity h. The remaining condition, containing E_0 , thus serves for the *final determination of the propagation constant h*, which here as for the single wire is of primary interest. The equation in question is obtained by comparison of the terms in (18a, b) which are independent of v and is

$$-E_0 u_0 = \sqrt{\frac{\varepsilon_0}{\varepsilon'}\frac{\mu}{\mu_0}}\frac{1}{a}\left(1 + 2\sum_{n=1}^{\infty} p^{2n}\right).$$

If here we substitute values from (7) and (9) for E_0 and u_0 and carry out the summation on the right, we obtain:

$$h = k + \frac{i}{2}\sqrt{\frac{\varepsilon_0}{\varepsilon'}\frac{\mu}{\mu_0}}\frac{1}{a}\frac{1 + p^2}{1 - p^2}\Big/\log\frac{1}{p}. \qquad (20)$$

We see that *h is given directly and by an elementary expression*, not, as for the single wire, by a *transcendental equation*. Furthermore for a perfectly conducting material, $\varepsilon_0/\varepsilon' \rightarrow 0$, $h = k$, as should be the case. For finite conductivity ε' is essentially positive imaginary, $\sqrt{\varepsilon'}$ hence of the type $\exp(i\pi/4)$ and $(\varepsilon_0/\varepsilon')^{\frac{1}{2}}$ of the type

$$e^{-i\pi/4} = \frac{1 - i}{\sqrt{2}}.$$

Hence the correction term in (20) is of the type $1 + i$. The real part of h thus becomes *greater* than k; this signifies a *propagation velocity less than c*. At the same time the real part of ih becomes negative. For our form $\exp\{i(hx - \omega t)\}$ this means *damping for propagation along the positive x-axis*. The correction term thus indicates a reasonable physical behavior in both its real and its imaginary parts.

Finally, to test the dependence of the correction term on the geometrical data of the Lecher system, i.e. the wire radius a and the wire separation $2b$, we may, for $2b \gg a$, replace F by $2b$. We then find from formula (11)

$$p = \frac{a}{2b} \ll 1, \qquad \frac{1 + p^2}{1 - p^2} \Big/ \log \frac{1}{p} \cong 1/\log \frac{2b}{a}. \qquad (20a)$$

The correction term then has only a logarithmic dependence on the separation of the wires and is inversely proportional to the wire radius.

If, conversely, b is only slightly greater than a, i.e. $b = a(1 + \alpha)$ with $\alpha \ll 1$, we obtain from (3b) and (11)

$$F = a(1 + \sqrt{2\alpha}), \qquad p = 1 - \sqrt{2\alpha}, \qquad \log \frac{1}{p} = \sqrt{2\alpha},$$

and hence

$$\frac{1}{a} \frac{1 + p^2}{1 - p^2} \Big/ \log \frac{1}{p} = \frac{1}{a} \left(\frac{1}{\sqrt{2\alpha}} - 1 \right) \Big/ \sqrt{2\alpha} = \frac{1}{2a\alpha} = \frac{1}{2(b - a)}. \qquad (20b)$$

We have carried out this short calculation to show that our final formula (20) also covers the case of slightly separated wires, where the mutual influence of the wires is very great and their skin effect must be enhanced unilaterally. We thus make it clear that our treatment is quite general with respect to the *geometric* circumstances. On the other hand, the limitation to wires of *high conductivity* already introduced in section A applies throughout.

F. Supplement Regarding the Remaining Boundary Conditions

To demonstrate the completeness of our solution and its freedom from contradiction we shall survey briefly the remaining boundary conditions. These are the continuity conditions for H_x on the one hand, for $E_\varphi = E_v$ on the other.

The representation of the interior field to be obtained from (23.7) by summation over n contained, in H_x, the coefficients D_n which till now we have suppressed. The boundary condition relates them to the constants H_n of the external field which occur in (6) and which, according to (8), are equal and opposite to the already known constants E_n. *The boundary condition for H_x thus serves the determination of the constants D_n of the internal field.* Without entering into the calculation, we find

$$| D_n | \ll | C_n |,$$

which we have already used in (17). Finally, no disposable coefficients are left over for the fulfilment of the boundary condition

$$E_\varphi \text{ (inside)} = E_v \text{ (outside)}.$$

On the other hand we know from our approximate calculation that everywhere outside, and in particular at the surfaces of the wires, $E_v = 0$. It is thus necessary to show that according to our representation (23.7) also the sum of all the E_φ (inside) is much smaller than that of the other field components. This is in fact so, by several orders of magnitude. The proof must however be omitted here.

G. Parallel and Push-Pull Operation

The approach already described is entirely adapted to push-pull excitation. In the parallel case the transversal components must be derived not from the potential (3), but from

$$w = \log (\zeta - \zeta_0) + \log (\zeta + \zeta_0), \tag{21}$$

and this only for sufficiently thin wires ($a \ll b$; otherwise the equipotential lines of (21) are not approximated by circles!). The bipolar coordinates then lose their usefulness, since they no longer coincide with the system of equipotentials of (21). In particular the electric lines of force no longer pass, as in Fig. 36, from Q_1 to Q_2, but repel each other and pass from Q_1 and Q_2 separately to infinity. We hence are now obliged to use ordinary cylindrical coordinates at the centers of the first and second wires, r, φ and $\bar{r}$, $\bar{\varphi}$, respectively, for the exterior of the wires as well. The field is now to be constructed by the superposition of the contributions of the two wires with *equal* sign. Furthermore, it is now *necessary*[1] to formulate these representations directly as solutions of the wave equation, which formerly, because of the limited range of the field, could be avoided. We thus return also for the exterior to Hondros' formulation (23.8) where, restricting ourselves to thin wires, we can limit our attention to the zero-order terms. In

---

[1] For the radiation condition for $r \to \infty$, which now becomes essential, see the discussion for the single wire in §22.

the representation of E_x we then obtain for the superposition of the two wires the sum:

$$\frac{\sqrt{k^2 - h^2}}{h}\left\{ H_0(\sqrt{k^2 - h^2}\, r) + H_0(\sqrt{k^2 - h^2}\, \bar{r}) \right\}. \tag{22}$$

On the periphery of the first wire $r = a$, $\bar{r} \cong 2b$ we obtain by (22.5), because of the smallness of $\sqrt{k^2 - h^2}$ (we of course continue to assume high conductivity),

$$\frac{\sqrt{k^2 - h^2}}{h}\frac{2i}{\pi}\left(\log\left\{ \frac{\gamma\sqrt{k^2 - h^2}}{2i}\, a \right\} + \log\left\{ \gamma\, \frac{\sqrt{k^2 - h^2}}{2i}\, 2b \right\} \right). \tag{22a}$$

This much about the exterior of the wires. Inside it is necessary to use the earlier perfectly general formulation (15) for E_x. The coefficient C_0 appearing here is determined with the aid of the continuity condition for H_φ. Herewith the longitudinal field E_x inside of the wire, more particularly its zero-order partial wave, is also determined. This must agree with the term (22) of the external field for $r = a$. We thus obtain an equation of the form

$$(k^2 - h^2)\left(\log\left\{ \frac{\gamma\sqrt{k^2 - h^2}}{2i}\, a \right\} + \log\left\{ \frac{\gamma\sqrt{k^2 - h^2}}{2i}\, 2b \right\} \right) = \text{const},$$

whose right side is known. It must be fulfilled by proper choice of h and is *transcendental* in character, as with the single wire. (It evidently can be made to correspond to the single-wire equation (22.15) by combining the two logarithms.) We now see the reason why the corresponding equation for h becomes elementary instead of transcendental in the Lecher case: The two logarithms are here superposed with the $-$ sign in place of the $+$ sign; in combining them the factor $\gamma\sqrt{k^2 - h^2}/(2i)$ under the log-sign cancels and only $\log\{a/(2b)\}$ remains, as in Eq. (20a). We also see that the calculation here outlined, without bipolar coordinates, would have been successful also in the Lecher case, but that it would have been much more involved than the earlier method, particularly without restriction to extremely thin wires.

Not only in mathematical formulation but also in physical structure the parallel wave resembles the single-wire wave. It decreases much more slowly outwards than the push-pull wave and is hence much more disturbed by the surroundings. It is obvious that at large distance the two similarly directed currents of the parallel wave must produce the same field as the alternating current of the single wire.

Experimentally a pure excitation of the push-pull wave is always desirable to avoid disturbances from the surroundings. However if the arrangement is not quite symmetrical parallel waves are also occasionally excited, which make the position of the nodal points, on which the wave-length determination rests, unsharp.

Hence even from a purely experimental standpoint it is important to keep in mind the possibility of the parallel processes and to take account of the theory of the single wire, which we have treated before the rest in §22, although as compared with the theory of the Lecher system it is of secondary practical importance.

PART III

THEORY OF RELATIVITY AND ELECTRON THEORY

§26. The Invariance of the Maxwell Equations in the Four-Dimensional World

The path taken by Einstein in 1905 in the discovery of the special theory of relativity was steep and difficult. It led through the analysis of the concepts of time and space and some ingenious imaginary experiments. The path which we shall take is wide and effortless. *It proceeds from the universal validity of the Maxwell equations* and the tremendous accumulation of experimental material on which they are based. It ends almost inadvertently at the *Lorentz transformation* and all its relativistic consequences.

A. The Four-Potential

We refer to the electrodynamic potentials in §19, which at that point still remained in the fog of an unsatisfactory formalism. I wish to create the impression in my readers that the true mathematical structure of these entities will appear only now, as in a mountain landscape when the fog lifts.

In the two differential equations (19.9) and (19.11), satisfied by $\mathbf{A}$ and Ψ, we had on the left the operator

$$\Delta - \frac{1}{c^2}\frac{\partial^2}{\partial t^2}. \tag{1}$$

We now introduce in place of x, y, z, t the new coordinates

$$x_1 = x, \qquad x_2 = y, \qquad x_3 = z, \qquad x_4 = ict, \tag{2}$$

where the proper remarks regarding the imaginary unit in x_4 will be made later. We call these x_i world coordinates since all events in the world are determined in space-time. The operator (1) then becomes the four-dimensional generalization of the Laplace operator and may be designated by[1]

$$\Box = \sum_{i=1}^{4} \frac{\partial^2}{\partial x_i^2}. \tag{3}$$

---

[1] Certain more advanced theories of Einstein and Kaluza employ also the five-dimensional symbol $\bigcirc = \sum_{i=1}^{5} \frac{\partial^2}{\partial x_i^2}$.

Like the independent coordinates x_i we combine the potentials **A**, Ψ in a four-dimensional entity, the four-potential $\mathbf{\Omega}$. Let its four components be

$$\Omega_1 = A_x, \qquad \Omega_2 = A_y, \qquad \Omega_3 = A_z, \qquad \Omega_4 = \frac{i}{c}\,\Psi. \tag{4}$$

The factor i in Ω_4 is reasonable in view of the definition of x_4 ; the factor $1/c$ gives the four components of $\mathbf{\Omega}$ the same dimension and will be justified below (7). The differential equations (19.9) and (19.11) then take the form

$$\Box\,\mathbf{\Omega} = -\,\mu_0\,\mathbf{\Gamma}. \tag{5}$$

The quantity $\mathbf{\Gamma}$ here introduced may be called the *four-current density*. It follows from (4), (19.9), and (19.11) that its four components are

$$\Gamma_1 = J_x, \qquad \Gamma_2 = J_y, \qquad \Gamma_3 = J_z, \qquad \Gamma_4 = ic\rho; \tag{6}$$

they all have the dimension Q/M^2S of a current density.

We now turn to the Eq. (19.10) relating the potentials **A** and Ψ. The second term on its left is, in view of (4),

$$\frac{1}{c^2}\,\dot{\Psi} = \frac{1}{ic}\,\frac{\partial\Omega_4}{\partial t} = \frac{\partial\Omega_4}{\partial x_4}.$$

Eq. (19.10) thus becomes

$$\frac{\partial\Omega_1}{\partial x_1} + \frac{\partial\Omega_2}{\partial x_2} + \frac{\partial\Omega_3}{\partial x_3} + \frac{\partial\Omega_4}{\partial x_4} = 0.$$

We write this more briefly

$$\mathrm{Div}\ \mathbf{\Omega} = 0, \qquad \mathrm{Div} = \sum_{i=1}^{4}\frac{\partial}{\partial x_i}. \tag{7}$$

We shall call the operator Div the *four-dimensional divergence*. Its four-dimensional symmetry indicates *isotropy* in space and time of world events. Our operator $\Box$ shows the same isotropy.

Thus we see the reason that we may regard our four-potential $\mathbf{\Omega}$ and our four-current density $\mathbf{\Gamma}$ as *vectors* in four-dimensional space, more briefly, as *four-vectors*. This notation contains a statement regarding the behavior of the quantities $\mathbf{\Omega}$ and $\mathbf{\Gamma}$ when the coordinates x_i are changed. As was shown in Vol. II, §2, an ordinary "three-vector" is a quantity which, for orthogonal transformation of the x, y, z, behaves just as the radius vector $\mathbf{r} = (x, y, z)$. Thus the four-vector attains a meaning in the four-dimensional world which is independent of the choice of the coordinate system. At the same point in Vol. II we defined a scalar as a quantity which is *invariant* with respect to orthogonal transformations; in particular, we showed by a simple calculation that the divergence of a three-vector possesses this property. This calculation may be transferred directly to four dimensions and shows that the divergence defined in (7), applied to any four-vector, yields a *scalar*, i.e. a four-dimensional invariant.

By reducing the Maxwell equations to the four-vector $\mathbf{\Omega}$ and the invariant operators $\square$, Div we have demonstrated at the same time their *general validity*, independent of the coordinate system. The isotropy of three-dimensional space found adequate expression in the vector calculus of parts I and II. It is now replaced, in view of the world isotropy, by the *four-dimensional vector calculus*. This states that, for a transition to a "primed" coordinate system x'_i, the Maxwell equations remain invariant, i.e. have the same form in the primed field components and coordinates as in the original "unprimed" ones. *This invariance is simply the principle of relativity in its electrodynamic formulation.* The Maxwell equations satisfy the relativity postulate from the very beginning. They need not be subsequently adapted to it, like the equations of mechanics (see §32).

B. The Six-Vectors of Field and Excitation

We now turn to the representation of the field component $\mathbf{E}$. By Eq. (19.7) we have e.g.

$$E_x = -\frac{\partial \Psi}{\partial x} - \frac{\partial A_x}{\partial t}.$$

According to Eqs. (2) and (4) this is equivalent to

$$\mathbf{E}_x = ic\left(\frac{\partial \Omega_4}{\partial x_1} - \frac{\partial \Omega_1}{\partial x_4}\right). \tag{8}$$

This relation suggests the introduction of the four-dimensional curl

$$\text{Curl}_{nm}\,\mathbf{\Omega} = \frac{\partial \Omega_m}{\partial x_n} - \frac{\partial \Omega_n}{\partial x_m}. \tag{9}$$

As a two-indices quantity it has *six* components (according to Vol. II, Eq. (2.17) it was preferable to give the three-dimensional curl as well two indices instead of one index). Evidently

$$\text{Curl}_{nn} = 0, \qquad \text{Curl}_{mn} = -\text{Curl}_{nm}. \tag{9a}$$

Curl $\mathbf{\Omega}$ is called a *six-vector* or, preferably, in view of the symmetry properties indicated by (9a), an *antisymmetric six-tensor*. The term "surface tensor" is also applied to it. The six components which differ from zero and from each other may be divided into three space-time and three space-space forms, corresponding to the arrangement of the indices

$$1\,4,\ 2\,4,\ 3\,4 \qquad \text{and} \qquad 2\,3,\ 3\,1,\ 1\,2. \tag{9b}$$

According to (8) the first three belong to the electric vector, the last three to the magnetic vector. However, we must not couple the entity of quantity $\mathbf{H}$ with the entity of intensity $\mathbf{E}$ to form a four-dimensional unit, but must employ for this the entity of intensity $\mathbf{B}$ or rather the quantity $c\mathbf{B}$, which has the same dimension as $\mathbf{E}$. We therefore write for example,

$$cB_x = c \, \mathrm{curl}_{yz} \, \mathbf{A} = c \left(\frac{\partial \Omega_3}{\partial x_2} - \frac{\partial \Omega_2}{\partial x_3} \right). \tag{10}$$

Combining (8) and (10) and extending them cyclically to the remaining field components we obtain the following representation of the *six-com-ponent field vector*, where ( ) is to indicate merely the combination of the two three-dimensional vectors into one four-dimensional quantity:

$$F = (c\mathbf{B}, -i\mathbf{E}) = c \, \mathrm{Curl} \, \boldsymbol{\Omega}. \tag{11}$$

This *field vector* F is to be given two indices in accord with the sequence (9b), like the six-vector Curl $\boldsymbol{\Omega}$.

We next ask about the *excitation vector*, which we shall denote by f. We form it from the two equally dimensioned quantities $\mathbf{H}$ and $c\mathbf{D}$. To pass from $\mathbf{H}$ and $\mathbf{D}$ to $\mathbf{B}$ and $\mathbf{E}$ we employ the constants for vacuum μ_0 and ε_0, since our part III will be limited throughout to space free from matter (e.g. a vacuum tube). Since

$$\mathbf{H} = \frac{\mathbf{B}}{\mu_0} = \sqrt{\frac{\varepsilon_0}{\mu_0}} \, c\mathbf{B} \quad \text{and} \quad c\mathbf{D} = c\varepsilon_0 \mathbf{E} = \sqrt{\frac{\varepsilon_0}{\mu_0}} \, \mathbf{E}$$

we obtain from (11) simply

$$f = (\mathbf{H}, -ic\mathbf{D}) = \sqrt{\frac{\varepsilon_0}{\mu_0}} \, c \, \mathrm{Curl} \, \boldsymbol{\Omega}. \tag{12}$$

Hence

$$f = \sqrt{\frac{\varepsilon_0}{\mu_0}} \, F. \tag{13}$$

To our satisfaction the geometric mean of the two three-dimensional constants of vacuum ε_0 and $\mu' = 1/\mu_0$ appears here. Already at the introduction of the permeability in §4, p. 21 we emphasized that the true magnetic analog of the dielectric constant ε is not μ, but its reciprocal μ'. In any case, Eq. (13) combines the three-dimensional relations between excitations and fields symmetrically, with a single constant of vacuum. At the same time our formulation (11) and (12) translates the earlier, highly heterogeneous representations (19.7) and (19.6) of $\mathbf{E}$ and $\mathbf{H}$ into an entirely symmetric and harmonious form.

To create a clear visual image of the structure of the antisymmetric tensor we write down the array of all the components of f in matrix form; the arrangement of the components of F is obtained herefrom by multiplication with $\sqrt{\mu_0/\varepsilon_0}$ and simultaneous exchange of $\mathbf{H}$ with $c\mathbf{B}$ and of $c\mathbf{D}$

with **E**. We distinguish the components of **H** and **D** in the customary three-dimensional fashion by the indices x, y, z; from our present standpoint it would be preferable to designate them by the double indices (9b), as in the matrix at the left. We point out specifically the order 12, 13, 14 in the first row, 21, 23, 24 in the second row, etc. as well as the change in sign for the converse order:

$$f = \begin{pmatrix} 0 & f_{12} & f_{13} & f_{14} \\ f_{21} & 0 & f_{23} & f_{24} \\ f_{31} & f_{32} & 0 & f_{34} \\ f_{41} & f_{42} & f_{43} & 0 \end{pmatrix} = \begin{pmatrix} 0 & H_z & -H_y & -icD_x \\ -H_z & 0 & H_x & -icD_y \\ H_y & -H_x & 0 & -icD_z \\ icD_x & icD_y & icD_z & 0 \end{pmatrix} \tag{14}$$

C. The Maxwell Equations in Four-Dimensional Form

We also want to write the original Maxwell equations with double indices. Proceeding from the equation

$$\dot{D}_x - \frac{\partial H_z}{\partial y} + \frac{\partial H_y}{\partial z} = -J_x$$

we obtain from the first row of the array (14), making use of the definition of $\boldsymbol{\Gamma}$ in (6),

$$-\frac{\partial f_{12}}{\partial x_2} - \frac{\partial f_{13}}{\partial x_3} - \frac{\partial f_{14}}{\partial x_4} = -\Gamma_1$$

and corresponding equations for the second and third of this triplet of Maxwell equations. We hence have in general form for $m = 1, 2, 3$

$$\sum_{n=1}^{4} \frac{\partial f_{mn}}{\partial x_n} = \Gamma_m. \tag{15}$$

If we extend this form to $m = 4$, we obtain by (14) and (6)

$$ic\left(\frac{\partial D_x}{\partial x} + \frac{\partial D_y}{\partial y} + \frac{\partial D_z}{\partial z}\right) = ic\rho = \Gamma_4.$$

Our original *definition of the charge density* ρ in (4.4b) thus proves to be the *four-dimensional completion* of the second *Maxwell triplet* (4.4).

The operation $\sum \dfrac{\partial}{\partial x_n}$ carried out in (15) bears the name "reduction" or "divergence" in general tensor analysis (see Vol. II, p. 60); it reduces a four-dimensional tensor to a four-vector. We symbolize it by **Div**, placing it parallel to the operation Div defined in (7), which reduces a four-vector to a scalar. We thus write in place of (15)

$$\mathbf{Div}_m f = \sum_{n=1}^{4} \frac{\partial f_{mn}}{\partial x_n} = \Gamma_m. \tag{16}$$

Generally for any antisymmetric tensor T

$$\text{Div } \mathbf{Div } T_{nm} = 0, \tag{16a}$$

which follows directly from $T_{nm} = -T_{mn}$. We conclude therefore that the *divergence of the four-current vanishes*:

$$\text{Div } \mathbf{\Gamma} = 0; \tag{16b}$$

this is simply the *continuity equation* (4.4c) in a refined, four-dimensional form.

What of the first triplet of the Maxwell equations (4.4)? Its x-component is

$$\dot{B}_x + \frac{\partial E_z}{\partial y} - \frac{\partial E_y}{\partial z} = 0.$$

In view of the meaning of our field tensor F, which is analogous to (14):

$$F = \begin{pmatrix} 0 & F_{12} & F_{13} & F_{14} \\ F_{21} & 0 & F_{23} & F_{24} \\ F_{31} & F_{32} & 0 & F_{34} \\ F_{41} & F_{42} & F_{43} & 0 \end{pmatrix} = \begin{pmatrix} 0 & cB_z & -cB_y & -iE_x \\ -cB_z & 0 & cB_x & -iE_y \\ cB_y & -cB_x & 0 & -iE_z \\ iE_x & iE_y & iE_z & 0 \end{pmatrix} \tag{17}$$

it may be rewritten in the form

$$i\left(\frac{\partial F_{23}}{\partial x_4} + \frac{\partial F_{34}}{\partial x_2} + \frac{\partial F_{42}}{\partial x_3}\right) = 0. \tag{17a}$$

This somewhat confusing distribution of subscripts becomes quite plain if we introduce the "dual" six-vector of F,

$$F^* = (-i\mathbf{E}, c\mathbf{B}), \tag{17b}$$

which is obtained from F by an exchange of the real and the imaginary constituents. The determination of the individual components of F^* is fixed by the rule

$$F^*_{mn} = F_{kl} \tag{17c}$$

with the prescription that the sequence of subscripts

$$k \, l \, m \, n \quad \text{arises from} \quad 1 \, 2 \, 3 \, 4 \tag{17d}$$

by an *even number of exchanges*. By this requirement we have uniquely

$$F_{23} = F^*_{14}, \qquad F_{34} = F^*_{12}, \qquad F_{42} = F^*_{13}, \tag{17e}$$

so that Eq. (17a) becomes

$$\frac{\partial F^*_{12}}{\partial x_2} + \frac{\partial F^*_{13}}{\partial x_3} + \frac{\partial F^*_{14}}{\partial x_4} = 0. \tag{17f}$$

It has thus become the first component of

$$\text{Div } F^* = 0. \tag{18}$$

The other two components of the Maxwell triplet in question take on similar forms.

But what is the meaning of the fourth component of (18)? It is

$$\frac{\partial F^*_{41}}{\partial x_1} + \frac{\partial F^*_{42}}{\partial x_2} + \frac{\partial F^*_{43}}{\partial x_3} = 0$$

and, by (17b) and (17) may be transformed into

$$-c \left(\frac{\partial B_x}{\partial x} + \frac{\partial B_y}{\partial y} + \frac{\partial B_z}{\partial z} \right) = 0.$$

It is thus identical with the familiar absence of *sources of the magnetic field intensity* **B**. This appears now, from the four-dimensional relativistic standpoint, as a *formally necessary completion* of our first Maxwell triplet, while originally, in Eq. (4.4a), it had to be *postulated* separately as an empirical fact.

We have thus a complete representation of Maxwell's theory for vacuum in the statements

$$\textbf{Div } f = \Gamma, \qquad \textbf{Div } F^* = 0, \qquad F = \sqrt{\frac{\mu_0}{\varepsilon_0}} f, \tag{19}$$

which parallel and are equivalent to the potential relations

$$\Box \, \Omega = -\mu_0 \Gamma, \qquad \text{Div } \Omega = 0, \qquad F = c \, \text{Curl} \, \Omega, \qquad \mu_0 f = \text{Curl} \, \Omega. \tag{20}$$

All quantities and operations appearing in both formulations have proper citizenship in the four-dimensional world and hence satisfy the principle of relativity.

It should be emphasized on this occasion that the theory of relativity leaves no doubt that the vectors **E** and **B** on the one hand, and **D** and **H** on the other, belong together, as parts of the higher entities F and f. This seemed clear to us from the beginning, partly for dimensional reasons, partly because of their different significance as entities of intensity and quantity. In particular, the theory of relativity leaves no doubt that this distinction is as necessary in vacuum as in any ponderable medium, i.e. that here also both six-vectors F and f (the four three-vectors **E**, **B**, **D**, **H**) have to be employed side by side.

D. On the Geometric Character of the Six-Vector and its Invariants

The *four-vector* is represented, as a matter of course, by a *straight line segment* in four-dimensional space (by an R_1 with sense of direction). It might appear appropriate to represent the *six-vector* by a two-dimensional *segment of a plane*, i.e. by its magnitude and position in four-dimensional space (an R_2, one of whose sides is designated as positive). However, this representation is too specialized. Such a segment of area has only 5 independent parameters, not 6, as a six-vector, i.e. *one* parameter indicating its size (shape is to be indifferent) and *four*[1] indicating its orientation (parallel displacements do not count). To obtain a geometrical interpretation of the general six-vector we note that every R_2 in four-dimensional space has uniquely correlated with it a second R_2 perpendicular[2] to it. If a segment of area is also prescribed in the second, a single further parameter is obtained (since the orientation in space is already determined by the orientation of the first segment of area), leading to the desired total number of six independent parameters. The geometric picture of the six-vector is thus not *one* segment of area, but *two mutually perpendicular segments of area of arbitrary size*. The components of the six-vector are equal to the sums of the projections of the *two* segments of area on the six coordinate planes (x_n, x_m). If the sizes of the two segments are interchanged, the original six-vector F passes over into the *dual* six-vector F^*; this confirms the relation (17c) between the components of F and F^*.

The four-vector has only *one* invariant, the square of its length, equal to the sum of the squares of its four components (the fourth of them taken, of course, with negative sign, in view of its imaginary character). On the other hand, every six-vector F has *two invariants*

$$F \cdot F \qquad \text{and} \qquad F \cdot F^*$$

both given, in accord with the rule for the scalar product, by summation over the six components with equal indices. We carry this out for the electrodynamic case. By (17) and (17a)

$$F \cdot F \ = F_{12}{}^2 + F_{23}{}^2 + F_{31}{}^2 + F_{14}{}^2 + F_{24}{}^2 + F_{34}{}^2 = c^2 \mathbf{B}^2 - \mathbf{E}^2$$
$$F \cdot F^* = F_{12}F_{34} + F_{23}F_{14} + F_{31}F_{24} + \cdots = -2ic\mathbf{B} \cdot \mathbf{E}. \tag{21}$$

Here $\cdots$ signifies repetition of the three preceding products with reversal of the sequence of the factors, i.e. simply the doubling of the sum. For the *vacuum light wave* both invariants (21) are zero. In fact, by §6, $\mathbf{B} \perp \mathbf{E}$ and $c \,|\, \mathbf{B} \,| = |\, \mathbf{E} \,|$. In view of our statement regarding invariance the

---

[1] If its orientation is thought of as defined by two four-vectors proceeding from the same point and lying in R_2, these are given by 3 quantities, e.g. the ratios of their four components. Both may however be rotated arbitrarily within R_2, so that the number $2 \cdot 3$ is reduced to $2 \cdot 3 - 2 = 4$.

[2] One R_2 may be designated as the "axis" of the other, since one can be rotated about the other arbitrarily within itself. The relation of the two is of course mutual or "dual". The axis of an R_2 in R_4 is thus a two-dimensional, not as in R_3 a one-dimensional manifold.

*

light wave retains this property in all reference systems of the four-dimensional world.

The excitation vector has of course the corresponding invariants

$$f \cdot f = \mathbf{H}^2 - c^2 \mathbf{D}^2,$$
$$f \cdot f^* = -2ic\mathbf{H} \cdot \mathbf{D}.$$
(21a)

The following mixed invariants differ from (21) and (21a) only by a factor:

$$f \cdot F = f_{12}F_{12} + f_{23}F_{23} + f_{31}F_{31} + f_{14}F_{14} + f_{24}F_{24} + f_{34}F_{34}$$
$$= c\mathbf{H} \cdot \mathbf{B} - c\mathbf{D} \cdot \mathbf{E},$$
(22)

$$f \cdot F^* = f_{12}F^*_{12} + f_{23}F^*_{23} + f_{31}F^*_{31} + f_{14}F^*_{14} + f_{24}F^*_{24} + f_{34}F^*_{34}$$
$$= -i\mathbf{H} \cdot \mathbf{E} - ic^2\mathbf{D} \cdot \mathbf{B} = -2i\mathbf{E} \cdot \mathbf{H}.$$
(23)

We define

$$\Lambda = \frac{1}{2c} f \cdot F = \frac{1}{2} \mathbf{H} \cdot \mathbf{B} - \frac{1}{2} \mathbf{D} \cdot \mathbf{E}$$
(24)

as the *Lagrange density* (Lagrange function of the moving electron per unit volume of the field). On the other hand the energy density W, given by $\frac{1}{2}\mathbf{H} \cdot \mathbf{B} + \frac{1}{2}\mathbf{D} \cdot \mathbf{E}$, will prove to be a *component* of a world tensor; by itself it has no meaning independent of the frame of reference.

Our invariants may be expressed as follows in terms of the areas a and b which are correlated in the six-vector:

$$F \cdot F = a^2 + b^2, \qquad F \cdot F^* = 2ab, \qquad \Lambda = \frac{\varepsilon_0}{2}(a^2 + b^2).$$
(25)

It follows from this that a particular six-vector ($b = 0$) is distinguished from the general one by the condition $F \cdot F^* = 0$.

E. Relativistically Invariant Three-Vectors

We now ask what properties a three-vector must have in order that it may exist legitimately in the four-dimensional world. For this purpose we consider a *six-vector which is dual to itself*. We may write it in the form

$$\bar{F} = F_{mn} + F^*_{mn} = F_{mn} + F_{kl}$$
(26)

This six-vector has in fact only *three* independent components. We can define them by

$$a_x = \bar{F}_{14} = \bar{F}_{23} = F_{14} + F_{23},$$
$$a_y = \bar{F}_{24} = \bar{F}_{31} = F_{24} + F_{31},$$
$$a_z = \bar{F}_{34} = \bar{F}_{12} = F_{34} + F_{12},$$
(26a)

and obtain as specific tensor arrangement for the three-vector **a** by (17):

$$\bar{F} = \begin{pmatrix} 0 & a_z & -a_y & a_x \\ -a_z & 0 & a_x & a_y \\ a_y & -a_x & 0 & a_z \\ -a_x & -a_y & -a_z & 0 \end{pmatrix}. \tag{26b}$$

In the special case of the electrodynamic tensor F we obtain as corresponding three-vector

$$a_x = -i(E_x + icB_x), \qquad a_y = -i(E_y + icB_y), \qquad a_z = -i(E_z + icB_z).$$

Thus the complex three-vector

$$\mathbf{E} + ic\mathbf{B} = i\mathbf{a} \tag{27}$$

may be regarded as a four-dimensional tensor of the form (26), which is dual to itself.

On the other hand, we may also write, in place of (26),

$$\bar{F} = F_{mn} - F^*_{mn} = F_{mn} - F_{kl}.$$

This six-vector is *oppositely dual* to itself and leads to the three-vector

$$\mathbf{E} - ic\mathbf{B} = -i\mathbf{b}. \tag{27a}$$

Even before the theory of relativity it was often noted that the complex combinations $\mathbf{E} \pm ic\mathbf{B}$ and the analogous $\mathbf{H} \pm ic\mathbf{D}$ have certain characteristic advantages for the integration of the Maxwell equations.

We can now also give the Hertzian vector $\mathbf{\Pi}$ its proper place in the four-dimensional world. It appeared in §19 as a *three-vector*, but it is in fact a disguised *six-vector*, with the structure of an electrostatic field vector F_{stat}. For it $\mathbf{B}_{\text{stat}} = 0$, $\mathbf{E}_{\text{stat}}$ equals a three-vector which, temporarily, we will denote by P_x, P_y, P_z. By (17) we then have

$$F_{\text{stat}} = \begin{pmatrix} 0 & 0 & 0 & -iP_x \\ 0 & 0 & 0 & -iP_y \\ 0 & 0 & 0 & -iP_z \\ iP_x & iP_y & iP_z & 0 \end{pmatrix}. \tag{28}$$

Reduction of this tensor leads to a four-vector which for the present will be denoted by $\mathbf{\Omega}$, like our four-potential:

$$\mathbf{\Omega} = \mathbf{Div}\, F_{\text{stat}}. \tag{28a}$$

It has the components

$$\Omega_{1,2,3} = -i\frac{\partial P_{x,y,z}}{\partial x_4} = -\frac{1}{c}\dot{\mathbf{P}}, \qquad \Omega_4 = i\,\mathrm{div}\,\mathbf{P}.$$

If, by Eq. (4), we pass from this $\Omega_{1,2,3,4}$ to the electrodynamic potential **A** and Ψ, we find

$$\mathbf{A} = -\frac{1}{c}\dot{\mathbf{P}}, \qquad \Psi = c \operatorname{div} \mathbf{P}. \qquad (28b)$$

If, finally, we set $\mathbf{P} = -\mu_0 c\mathbf{\Pi}$, our Eqs. (28b) pass over exactly into Eqs. (19.15) and (19.16a), by which we had originally defined the Hertzian vector. Thus, like **P**, the three-vector **Π** has been reduced to the six-vector (28). This six-vector is here referred to a coordinate system in which the Hertzian dipole *rests*. The transition from this "system at rest" to an arbitrary reference system can be carried out by the rules of the next section.

§27. The Group of the Lorentz Transformations and the Kinematics of the Theory of Relativity

In his "Erlangen Program" Felix Klein[1] has classified the several geometric disciplines on the basis of the group of transformations permitted in them. *Projective geometry* regards all figures as the same, which pass over into each other by *central projections* in three-dimensional space. *Affine geometry* keeps the infinitely distant plane fixed, and hence permits only *parallel projections*. For *elementary geometry* also the shapes of figures, their angles and ratios of linear dimensions, are of importance. Its group is that of the *orthogonal transformations in three-dimensional space*, extended by the similarity transformations. Here the imaginary sphere circle contained in the infinitely distant plane is kept fixed in addition to this plane itself. The geometry of the general *point transformations* can transform any surface into any other, but subjects any small region of space only to linear (projective) changes. The group of the *contact transformations* dissolves even the content of surfaces in space and leaves only the combined position of surface point and tangential plane untouched.

The fact that we have spoken here not of individual transformations, but only of transformation *groups* evidently derives from the necessity of regarding transformations resulting from a sequence or combination of transformations as equally valid. If attention is focused on the unaltered properties, rather than the changes of geometric structures, we speak of the *theory of invariants* belonging to the transformation group in question.

The transformation group of *classical mechanics* is that of the *Galilei* transformations (see Vol. I, p. 10). It may be divided into the group of the orthogonal transformations of space and the displacements of the time scale, corresponding to Newton's idea of absolute space and absolute time. Its invariants are the square of the separation in space and the time difference. The group of *Maxwell's electrodynamics* is, as we saw in the last

---

[1] Comparative Study of More Recent Researches in Geometry, Erlangen, 1872.

section, that of the *orthogonal transformations in space-time*. In honor of the great Dutch physicist Hendrick Antoon Lorentz, Poincaré has called them the *Lorentz transformations*. Just as the nature of the several geometries is characterized by their particular group, the essence of Maxwell's theory rests in its *invariance within the Lorentz group*. Its fundamental invariant is the *separation of two world points*, in particular the *four-dimensional line element*, i.e. the separation of two neighboring points in space-time.

A. The General and the Special Lorentz Transformation

We have recorded the general pattern of the Lorentz transformation already in (2.10) of Vol. I. It is contained in the formulas

$$x'_i = \sum \alpha_{ik} x_k, \qquad x_k = \sum \alpha_{ik} x'_i, \qquad \left.\begin{matrix} i \\ k \end{matrix}\right\} = 1, 2, 3, 4, \qquad (1)$$

$$\sum_{j=1}^{4} \alpha_{ij}\alpha_{kj} = \sum_{j=1}^{4} \alpha_{ji}\alpha_{jk} = \delta_{ik} = \begin{cases} 0 & k \neq i \\ 1 & k = i \end{cases}. \qquad (2)$$

The orthogonal transformations of three-dimensional space, or its rotations within itself, form a subgroup. Correspondingly the general Lorentz transformation may be designated as a *rotation in space-time*. We have also already derived, in Vol. I, Eq. (2.14), the *special Lorentz transformation* in which two coordinates remain unchanged. If we choose for the latter the y- and z-coordinates, the transformation matrix reduces to

	x_1	x_2	x_3	x_4
x'_1	α_{11}	0	0	α_{14}
x'_2	0	1	0	0
x'_3	0	0	1	0
x'_4	α_{41}	0	0	α_{44}

(2a)

According to conditions (2) we must have

$$\alpha_{11}^2 + \alpha_{14}^2 = \alpha_{41}^2 + \alpha_{44}^2 = \alpha_{11}^2 + \alpha_{41}^2 = \alpha_{14}^2 + \alpha_{44}^2 = 1, \qquad (3)$$

so that

$$\alpha_{11}^2 = \alpha_{44}^2, \qquad \alpha_{14}^2 = \alpha_{41}^2. \qquad (4)$$

we put

$$\alpha_{11} = \alpha_{44} = \alpha \qquad (5)$$

and find from (2)

$$\alpha_{11}\alpha_{14} + \alpha_{41}\alpha_{44} = \alpha(\alpha_{14} + \alpha_{41}) = 0. \tag{6}$$

We thus may write, introducing a new constant β:

$$\alpha_{14} = -\alpha_{41} = i\alpha\beta. \tag{7}$$

The added factor i is necessary because of the imaginary character of x_4 and x_4' if real values are to be obtained for x_1' and x_1 in the equations of our system

$$x_1' = \alpha x_1 + \alpha_{14}x_4 \quad \text{and} \quad x_1 = \alpha x_1' + \alpha_{41}x_4' .$$

Substitution of (5) and (7) in (3) finally leads to

$$\alpha^2(1 - \beta^2) = 1, \qquad \alpha = \frac{1}{\sqrt{1 - \beta^2}}. \tag{8}$$

Thus our system (2a) becomes

$$x_1' = \frac{1}{\sqrt{1 - \beta^2}} (x_1 + i\beta x_4), \qquad x_2' = x_2, \qquad x_3' = x_3,$$

$$x_4' = \frac{1}{\sqrt{1 - \beta^2}} (- i\beta x_1 + x_4) \tag{9}$$

or, in real terms,

$$x' = \frac{1}{\sqrt{1 - \beta^2}} (x - \beta ct), \qquad y' = y, \qquad z' = z,$$

$$t' = \frac{1}{\sqrt{1 - \beta^2}} \left(t - \frac{\beta}{c} x \right). \tag{10}$$

If, in (10), we carry out the transition to the limit

$$c \to \infty, \qquad \beta \to 0, \qquad \text{but} \qquad \beta c = v = \text{finite},$$

we obtain

$$x' = x - vt, \qquad y' = y, \qquad z' = z, \qquad t' = t. \tag{10a}$$

We follow Ph. Frank in calling these equations the *Galilei transformation*. From the standpoint of this transformation group time and space have become "absolute". It takes the place of the Lorentz group only when

$$v \ll c, \qquad \text{i.e.} \qquad \beta \ll 1. \tag{10b}$$

The universally accepted notation $\beta = v/c$ may recall the β-rays, which have a velocity comparable with c, so that the Galilei group is not applicable to them.

Eqs. (10) signify that the two systems (x, t) and (x', t') move with respect to each other with the velocity $v = \beta c$. If we consider a particular point $x' = $ const, we find for it from (10)

$$x - \beta ct = x - vt = \text{const}.$$

The primed system thus progresses along the positive x-axis, which coincides with the x'-axis, with the velocity v. The two other axes y' and z' displace themselves, in space, with the same velocity v, remaining parallel to the axes y and z. For $t = 0$ the "moving system" and the "system at rest" coincide.[1]

Problem III.1 will treat the somewhat more general case that the relative motion of the two systems will not be along the x-axis, but for example in some other direction lying in the xy-plane.

B. The Relative Nature of Time

From Eqs. (10) and (10a) we see that the course of time is absolute only in the limit $c \to \infty$, whereas for finite c it depends on the frame of reference of the observer: The "primed" observer measures a different time than the "unprimed" observer.

This becomes obvious if we return from the real representation (10) to the complex representation (9), but nevertheless plot the entities occurring

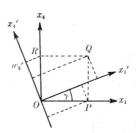

FIG. 37. The system x_1, x_4 is transformed into the system x_1', x_4' by the imaginary angle of rotation γ. The two events R, Q which are simultaneous in t then receive different coordinates x_4', just as the two events P, Q with the same x-coordinate have different coordinates x_1'.

there as real quantities (Fig. 37). We may then write, as in plane analytic geometry,

$$x_1' = x_1 \cos \gamma + x_4 \sin \gamma, \qquad x_4' = -x_1 \sin \gamma + x_4 \cos \gamma \qquad (11)$$

with

$$\cos \gamma = \frac{1}{\sqrt{1 - \beta^2}}, \qquad \sin \gamma = \frac{i\beta}{\sqrt{1 - \beta^2}}, \qquad \tan \gamma = i\beta. \quad (11a)$$

The first Eq. (11), as is well known, signifies the projection of the broken line OPQ on the x_1'-axis, the second that on the x_4'-axis. However, the relative angle of rotation of the systems is here *imaginary* and similarly also $\sin \gamma$ and $\tan \gamma$; $\cos \gamma$ is actually a hyperbolic cosine and hence >1, since $\beta < 1$.

Fig. 37 also shows directly that two "events" (points in space-time) Q

---

[1] It is of course equally permissible to regard the primed system as "system at rest," relative to which the unprimed system moves with the velocity v in the direction of the negative x-axis. This will occur occasionally in §§28 and 33.

and R, which are simultaneous in the unprimed system, are no longer simultaneous in the primed system. This *removal of the sameness in time* (simultaneity) now surprises us no more than the *removal of the sameness of the x-value* of the events Q and P.

Our pseudo-real representation in Fig. 37 will also prove useful in the future and can scarcely lead to misunderstandings. It is true that we deviate herein from the great example of Hermann Minkowski. In his classic lecture "Space and Time" before the Kölner Naturforscher-Gesellschaft in 1908 he calculates throughout with real quantities. What we would call the unit circle $x_1^2 + x_4^2 = 1$ is for him the hyperbola $x^2 - c^2t^2 = 1$; two straight lines which, to us, are perpendicular to each other, then become conjugate diameters of this hyperbola. It need scarcely be emphasized that, in spite of this (only superficial) difference, we have stood on Minkowski's shoulders even in the preceding paragraph and will continue to follow his conception of the theory of relativity.

C. The Lorentz Contraction

This was proposed by H. A. Lorentz even before the theory of relativity as an ad hoc hypothesis to explain the negative result of the Michelson experiment. Deferring discussion of this experiment to Vol. IV, we state Lorentz's hypothesis in the following manner: *To an observer at rest a rod of "intrinsic length" l_0 appears, if moving with uniform velocity v in the direction of its length, shortened to $l = l_0 \sqrt{1 - \beta^2}$.*

Let the rod rest in the moving system x', t' and let its endpoints in this system have the coordinates x'_a, x'_b; their difference is the intrinsic length $l_0 = x'_b - x'_a$. We are here not concerned with the times t'_a, t'_b of the measurements. The situation is different for the observer at rest. He must arrange the measurements of the two ends of the rod so that, from his standpoint, they occur *simultaneously*, i.e. at the same instant $t_a = t_b$. He thus finds the points x_a and x_b and regards their difference, $x_b - x_a$, as the length of the rod l. From the first equation (10) it follows that

$$x'_a = \frac{1}{\sqrt{1 - \beta^2}}(x_a - \beta c t_a), \qquad x'_b = \frac{1}{\sqrt{1 - \beta^2}}(x_b - \beta c t_b),$$

and hence, since $t_a = t_b$, $x_b - x_a = l$, $x'_b - x'_a = l_0$,

$$l_0 = \frac{l}{\sqrt{1 - \beta^2}}, \qquad l = l_0\sqrt{1 - \beta^2}. \tag{12}$$

The hypothetical Lorentz contraction is thus a direct consequence of the Lorentz transformation.

It is not superfluous to interpret this result graphically. The location of the rod in the moving system is represented in Fig. 38 by the strip which is shaded parallel to the x_1'-axis. The observer at rest makes a cut of this strip parallel to the x_1-axis. From the figure its length is

$$l = \frac{l_0}{\cos \gamma}.$$

In our pseudo-real representation it appears longer than the intrinsic length l_0, although in fact it is shorter since $\cos \gamma = (1 - \beta^2)^{-\frac{1}{2}}$.

As a result of this contraction a moving sphere of radius a is flattened into an oblate spheroid with the small axis $b = a\sqrt{1 - \beta^2}$ and with the large axis a. Lorentz based on this his hypothesis of the *deformable electron*, first stated in 1903 as final result of his great paper on electron theory.[1]

The frequently raised question whether the Lorentz contraction is "real" or "apparent" is of course just as idle as the question whether a body "actually" moves. The distinction between a moving system and a system

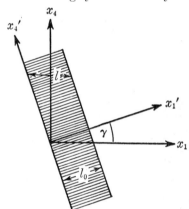

FIG. 38. A rod at rest in the primed system, moving with respect to the unprimed system, is represented by the shaded strip. Its length l in the unprimed system appears longer than its intrinsic length l_0 in the figure, but is in fact shorter, in view of the fact that γ is imaginary: *Lorentz contraction*.

at rest, which we permitted in the preceding for the sake of simplicity of expression, is equally meaningless and arbitrary.

D. The Einstein Dilatation of Time

Let a clock[2] rest in the primed system and mark, by its pendulum swings, the times and successive time differences

$$t_1', t_2', t_3', \cdots, t_2' - t_1' = t_3' - t_2' = \cdots = \tau'.$$

They project themselves in the unprimed system into

$$t_1, t_2, t_3, \cdots, t_2 - t_1 = t_3 - t_2 = \cdots = \tau.$$

In Fig. 39 τ seems shortened as compared with τ', but is actually expanded in view of

[1] Enzyklopaedie der Math. Wiss., Vol. V_2 . Pages 277–279 of this article are forerunners of the theory of relativity.

[2] Instead of speaking, with Einstein, of a clock we may adhere to electromagnetic patterns by thinking of a tuned circuit and its natural period.

$$\tau = \tau' \cos \gamma = \frac{\tau'}{\sqrt{1 - \beta^2}}. \tag{13}$$

This is realized most simply analytically if Eq. (10) is inverted, in which v simply changes sign, as may be verified by calculation. Thus

$$x = \frac{1}{\sqrt{1 - \beta^2}} (x' + \beta ct'), \qquad t = \frac{1}{\sqrt{1 - \beta^2}} \left(t' + \frac{\beta x'}{c} \right). \tag{14}$$

Furthermore, since $x' = $ const, the second of these equations yields for the successive differences $t_2 - t_1$, $t_3 - t_2$, $\cdots$; $t_2' - t_1'$, $\cdots$

$$\tau = \frac{\tau'}{\sqrt{1 - \beta^2}}.$$

Such a clock is realized in a rapidly moving atom which emits a mono-chromatic spectral line, e.g. a hydrogen canal ray. Einstein regarded the expected red shift of the spectral line as the crucial experiment of the

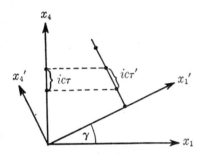

FIG. 39. The period of oscillation of the moving clock τ' appears shortened in the figure for an observer at rest in the x_1, x_4-system to τ, is however actually lengthened in view of the imaginary character of γ: *Einstein's time dilatation.*

theory of relativity and spoke of a "transversal Doppler effect," considering observation at 90° relative to the canal rays. Ives has shown, however, that the observation may as well be carried out at an arbitrary angle, preferably a small angle, where the primary light of the canal rays can be compared with light reflected by a mirror from the (oppositely directed) canal rays. We then observe, e.g. for the H_α line of the hydrogen canal rays, in addition to the primary light, which is shifted toward the blue, the reflected light, which is shifted toward the red. The arithmetic mean of the two wave-lengths does not, however, coincide with the spectral line H_α of the atom at rest, but is displaced from it by the relativistic red shift independently of the direction of viewing. The experiment[1] fully confirms Einstein's expectation.

[1] H. E. Ives and G. R. Stillwell, J. Optical Soc. Am. *28*, 215, 1938; H. E. Ives, J. Optical Soc. Am. *29*, 183 and 294, 1939. G. Otting, Munich thesis, Phys. Z. *40*, 681, 1939. There is a difference in the theoretical interpretation of the American papers and the simultaneous German thesis which is notable in view of the times (1939!):

A *radioactive sample* also has an intrinsic time τ' in the form of its mean life. Hence, observed in the canal ray, it should have a *longer life* than at rest. This experiment is realized under the most favorable circumstances (β nearly equal to 1) in the *meson* disintegration of cosmic rays. Rasetti found for the life of mesons which had become trapped in an absorber and hence were practically at rest the value $\tau' \cong 1.5 \cdot 10^{-6}$ sec, determining the time difference between the incidence of the meson and the appearance of the secondary electron produced in the disintegration. On the other hand, absorption measurements on the mesons of cosmic rays lead to a most probable range of the order of 20 km. In the (unprimed) time measure of the terrestrial observer this corresponds to a mean life $\tau \cong 20$ km/$c \cong 7 \cdot 10^{-5}$ sec. The time expansion thus has here the enormous value[1]

$$\frac{\tau}{\tau'} \cong \frac{7 \cdot 10^{-5}}{1.5 \cdot 10^{-6}} \cong 50.$$

We hence find for the velocity of the mesons, by (13),

$$\frac{1}{\sqrt{1 - \beta^2}} = 50, \qquad v = c\left(1 - \frac{1}{5000}\right).$$

This consideration is confirmed by the experimental determination of the energy of the mesons; for the most commonly occurring mesons approximately 50 times the rest energy is found, which fully agrees with the dependence of the kinetic energy on the velocity (see §32).

E. The Addition Theorem for the Velocity

Two velocities v_1 and v_2 having the same direction do not combine relativistically according to the rule

$$v = v_1 + v_2 .$$

We have instead

$$v = \frac{v_1 + v_2}{1 + \dfrac{v_1 v_2}{c^2}}. \tag{15}$$

Here v_2 is the velocity with which a point 2 moves relative to a body 1, which itself moves in the same direction with the velocity v_1. When Einstein in 1905 proposed this formula, it naturally aroused surprise. It becomes entirely reasonable, however, when we note that we are here dealing with the composition of two Lorentz transformations, and that each of them, according to Fig. 37, denotes a rotation.[2] Let γ_2 be the angle of rota-

---

the American papers seek to retain the concept of the absolute ether, while the German paper assumes the relativistic standpoint from the very beginning.

[1] W. Heisenberg, Vorträge über kosmische Strahlen, Springer, 1943, pp. 78 ff.

[2] The two rotations γ_1 and γ_2 as well as their resultant γ take place about the same "axis", i.e. the R_2 perpendicular to the $x_1 x_4$-plane (see p. 219, footnote 2).

tion which, by Eq. (11a), pertains to v_2, and γ_1 that pertaining to v_1. The result of the composition of the two rotations is a rotation through the angle

$$\gamma = \gamma_1 + \gamma_2. \tag{15a}$$

Thus the *angles* of rotation are added, not their *tangents*. For the latter we have instead

$$\tan \gamma = \frac{\tan \gamma_1 + \tan \gamma_2}{1 - \tan \gamma_1 \tan \gamma_2}.$$

By (11a) this leads to

$$\beta = \frac{\beta_1 + \beta_2}{1 + \beta_1 \beta_2}, \tag{15b}$$

which agrees with (15). The addition theorem for the velocities is hence in essence merely the addition formula for the tangent function.

The same formula may be obtained quite readily, though more indirectly, by superposing the two Lorentz transformations. They may be written, e.g., in the form of Eqs. (14):

$$\sqrt{1 - \beta_1^2}\, x = x_1 + \beta_1 c t_1, \qquad \sqrt{1 - \beta_2^2}\, x_1 = x_2 + \beta_2 c t_2,$$

$$\sqrt{1 - \beta_1^2}\, t = t_1 + \frac{\beta_1}{c} x_1, \qquad \sqrt{1 - \beta_2^2}\, t_1 = t_2 + \frac{\beta_2}{c} x_2. \tag{15b}$$

Elimination of x_1, t_1 then yields for the direct transition from x, t to x_2, t_2

$$\frac{\sqrt{1 - \beta_1^2}\,\sqrt{1 - \beta_2^2}}{1 + \beta_1 \beta_2} x = x_2 + \frac{\beta_1 + \beta_2}{1 + \beta_1 \beta_2} c t_2,$$

$$\frac{\sqrt{1 - \beta_1^2}\,\sqrt{1 - \beta_2^2}}{1 + \beta_1 \beta_2} t = t_2 + \frac{\beta_1 + \beta_2}{1 + \beta_1 \beta_2} \frac{x_2}{c}. \tag{15c}$$

This is again a Lorentz transformation of the form (14) if we put

$$\beta = \frac{\beta_1 + \beta_2}{1 + \beta_1 \beta_2}, \qquad \sqrt{1 - \beta^2} = \frac{\sqrt{1 - \beta_1^2}\,\sqrt{1 - \beta_2^2}}{1 + \beta_1 \beta_2}. \tag{15d}$$

The first of these formulas agrees with (15a); the second follows from it as may be verified by a simple calculation.

For small velocities ($v_1 \ll c$ and $v_2 \ll c$) (15) of course passes over into the elementary superposition formula.

F. c as Upper Limit for All Velocities

If by repeated superposition of velocities the resultant approaches the velocity of light, the further addition of any arbitrary velocity is without effect. In fact we have, by (15b), for $\beta_1 \cong 1$:

$$\beta \cong \frac{1 + \beta_2}{1 + \beta_2} \cong 1.$$

The velocity of light c can only be approached, never exceeded. Even a cyclotron or betatron, which operates with continuous increases in velocity, cannot yield velocities greater than that of light.

We shall define our statement more precisely. To begin with, we obviously mean by "velocity" "relative velocity". But that does not suffice. Consider sample of radium. It emits electrons with almost the velocity of light. Two electrons which fly off simultaneously in opposite directions have very nearly the relative velocity $2c$, viewed from the laboratory in which the sample of radium is at rest. However, in order to properly define relative velocity as used in our statement we must view one electron from the other. Then and then only the seemingly paradoxical equation $c + c = c$ applies. We are thus concerned, in our statement, with the *relative velocity of a moving point with respect to a reference system which is transformed to a state of rest.*

The moving point need not be a material point; it may also be a *process resulting in material changes.* Such a process is called a *signal* and we then speak of the *signal velocity.* If this should ever exceed c, the whole time sequence would be disturbed (see below). In wireless telegraphy and radar a bundle of electromagnetic waves serves as signal; a *monochromatic* wave, on the other hand, constitutes no signal, since a purely periodic wave has neither beginning nor end. Its velocity of propagation is hence not governed by our statement. In fact, we found for wave guides, in §24, phase velocities $/h$ which were greater than c. Similarly, we will see in Vol. IV that phase velocities greater than c may occur in the anomalous dispersion of light waves. There are also quite trivial processes with velocities exceeding that of light, which, then, obviously cannot serve as signals. An example is the intersection of the edge of a ruler with a straight line with a very acute angle. If we displace the ruler at right angles to itself even with only moderate velocity, the intersection will move along the straight line with a velocity exceeding that of light, provided only that the angle has been chosen small enough.

A formal indication of the prohibition of $v > c$ is evidently given already by the Lorentz transformation in the form (10), since here $\sqrt{1 - \beta^2}$ and consequently also x' and t' would become imaginary.

. Light Cone; Space-Like Vectors and Time-Like Vectors; Intrinsic Time

The four-dimensional form

$$\sum_{i=1}^{4} x_i^2 = 0, \quad \text{in real terms } r^2 - c^2 t^2 = 0, \tag{16}$$

is characteristic for the metric of the Lorentz transformations. It represents, in three dimensions, a sphere expanding with the velocity of light, in four dimensions, a conic R_3 with rotational symmetry about the t-axis.

We call it, with Minkowski, the *light cone*. The interior is called the *forecon* and the *aftercone,* depending on whether $t < 0$ or $t > 0$.

All four-vectors leaving the origin which lie outside of the light con are called *space-like,* those which lie inside of it, *time-like.* Thus the vector in the equatorial plane of the light cone is space-like, whereas all permitte velocities leaving the origin are time-like.

The sequence of all four-dimensional positions assumed by a movin material point is called its *world line.* The world line of a point at rest parallel to the t-axis. All world lines passing through the origin lie in th aftercone for $t > 0$, in the forecone for $t < 0$.

We consider the *element of a world line*

$$ds = \sqrt{\sum_{i=1}^{4} dx_i^2}.$$

As the distance between two neighboring world points, it is Lorentz-in variant (see p. 213). The same applies to *Minkowski's intrinsic time*

$$d\tau = \frac{ds}{ic} = \sqrt{dt^2 - \frac{1}{c^2}(dx^2 + dy^2 + dz^2)}$$

$$= dt\sqrt{1 - \frac{v^2}{c^2}} = dt\sqrt{1 - \beta^2}.$$ (17

We will now define the *four-vector of the velocity* along a world line. Th form

$$\frac{dx}{dt}, \frac{dy}{dt}, \frac{dz}{dt}, \quad ic = (\mathbf{v}, ic)$$

would not be a permissible definition, since dt has no invariant meanin This does not apply, however, to

$$\mathbf{V} = \frac{dz}{d\tau}, \frac{dy}{d\tau}, \frac{dz}{d\tau}, \quad ic\frac{dt}{d\tau} = \frac{dt}{d\tau}(\mathbf{v}, ic).$$ (18

The square of its length is

$$\mathbf{V}\cdot\mathbf{V} = \frac{dx^2 + dy^2 + dz^2 - c^2 dt^2}{d\tau^2} = -c^2,$$ (18a

i.e. in fact an invariant which, furthermore, has the same value for al velocity vectors $\mathbf{V}$. The *four-vector of the acceleration* should be defined corre spondingly by

$$\mathbf{W} = \frac{d\mathbf{V}}{d\tau} = \frac{d^2x}{d\tau^2}, \frac{d^2y}{d\tau^2}, \frac{d^2z}{d\tau^2}, \quad ic\frac{d^2t}{d\tau^2}.$$ (18b

t is, in the four-dimensional sense, perpendicular to the four-vector of the
elocity; differentiation of (18a) with respect to τ leads to:

$$\mathbf{V} \cdot \mathbf{W} = 0. \tag{18c}$$

H. The Addition Theorem for Velocities of Different Directions

Velocities of the *same* direction were to be combined in such fashion that
their *angles* were added. Since in their elementary meaning angles denote
rcs on the unit circle, their addition is equivalent to the joining of arcs
n a circle which in our case, it is true, has the radius i instead of 1. The
ormulas of *plane* trigonometry could be applied to the joining of these
rcs.

In order to combine velocities with *different* directions it is necessary
o pass from the circle to the sphere, i.e. from the formulas of plane to those
f *spherical* trigonometry, and for a sphere of radius i, not of radius 1.
Combination of the velocities v_1 and v_2 to form the resultant v is therefore
quivalent to the combination of the angles γ_1 and γ_2 to the resultant angle
, i.e. the construction of a spherical triangle with the sides γ_1, γ_2, and γ.
f α is the angle of inclination of v_2 relative to v_1, then α appears as the
xternal angle between the sides γ_1 and γ_2 in the spherical triangle. We then
ave by the cosine law (see Problem I.4):

$$\cos \gamma = \cos \gamma_1 \cos \gamma_2 - \sin \gamma_1 \sin \gamma_2 \cos \alpha. \tag{19}$$

This is the desired *generalized addition theorem*. For $\alpha = 0$ we obtain $\cos \gamma$
$= \cos(\gamma_1 + \gamma_2)$; $\gamma = \gamma_1 + \gamma_2$, i.e. the earlier Eq. (15a). In view of the rela-
ion $\cos \gamma = (1 - \beta^2)^{-\frac{1}{2}}$ etc. (19) is equivalent with the rather untransparent
ormula

$$\beta^2 = \frac{\beta_1^2 + \beta_2^2 + 2\beta_1\beta_2 \cos \alpha - \beta_1^2 \beta_2^2 \sin^2 \alpha}{(1 + \beta_1 \beta_2 \cos \alpha)^2}. \tag{19a}$$

which was given already by Einstein. In Problem III.2 it will be proved
analytically by application of the Lorentz transformation.

The introduction of our sphere of radius i may seem an arbitrary trick;
actually, it is merely an expression of the fact that the arcs γ_1, γ_2 which
we must combine are imaginary, according to (11a).

We shall mention one more interesting result, which may be read off
directly on Fig. 40: In the theory of relativity the sequence of differently
oriented velocities is *not exchangeable*; the result of the combination of
v_1 and v_2 *differs* from that of the combination of v_2 and v_1. Though the
magnitude of the resultant is the same, the *direction* differs. The difference
n the two directions increases as the velocities increase; in fact, as we
shall know, it is equal to the spherical excess of the spherical triangle formed
n our construction.

In Fig. 40 the angle α between v_1 and v_2 has been chosen equal to $\pi/2$

to simplify the drawing and the arc γ_1 corresponding to v_1 has been placed on the equator of the sphere. The extension of the arc γ_2 then passes through the northpole N. If, on the other hand, starting from the same point A we first record γ_2 (denoted by γ_2' on the figure), perpendicular to the equator and with its extension also passing through N, we must draw through the endpoint B' of γ_2' a great circle perpendicular to the meridian AB' and measure off on it $\gamma_1' = \gamma_1 = AB$. The point A' located in this manner does not coincide with C; instead, the connecting arcs AC and $C'A'$ enclose a certain angle ε. In view of the equality of the two triangles ABC and $A'B'C'$ we have here $\sphericalangle BAC = \sphericalangle B'A'C'$ and $\sphericalangle ACB = \sphericalangle A'C'B'$. If we call

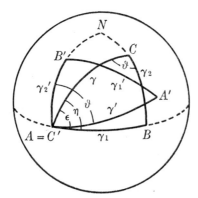

Fig. 40. Combination of two different directed velocities v_1 and v_2 to form the resultant v, corresponding to the circular arcs γ_1, γ_2, and γ on a sphere of radius i. For convenience in representation the angle between v_1 and v_2 has been set equal to $\pi/2$. The figure shows the non-commutative character of the components: v_1, $v_2 = AB$ $\neq v_2$, $v_1 = C'B'A$; the angle ε between A and $C'A'$ is equal to the spherical excess of the triangle ABC (and that of the triangle $A'B'C'$ which is congruent to it).

these two angles η and ϑ, we see that the right angle at A is formed by η, ϑ, and ε in the following manner:

$$\frac{\pi}{2} = \eta + \vartheta - \varepsilon.$$

Hence

$$\varepsilon = \eta + \vartheta - \pi/2 = \eta + \vartheta + \pi/2 - \pi. \tag{20}$$

ε thus is in fact the spherical excess of our right spherical triangle ABC and the congruent triangle $A'B'C'$. (The same applies for a general spherical triangle.)

The limiting case $\gamma_1 = \gamma_2 = \pi/2$, where the two triangles ABC and $A'B'C'$ become equal to the same spherical octant, is particularly simple. Here the resultants are evidently perpendicular to each other and, in view of $\eta = \vartheta = \pi/2$, the spherical excess is also $\pi/2$.

J. The Principles of the Constancy of the Velocity of Light and of Charge

Einstein in 1905 expressly added the first of these principles to the principle of relativity as an empirical postulate. It states that the velocity

propagation of light is independent of the state of rest or motion of the
emitting body. This principle is already included in the original formulation
of our world geometry insofar as we have demanded the universal validity
of the Maxwell equations. Like the velocity, the spherical propagation of
the light is invariant in the transition from $x_1 \cdots x_4$ to $x_1' \cdots x_4'$. The
Lorentz transformation does not change the light sphere into a light ellipsoid, but leaves it a light *sphere*. (This does not apply to the *wave-length*
of the light, which is not invariant but is known to depend on the frame of
reference of the observer: Doppler effect.)

In the earlier but long since discarded theory of the universal *ether*, the
dependence of the light wave from the state of motion of the emitting
body was readily understood: once transferred to the ether, it propagates
itself in accord with the (elastic or electromagnetic) properties of this
medium. Constancy of the velocity of light was here equivalent with field
action. The same does not apply for a mechanical emission theory such as
that surmised by Newton. Here a transfer of the velocity of the emitting
body to the emitted light particles seems almost unavoidable.[1] We may say:
the constancy of the velocity of light is today the only valid *remnant of
the ether concept*. If at present we should speak of an ether, we would have
to assign a separate ether to every frame of reference, i.e. speak e.g. of a
primed and an unprimed ether. We now regard Lenard's "absolute ether
(Uräther)" merely as a freak and the Aristotelian and scholastic "quintes-
sence" (the fifth element, added to fire, water, air, and earth) as an historical
curiosity. Thus in parts I and II, we have almost never spoken of the
ether", but used instead the not readily misinterpreted word "vacuum".

The principle of *constant charge* is as important as that of the constancy
of the velocity of light. The charge is the same for every frame of reference.
This is not obvious, but follows from the Maxwell equations if we can claim
their universal validity for all frames of reference. On the other hand the
principle of the constancy of mass with change of the system of reference,
formerly regarded as obvious, cannot be upheld, as we shall see presently.
*The charge is an absolute invariant with respect to Lorentz transformations;
mass and*, as we shall also see, *energy, are not*.

Summarizing the content of this and the preceding section we may say:
From the standpoint of the Maxwell equations the theory of relativity is
obvious. A mathematician whose eyes had been trained by Klein's Erlangen
program could have read from the form of the Maxwell equations its transformation group along with all its kinematic and optical consequences.

---

[1] The fact that Newton's emission theory could in a sense, experience a resurrec-
tion in the present theory of the light quanta rests solely on the addition theorem of
the theory of relativity according to which effectively $c + v = c$ (c = velocity of light
quanta, v = velocity of the emitting body).

§28. Preparation for the Electron Theory

Maxwell had directed attention away from the charges and toward th lines of force. Since the discovery of the electron and Helmholtz's earlie remarks[1] on the atomism of electricity, interest has once more returne to the sources of the lines of force, the electrons and ions. H. A. Lorent has created the secure mathematical basis for this new electrodynamic (which might be called *electron dynamics*). The judgment exhibited by hir here is remarkable; he introduced only concepts which retained thei substance in the later theory of relativity. We will abbreviate our treat ment by inverting the historical development and basing the electro theory on the theory of relativity.

Unlike Maxwell, Lorentz does not recognize a host of media differin electrically and magnetically; all events take place in a single uniforn medium, the vacuum. The different properties of matter arise simply fron the varied binding and state of motion of the electrons and ions. In di electrics the electrons are bound to ions, in conductors they are more o less freely mobile, and in magnetic materials we are dealing with electron which, as the result of their spin, are aligned in the magnetic field.

In this explanation of the electromagnetic properties of matter we hav the simultaneous action of great numbers of electrons, i.e. a *statistics o electrons*. We shall treat this subject in greater detail in Vol. V. In th present volume we must limit ourselves to the *theory of the individua electron*. It is true that the basic question regarding the *nature* of the electro will remain unclarified. *The electron is a stranger in electrodynamics*, a Einstein has said on occasion. We cannot comprehend, from the electro dynamic standpoint, how the finite electron charge e, concentrated in point or in a very small volume, can cohere stably in spite of the Couloml forces between parts of the charge. For a solution of this problem we mus look to a *general theory of the elementary particles*, the electron, proton neutron, neutrino, positron, meson (and other elementary particles whicl are yet to be discovered). It is clear, however, that such a theory is at th moment still remote.

[1] In his Faraday Lecture in 1881: "If we accept atoms for the chemical element we cannot avoid concluding that also both positive and negative electricity is sub divided into certain elementary quanta which behave like atoms of electricity."

[2] In his book "Versuch einer Theorie der elektrischen und optischen Erschein ungen in bewegten Körpern," Leyden, 1895; unaltered reprinting, Teubner, 1906 See also the later "Theory of Electrons," Teubner, 1909. Emil Wiechert reached th same conclusions and formulas independently of Lorentz at almost the same time ir "The Theory of Electrodynamics and Röntgen's Discovery," Abh. der Physikalisch ökonomischen Gesellschaft zu Königsberg.

A. The Transformation of the Electric Field. Introduction to The Lorentz Force

In order to calculate in the most elementary fashion, i.e. only with four-vectors, we return to the four-potential Ω, which transforms itself like the coordinate vector. We employ Eqs. (27.11) which, applied to Ω, take the form

$$\Omega_1' = \cos\gamma\,\Omega_1 + \sin\gamma\,\Omega_4\,, \qquad \Omega_2' = \Omega_2\,, \qquad \Omega_3' = \Omega_3\,,$$

$$\Omega_4' = -\sin\gamma\,\Omega_1 + \cos\gamma\,\Omega_4\,. \tag{1}$$

The same equations (27.11), solved for x, yield

$$x_1 = \cos\gamma\,x_1' - \sin\gamma\,x_4'\,, \qquad x_2 = x_2'\,, \qquad x_3 = x_3'\,,$$

$$x_4 = \sin\gamma\,x_1' + \cos\gamma\,x_4'\,. \tag{2}$$

We concern ourselves first with E_y' and corresponding to (26.9), form

$$\text{Curl}_{24}'\,\Omega' = \frac{\partial\Omega_4'}{\partial x_2'} - \frac{\partial\Omega_2'}{\partial x_4'}. \tag{3}$$

From (1) and (2) we obtain

$$\frac{\partial\Omega_4'}{\partial x_2'} = \frac{\partial\Omega_4'}{\partial x_2} = -\sin\gamma\frac{\partial\Omega_1}{\partial x_2} + \cos\gamma\frac{\partial\Omega_4}{\partial x_2}, \tag{3a}$$

$$\frac{\partial\Omega_2'}{\partial x_4'} = \frac{\partial\Omega_2}{\partial x_1}\frac{\partial x_1}{\partial x_4'} + \frac{\partial\Omega_2}{\partial x_4}\frac{\partial x_4}{\partial x_4'} = -\sin\gamma\frac{\partial\Omega_2}{\partial x_1} + \cos\gamma\frac{\partial\Omega_2}{\partial x_4} \tag{3b}$$

and as the difference of the right sides of (3a, b)

$$\cos\gamma\left(\frac{\partial\Omega_4}{\partial x_2} - \frac{\partial\Omega_2}{\partial x_4}\right) + \sin\gamma\left(\frac{\partial\Omega_2}{\partial x_1} - \frac{\partial\Omega_1}{\partial x_2}\right)$$

$$= \cos\gamma\,\text{Curl}_{24}\,\Omega + \sin\gamma\,\text{Curl}_{12}\,\Omega. \tag{4}$$

This is at the same time the right side of (3). We hence have

$$\text{Curl}_{24}'\,\Omega' = \cos\gamma\,\text{Curl}_{24}\,\Omega + \sin\gamma\,\text{Curl}_{12}\,\Omega. \tag{5}$$

By Eq. (26.11) we conclude therefore

$$-iE_y' = \cos\gamma\,(-iE_y) + \sin\gamma\,(cB_z). \tag{6}$$

In view of (27.11a) we may write instead

$$E_y' = \frac{E_y - \beta cB_z}{\sqrt{1-\beta^2}}. \tag{6a}$$

A corresponding calculation yields

$$E_z' = \frac{E_z + \beta cB_y}{\sqrt{1-\beta^2}}. \tag{6b}$$

The calculation for the x-component is somewhat more complicated insofar as it leads first to 8 terms, of which 4 are multiplied with $\sin \gamma \cos \gamma$, and two each with $\sin^2 \gamma$ and $\cos^2 \gamma$, respectively. The first cancel each other, whereas the remaining ones may be reduced to

$$\frac{\partial \Omega_1}{\partial x_4} - \frac{\partial \Omega_4}{\partial x_1}$$

and yield simply

$$E'_x = E_x . \tag{7}$$

So as to remove the distinctive treatment of the x-axis we indicate by the subscripts $\|$ and $\perp$ the direction parallel and perpendicular to the relative motion of the two systems. The Eqs. (7) and (6a, b) then become

$$E'_{||} = E_{||}, \qquad E'_\perp = \left(\frac{\mathbf{E} + \mathbf{v} \times \mathbf{B}}{\sqrt{1 - \beta^2}} \right)_\perp . \tag{8}$$

Since $(\mathbf{v} \times \mathbf{B})_{||} = 0$ we may write instead

$$E'_{||} = (\mathbf{E} + \mathbf{v} \times \mathbf{B})_{||}, \qquad E'_\perp = \left(\frac{\mathbf{E} + \mathbf{v} \times \mathbf{B}}{\sqrt{1 - \beta^2}} \right)_\perp . \tag{8a}$$

The quantity $\mathbf{E} + \mathbf{v} \times \mathbf{B}$, which appears here automatically, when multiplied with e, has the dimension "newton" and is called the *Lorentz force*

$$\mathbf{K} = e(\mathbf{E} + \mathbf{v} \times \mathbf{B}). \tag{9}$$

Through its formulation (more precisely, the formulation of the force density $\mathbf{k}$ to be introduced presently) Lorentz put an end to the fruitless discussions of the older theory with regard to the ponderomotive forces on moving charges. In spite of its amazing simplicity Eq. (9) represents the sum total of the forces acting in arbitrary electromagnetic fields. An experiment of W. Wien on hydrogen canal rays confirms this directly.[1] After J. Stark had demonstrated the splitting of the Balmer lines in an electric field, Wien could produce qualitatively the same effect by letting a magnetic field corresponding to the electric field act on the rays. He thus replaced $\mathbf{E}$ by the $\mathbf{v} \times \mathbf{B}$ which is equivalent to it.

It may incidentally be noted that $\mathbf{v} \times \mathbf{H}$ is commonly written in place of $\mathbf{v} \times \mathbf{B}$ in (9); from our dimensional standpoint this is an absurdity.

B. The Magnetic Analog to the Lorentz Force

We must now calculate $\mathbf{B}'$, i.e. the space-space components of the curl of Ω', instead of the preceding space-time components. This becomes very

---

[1] Preuss. Akad., January 1914.

simple for the component in the direction of motion. In view of (1) and (2) it becomes

$$\text{Curl}'_{23}\ \Omega' = \frac{\partial \Omega'_3}{\partial x'_2} - \frac{\partial \Omega'_2}{\partial x'_3} = \frac{\partial \Omega_3}{\partial x_2} - \frac{\partial \Omega_2}{\partial x_3} = \text{Curl}_{23}\ \Omega.$$

By Eq. (26.11) this leads directly to

$$B'_x = B_x. \tag{10}$$

For B'_y we proceed as before for E'_y, noting that

$$\text{Curl}'_{31}\ \Omega' = \frac{\partial \Omega'_1}{\partial x'_3} - \frac{\partial \Omega'_3}{\partial x'_1} = \frac{\partial}{\partial x_3}\left\{\cos\gamma\ \Omega_1 + \sin\gamma\ \Omega_4\right\} - \frac{\partial \Omega_3}{\partial x'_1},$$

$$\frac{\partial \Omega_3}{\partial x'_1} = \left\{\cos\gamma\ \frac{\partial}{\partial x_1} + \sin\gamma\ \frac{\partial}{\partial x_4}\right\}\Omega_3, \quad \text{so that}$$

$$\text{Curl}'_{31}\ \Omega = \cos\gamma\left(\frac{\partial \Omega_1}{\partial x_3} - \frac{\partial \Omega_3}{\partial x_1}\right) + \sin\gamma\left(\frac{\partial \Omega_4}{\partial x_3} - \frac{\partial \Omega_3}{\partial x_4}\right)$$

$$= \cos\gamma\ \text{Curl}_{31}\ \Omega + \sin\gamma\ \text{Curl}_{34}\ \Omega.$$

From this we obtain by Eqs. (26.11) and (27.11a):

$$B'_y = \frac{B_y + \beta E_z/c}{\sqrt{1 - \beta^2}}. \tag{10a}$$

Similarly we find

$$B'_z = \frac{B_z - \beta E_y/c}{\sqrt{1 - \beta^2}}. \tag{10b}$$

These formulas (10) may be generalized vectorially to

$$B'_{||} = (\mathbf{B} - \mathbf{v} \times \mathbf{E}/c^2)_{||}, \qquad B'_{\perp} = \left(\frac{\mathbf{B} - \mathbf{v} \times \mathbf{E}/c^2}{\sqrt{1 - \beta^2}}\right)_{\perp}.$$

The quantity $\mathbf{B} - \mathbf{v} \times \mathbf{E}/c^2$ appearing in (11) is at the same time the ponderomotive force on a magnetic pole of strength 1, i.e. the *magnetic analog* of the *Lorentz force* exerted by the field on the charge 1.

C. The Intrinsic Field of an Electron in Uniform Motion

In a frame of reference x, y, z which moves with the electron the intrinsic field of the electron is electrostatic in character. Thus, for $r = \sqrt{x^2 + y^2 + z^2}$:

$$\mathbf{E} = -\frac{e}{4\pi\varepsilon_0}\ \text{grad}\ \frac{1}{r} \quad \text{and}\quad \mathbf{B} = 0. \tag{12}$$

For an observer at rest, with respect to whom the electron moves in the direction of the negative x-axis (see footnote at the end of §27A), we then have in view of (6a, b), (7), (10), and (10a, b)

$$E'_x = E_x \qquad E'_y = \frac{1}{\sqrt{1 - \beta^2}} E_y, \qquad E'_z = \frac{1}{\sqrt{1 - \beta^2}} E_z.$$

$$B'_x = 0, \qquad B'_y = \frac{v}{c^2\sqrt{1 - \beta^2}} E_z, \qquad B'_z = - \frac{v}{c^2\sqrt{1 - \beta^2}} E_y. \tag{12a}$$

We express these primed fields in terms of the primed coordinates x', y', z' of the point of the field considered, which, like the x, y, x, we shall measure from the momentary position of the electron and consider the Lorentz contraction along the x-coordinate:

$$x' = \sqrt{1 - \beta^2}x, \qquad y' = y, \qquad z' = z. \tag{13}$$

Simultaneously we set $s(x', y', z') \equiv r(x, y, z)$ or

$$s = \sqrt{\frac{x'^2}{1 - \beta^2} + y'^2 + z'^2}. \tag{13a}$$

We then obtain from (12) and (12a)

$$E'_x, \quad E'_y, \quad E'_z = \frac{e}{4\pi\varepsilon_0\sqrt{1 - \beta^2}} \frac{x', y', z'}{s^3}, \tag{14}$$

$$B'_x, \quad B'_y, \quad B'_z = \frac{ev}{4\pi\varepsilon_0 c^2\sqrt{1 - \beta^2}} \frac{0, z', -y'}{s^3}. \tag{14a}$$

Thus our primed observer, unlike one moving with the electron, is aware of a *magnetic* field in addition to the electric field. By (14a) its lines of force are circles about the direction of motion; its intensity is, if we replace $\varepsilon_0 c^2$ in the denominator of (14a) by μ_0 in the numerator and pass over from the field strength $\mathbf{B}$ to the excitation $\mathbf{H} = \mathbf{B}/\mu_0$,

$$|\mathbf{H}| = \frac{ev}{4\pi\sqrt{1 - \beta^2}} \frac{\sin \vartheta}{s^2}, \qquad \sin \vartheta = \frac{\sqrt{y'^2 + z'^2}}{s}. \tag{14b}$$

This expression should be compared with the expression (15.12) for the Biot-Savart force, from which (14b) differs only by relativistic corrections of the second order in β. Thus in a sense a moving electron in a cathode ray realizes the commonly mentioned, but unreal, current element of the earlier theory; ev here takes the place of I ds.

The electric lines of force, on the other hand, are according to (14) straight lines diverging from the instantaneous position of the electron in the primed system (in view of the proportionality of the components of $\mathbf{E}'$ in (14) with x', y', z') as well as in the unprimed system; however, they do not have the same density in all directions in the former as in the latter case. Rather, they are squeezed together in the equatorial plane $x' = 0$. Because of the meaning of s in (13a), $s \to \infty$ and $\mathbf{E}' \to 0$ for $\beta \to 1$ *unless* $x' = 0$. In this limiting case the electric field would be concentrated entirely in the equatorial plane. Thus the electron is *flattened* in the limit $v \to c$ not only in respect to its shape (with which we are not concerned here), but also in respect to its field.

In the preceding we have convinced ourselves that the determination of the field for uniform motion is merely a matter of *algebraic transformation*, while in the older electrodynamics it involved at least some integration.[1] §30 will deal with the field of accelerated motion.

We emphasize in general: The electric and magnetic fields form a single *unit* and can be distinguished only with reference to the particular reference system employed. Together they form a six-vector. In changing the frame of reference its electric components contribute to the magnetic ones and vice-versa. We are here dealing with an *effect of perspective in four dimensions*. The aspect of a cube furnishes the three-dimensional analog: For a particular choice of the viewing direction we see only the ("electric") front face, for other, oblique, directions the ("magnetic") lateral faces as well.

D. An Invariant Approach to the Lorentz Force; the Four-Vector of the Force Density

From the four-vector $x_1 \cdots x_4$ we obtain as the difference in position of two neighboring world points the *four-vector*

$$dx_1, \, dx_2, \, dx_3, \, ic \, dt. \tag{15}$$

Furthermore the four-dimensional volume element

$$dx_1 \, dx_2 \, dx_3 \cdot ic \, dt \tag{15a}$$

is also independent of the choice of coordinates, just like the volume element $dx_1 dx_2 dx_3$ in three dimensions. Since the corresponding charge Δe (the number of electrons contained in the element of volume), just like e itself, is also *invariant*, division of Δe by (15a) leads to another invariant scalar and multiplication of this scalar by (15) to another four-vector. As in (26.6) we call it the four-current density $\boldsymbol{\Gamma}$:

$$\boldsymbol{\Gamma} = \frac{\Delta e}{dx_1 \, dx_2 \, dx_3} \left(\frac{dx_1}{dt}, \frac{dx_2}{dt}, \frac{dx_3}{dt}, ic \right) = \rho(\mathbf{v}, ic). \tag{16}$$

ρ is the usual three-dimensional charge density. A comparison of the preceding definition of $\boldsymbol{\Gamma}$ with that in (26.6) shows that the current density $\mathbf{J}$

---

[1] See Oliver Heaviside, Phil. Mag. 1889. The surface $s = $ const (Eq. (13a)) is known as the Heaviside ellipsoid; see Problem III.3.

of electrodynamics passes over into the *convection current* density $\rho\mathbf{v}$ in the electron theory and, furthermore, that the four-vector $\boldsymbol{\Gamma}$ has the same direction as the velocity four-vector $\mathbf{V}$ defined in (27.18), in view of the relation

$$\boldsymbol{\Gamma} = \rho\frac{d\tau}{dt}\mathbf{V} = \rho\sqrt{1 - \beta^2}\mathbf{V}. \tag{16a}$$

We now multiply the four-vector $\boldsymbol{\Gamma}$ with the six-vector $\mathbf{F}$ of the field. This results, by the process of "reduction", again in a four-vector, just as for the operation **Div** in (26.16). After having divided it by c, for dimensional reasons, we call it *force density* and denote it by $\mathbf{k}$, its nth component by k_n :

$$\mathbf{k} = \frac{1}{c}\boldsymbol{\Gamma}\cdot\mathbf{F}, \qquad ck_n = \sum_{m=1}^{4}\Gamma_m F_{nm}, \qquad n = 1, 2, 3, 4. \tag{17}$$

Written term by term this becomes

$$\begin{aligned}
ck_1 &= & \Gamma_2 F_{12} + \Gamma_3 F_{13} + \Gamma_4 F_{14} , \\
ck_2 &= \Gamma_1 F_{21} & + \Gamma_3 F_{23} + \Gamma_4 F_{24} , \\
ck_3 &= \Gamma_1 F_{31} + \Gamma_2 F_{32} & + \Gamma_4 F_{34} , \\
ck_4 &= \Gamma_1 F_{41} + \Gamma_2 F_{42} + \Gamma_3 F_{43} & ;
\end{aligned} \tag{17a}$$

or, in three-dimensional coordinates,

$$\left.\begin{aligned}
k_1 &= k_x = \rho(v_y B_z - v_z B_y + E_x) \\
k_2 &= k_y = \rho(v_z B_x - v_x B_z + E_y) \\
k_3 &= k_z = \rho(v_x B_y - v_y B_x + E_z)
\end{aligned}\right\} = \rho(\mathbf{E} + \mathbf{v} \times \mathbf{B}). \tag{17b}$$

In his original theory Lorentz operates primarily with the three-dimensional vector on the right.

We are of course also interested in the fourth component. It is, by (17a),

$$ck_4 = i\rho(v_x E_x + v_y E_y + v_z E_z) = i\rho\mathbf{v}\cdot\mathbf{E} = i\rho L. \tag{17c}$$

L is here the *power* expended by the electric field strength on a unit charge moving with velocity $\mathbf{v}$.

It may be shown readily from the representation (17a) that the four-vector $\mathbf{k}$ is perpendicular to the world line of the charge. In view of the proportionality of $\boldsymbol{\Gamma}$ and $\mathbf{V}$ in (16a) and the antisymmetric character of F we obtain for the scalar product of $\mathbf{k}$ and $\mathbf{V}$:

$$\mathbf{V} \cdot \mathbf{k} = 0. \tag{17d}$$

We pass from the *force density* to the *force* itself. It is not permissible here, however, to simply change ρ into the electronic charge e by integration over space, since the three-dimensional volume element is not a relativistic invariant, but an arbitrary section through the "world tube" described by the electron (perpendicular to the also arbitrarily chosen time axis). It is much more appropriate to place the section perpendicular to the *world line* of the electron, which is independent of the orientation of the t-axis, or, what is the same thing, perpendicular to the generatrices of the mantel surface of the world tube. If we denote the angle between the world line and t-axis by γ, the three-dimensional $dx\,dy\,dz$ projects itself into the "world tube cross section"

$$dx\,dy\,dz \cos \gamma = \frac{dx\,dy\,dz}{\sqrt{1 - \beta^2}} \tag{18}$$

with the general meaning of γ given by Eq. (27.11a). By integration over this cross section we obtain

$$\int \rho\,dx\,dy\,dz \cos \gamma = \frac{e}{\sqrt{1 - \beta^2}}. \tag{18a}$$

From the representation (17b) for the force density we thus find directly

$$\int \frac{\mathbf{k}\,dx\,dy\,dz}{\sqrt{1 - \beta^2}} = \frac{e}{\sqrt{1 - \beta^2}}\,(\mathbf{E} + \mathbf{v} \times \mathbf{B}) = \mathbf{K}/\sqrt{1 - \beta^2}. \tag{19}$$

$\mathbf{K}$ is the *Lorentz force* of Eq. (9). *It is not directly a part of a four-vector, but becomes one after division by $\sqrt{1 - \beta^2}$.*

The corresponding fourth *energetic component* of this four-vector is according to (17c)

$$\int \frac{k_4\,dx\,dy\,dz}{\sqrt{1 - \beta^2}} \frac{ei}{c\sqrt{1 - \beta^2}}\,\mathbf{v}\cdot\mathbf{E} = \frac{eiL}{c\sqrt{1 - \beta^2}}. \tag{19a}$$

We call the *four-vector of the force*, completed in this manner, $\mathbf{F}$; Its four components may be expressed collectively by

$$\mathbf{F} = \left\{ \frac{\mathbf{K}}{\sqrt{1 - \beta^2}},\quad \frac{eiL}{\sqrt{1 - \beta^2}} \right\}. \tag{19b}$$

In view of its derivation from the Lorentz force density $\mathbf{k}$ and of the relation (17d) it is *perpendicular* to the world line of the electron:

$$\mathbf{V}\cdot\mathbf{F} = 0. \tag{19c}$$

E. The General Orthogonal Transformation of a Tensor of the Second Rank

As generalization of the antisymmetric field tensor F we now consider an arbitrary (symmetric or asymmetric) tensor of the second rank T_{nm}, whose components T_{nn} need not vanish and for which we do not necessarily have $T_{nm} = T_{mn}$. Here we define as tensor a quantity whose components T_{nn} and T_{nm} behave like the squares and products x_n^2 and $x_n x_m$ of the four-dimensional coordinates in the orthogonal transformation (27.1). The formula which, by (27.1), applies for the product $x_n x_m$:

$$x_n' x_m' = \sum_{i=1}^{4} \sum_{k=1}^{4} \alpha_{ni} \alpha_{mk} x_i x_k$$

may be transferred to T in the following manner:

$$T_{nm}' = \sum_{i=1}^{4} \sum_{k=1}^{4} \alpha_{ni} \alpha_{mk} T_{ik} . \tag{20}$$

We shall encounter a symmetric tensor in §31. It remains symmetric in the transformation.

For an *antisymmetric* tensor (20) may be written

$$T_{nm}' = \sum_{i>k} \sum (\alpha_{ni} \alpha_{mk} - \alpha_{mi} \alpha_{nk}) T_{ik} = \sum_{i>k} \sum \begin{vmatrix} \alpha_{ni} & \alpha_{nk} \\ \alpha_{mi} & \alpha_{mk} \end{vmatrix} T_{ik} . \tag{20a}$$

Here the components $T_{ki} = -T_{ik}$ are already accounted for with the components T_{ik} . Hence the double sum in (20a) must be carried out in such fashion that, whereas i traverses all values from 1 to 4, k only assumes the values $k < i$. The value $k = i$ evidently need not be considered since $T_{ii} = 0$ and, in addition, since all the determinants in (20a) vanish in this case. An antisymmetric tensor remains antisymmetric in the transformation since the interchange of n and m reverses the sign of all the determinants.

After these general considerations we return once more to the behavior of the six-vector in the special Lorentz transformation (27.2a). We convince ourselves that all the subdeterminants of this matrix vanish with the exception of

$$\begin{vmatrix} \alpha_{11} & \alpha_{12} \\ \alpha_{21} & \alpha_{22} \end{vmatrix} = \begin{vmatrix} \alpha_{11} & \alpha_{13} \\ \alpha_{31} & \alpha_{33} \end{vmatrix} = \begin{vmatrix} \alpha_{22} & \alpha_{24} \\ \alpha_{42} & \alpha_{44} \end{vmatrix} = \begin{vmatrix} \alpha_{33} & \alpha_{34} \\ \alpha_{43} & \alpha_{44} \end{vmatrix} = \frac{1}{\sqrt{1 - \beta^2}},$$

$$\begin{vmatrix} \alpha_{22} & \alpha_{23} \\ \alpha_{32} & \alpha_{33} \end{vmatrix} = \begin{vmatrix} \alpha_{11} & \alpha_{14} \\ \alpha_{41} & \alpha_{44} \end{vmatrix} = 1,$$

$$\begin{vmatrix} \alpha_{12} & \alpha_{14} \\ \alpha_{22} & \alpha_{24} \end{vmatrix} = \begin{vmatrix} \alpha_{13} & \alpha_{14} \\ \alpha_{33} & \alpha_{34} \end{vmatrix} = - \begin{vmatrix} \alpha_{21} & \alpha_{22} \\ \alpha_{41} & \alpha_{42} \end{vmatrix} = - \begin{vmatrix} \alpha_{31} & \alpha_{33} \\ \alpha_{41} & \alpha_{43} \end{vmatrix} = \frac{-i\beta}{\sqrt{1 - \beta^2}}$$

in view of the values of the α_{ik} given by Eqs. (27.5, 7, 8). On this basis the sums of six terms in (20a) are reduced to one or two terms. We find specifically

$$T'_{12} = \frac{1}{\sqrt{1 - \beta^2}}\,(T_{12} - i\beta T_{24}). \qquad T'_{23} = T_{23},$$

$$T'_{31} = \frac{1}{\sqrt{1 - \beta^2}}\,(T_{31} + i\beta T_{34}),$$

$$T'_{14} = T_{14}, \qquad T'_{24} = \frac{1}{\sqrt{1 - \beta^2}}\,(T_{24} + i\beta T_{12}),$$

$$T'_{34} = \frac{1}{\sqrt{1 - \beta^2}}\,(T_{34} - i\beta T_{31}).$$

If here we substitute for the T_{ik} the electromagnetic equivalent of the F_{ik} given by the matrix arrangement (26.17) we see readily that the preceding transformation formulas correspond to Eqs. (8) and (11). The present procedure for the derivation of these equations may be slower than the earlier one, but it is fundamentally more elementary and certainly more general, since it covers any tensors of the second rank.

§*29. Integration of the Differential Equation of the Four-Potential*

We now turn to the differential equation (26.5),

$$\Box \boldsymbol{\Omega} = -\mu_0 \boldsymbol{\Gamma}, \tag{1}$$

where we are dealing with a problem of *four-dimensional potential theory*. The three-dimensional potential theory, Eqs. (7.4a) and (7.5), may serve us as example. To begin with, we require the four-dimensional analog to Newton's potential $1/r$. It is given by

$$\mathbf{U} = \frac{1}{R^2}, \qquad R^2 = (\xi_1 - x_1)^2 + (\xi_2 - x_2)^2 + (\xi_3 - x_3)^2 + (\xi_4 - x_4)^2. \tag{2}$$

As proof we calculate

$$\frac{\partial}{\partial \xi_i} \frac{1}{R^2} = -2\,\frac{\xi_i - x_i}{R^4}, \qquad \frac{\partial^2}{\partial \xi_i^2} \frac{1}{R^2} = -\frac{2}{R^4} + \frac{8}{R^6}\,(\xi_i - x_i)^2;$$

From this we deduce

$$\Box \mathbf{U} = \sum_{i=1}^{4} \frac{\partial^2}{\partial \xi_i^2} \frac{1}{R^2} = -\frac{8}{R^4} + \frac{8R^2}{R^6} = 0, \tag{3}$$

which is valid for all points except the "source point" $\xi_i = x_i$, $i = 1, \cdots 4$. It may be proved similarly that in a space of $p + 2$ dimensions the centrally symmetric potential is represented by $\mathbf{U} = R^{-p}$ if the meaning of R^2 is generalized correspondingly.

Furthermore, as a preliminary, we shall determine the "surface" of the sphere $R = $ const, i.e. a three-dimensional structure in four-dimensional space. If ω signifies the surface area of the unit sphere, it is

$$\omega R^3 \quad \text{with} \quad \omega = 2\pi^2. \tag{4}$$

In proof we consider the integral

$$J = \iiiint_{-\infty}^{\infty} \exp\left\{ -\sum_{i=1}^{4} \xi_i^2 \right\} d\xi_1 \, d\xi_2 \, d\xi_3 \, d\xi_4. \tag{4a}$$

By carrying out the integral for each coordinate separately and utilizing the familiar value of the Laplace integral we obtain the fourth power of $\sqrt{\pi}$ or π^2. On the other hand, if we introduce polar coordinates, with $r^2 = \sum \xi_i^2$, we find

$$J = \omega \int_0^{\infty} e^{-r^2} r^3 \, dr = \frac{\omega}{2}. \tag{4b}$$

A comparison of (4a) and (4b) proves Eq. (4). Similarly, for space with $p + 2$ dimensions

$$\omega = 2\pi^{p/2+1}/\Gamma(p/2 + 1). \tag{4c}$$

(The reader may check the validity of this formula in the three-dimensional and two dimensional cases, $p = 1$ and $p = 0$).

A. Four-Dimensional Form of the Potential Ω

We now apply Green's theorem to our two potentials Ω and $\mathbf{U}$:

$$\int (\Omega \ \square \ \mathbf{U} - \mathbf{U} \ \square \ \Omega) \, d\xi_1 \cdots d\xi_4 = \int \left(\Omega \frac{\partial \mathbf{U}}{\partial n} - \mathbf{U} \frac{\partial \Omega}{\partial n} \right) d\sigma. \tag{5}$$

The integration at the left is to be extended over infinite four-dimensional space, with the exclusion of the source point $x_i = \xi_i$ by means of a sphere K of radius $R \to 0$. The integration on the right is to be carried out over this sphere K and a sphere $R \to \infty$, which however, as in the three-dimensional case (see p. 39), does not contribute to the integral. Substitution from the differential equations (1) and (3) then yields for (5) ($\mathbf{\Gamma}$ now denotes the four-current, not the Γ-function):

$$\mu_0 \int \mathbf{\Gamma} \frac{d\xi_1 \cdots d\xi_4}{R^2} = \int_K \Omega \frac{\partial}{\partial n} \frac{1}{R^2} \, d\sigma - \int_K \frac{\partial \Omega}{\partial n} \frac{d\sigma}{R^2}. \tag{5a}$$

Since by Eq. (4) $\int d\sigma = 2\pi^2 R^3$, the second integral on the right vanishes for $R \to 0$. Since furthermore $dn = -dR$ (the normal is to be taken positive if it points outward from the space of integration, i.e. inward into the sphere K),

$$\int_K \Omega \frac{2}{R^3} \, d\sigma = 4\pi^2 \Omega.$$

Ω here denotes the value of our potential at $R = 0$, i.e. for $\xi_i = x_i$. Hence we find from (5a)

$$4\pi^2\,\Omega(x_1,\,x_2,\,x_3,\,x_4)/\mu_0 \;=\; \int \mathbf{\Gamma}\,\frac{d\xi_1 \cdot\cdot\; d\xi_4}{R^2}, \tag{6}$$

in perfect analogy to (7.5). (6) *represents the four-potential in an arbitrary world point* $x_1 \cdots x_4$ *by a four-dimensional integration over the four-current density* $\mathbf{\Gamma}$, *which is assumed to be known.*

However, $\mathbf{\Gamma}$ is known to us only *for the real times* $\tau < t$, *which precede the time of observation* t; we might also say, for the times $\tau < 0$, if, without loss of generality, we put the time of observation t temporarily equal to 0.

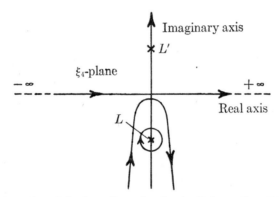

FIG. 41. Integration of the four-dimensional potential equation $\Box\Omega = \mu_0\mathbf{\Gamma}$. Deformation of the original path of integration along the real ξ_4-axis into a loop about the "light point" L on the negative imaginary axis.

Accordingly $\mathbf{\Gamma}$ is not known to us along the real ξ_4 axis, as we have implicitly assumed till now, but only for the negative imaginary values

$$\xi_4 = ic\tau = -ic\,|\,\tau\,| \,. \tag{6a}$$

Accordingly we shall distort the path of integration for ξ_4 along the real axis, $-\infty < \xi_4 < \infty$ into a loop about the negative imaginary half-axis, which leads from $-i\infty$ by way of the neighborhood of the origin of the complex ξ_4-plane back to $-i\infty$, as shown in Fig. 41. This does not alter our representation (6) or the fact that (6) satisfies our differential equation (1).

We still want to convince ourselves that (6) satisfies also the auxiliary condition (26.7), Div $\Omega = 0$. This follows from the fact that $\mathbf{\Gamma}$ satisfies the continuity equation (26.16b) Div $\mathbf{\Gamma} = 0$. In fact, if we indicate the differentiation with respect to x_i and ξ_i by subscripts, and carry it out under the fourfold integral sign we find

$$4\pi^2 \text{ Div } \mathbf{\Omega}/\mu_0 = \int \left(\mathbf{\Gamma} \text{ Grad}_x \frac{1}{R^2} \right) d\xi_1 \cdots d\xi_4$$

$$= -\int \left(\mathbf{\Gamma} \text{ Grad}_\xi \frac{1}{R^2} \right) d\xi_1 \cdots d\xi_4, \tag{7}$$

i.e., after carrying out an integration by parts,

$$4\pi^2 \text{ Div } \mathbf{\Omega}/\mu_0 = \int \text{ Div}_\xi \mathbf{\Gamma} \frac{d\xi_1 \cdots d\xi_4}{R^2} = 0, \tag{7a}$$

which was to be proved.

For the further treatment of the representation (6) we can carry out first either the integration with respect to ξ_4 or that with respect to ξ_1, ξ_2, and ξ_3. For the present we shall follow the first course.

B. Retarded Potentials

With reference to Fig. 41, we look for those points of the complex ξ_4-plane at which the denominator R^2 vanishes. We write

$$R^2 = r^2 + (x_4 - \xi_4)^2, \tag{8}$$

where r signifies the three-dimensional distance between the point of integration ξ_1, ξ_2, ξ_3 and the reference point x_1, x_2, x_3, and where we have dropped our temporary convention $x_4 = 0$, which merely served the more convenient description of Fig. 41.

There are two points at which $R^2 = 0$, i.e.

$$x_4 - \xi_4 = +ir \tag{8a}$$

and

$$x_4 - \xi_4 = -ir. \tag{8b}$$

We call the first, with Minkowski, the "light point" L; the second[1] is designated with L' in the figure.

In the neighborhood of L we have according to (8) and (8a)

$$R^2 = (x_4 - \xi_4 - ir)(x_4 - \xi_4 + ir) \cong 2ir(x_4 - \xi_4 - ir). \tag{8c}$$

By Cauchy's theorem we can now distort the path of integration in Fig. 41 into a circuit about L, yielding by the method of residues

$$\oint \mathbf{\Gamma} \frac{d\xi_4}{R^2} = \frac{\mathbf{\Gamma}_L}{2ir} \oint \frac{d\xi_4}{x_4 - \xi_4 - ir} = \frac{\mathbf{\Gamma}_L}{2ir} (+2i\pi) = \frac{\pi \mathbf{\Gamma}_L}{r}. \tag{9}$$

$\mathbf{\Gamma}_L$ is the value of $\mathbf{\Gamma}$ at the light point. The factor $(+2\pi i)$ results from the fact that, on the one hand, $-\xi_4$ occurs in the denominator, on the other,

---

[1] The distortion of the original real path of integration into a loop about L' would lead to the "advanced" instead of the retarded potentials (see p. 148).

the integration path is traversed clockwise about L, i.e. in the negative direction from a function-theoretical standpoint.

Substituting (9) in (6) we find

$$4\pi\Omega/\mu_0 = \int \frac{\Gamma_L}{r}\, d\xi_1\, d\xi_2\, d\xi_3. \tag{10}$$

Resolved into components this yields, by (26.4) and (26.6)

$$4\pi\mathbf{A}/\mu_0 = \int \frac{\mathbf{J}_L}{r}\, d\xi_1\, d\xi_2\, d\xi_3, \qquad 4\pi\varepsilon_0\Psi = \int \frac{\rho_L}{r}\, d\xi_1\, d\xi_2\, d\xi_3. \tag{10a}$$

These are, however, exactly the representations of the *retarded potentials* in Eq. (19.13). In fact our present $\mathbf{J}_L$ and ρ_L have the same meaning as our earlier $[\mathbf{J}]$ and $[\rho]$ in §19. For if we designate the time of the light point, which precedes that of the observation, by τ, as in (6a), we find from (8a)

$$ict = ic\tau + ir, \qquad \tau = t - r/c. \tag{10b}$$

This is however exactly the time defined in (19.13c), for which $[\mathbf{J}]$ and $[\rho]$ were to be calculated. In this manner the formerly suppressed proof of (19.13) has been brought in a mathematically particularly appropriate fashion. It should be noted that G. Herglotz had devised the method given here even before the theory of relativity, just on the basis of mathematical symmetry and elegance.[1]

C. The Lienard-Wiechert Approximation

We now take the second course mentioned above and carry out the integration over ξ_1, ξ_2, ξ_3. We here imagine current and charge to be concentrated in a single point, the *electron*, rather than spacially distributed as up to now. We use for Γ its electron-theory value (28.16), by which the conduction current $\mathbf{J}$ was interpreted as convection current, and with e as electron charge, obtain from it

$$\int \Gamma\, d\xi_1\, d\xi_2\, d\xi_3 = e(\mathbf{v}, ic) = -e\dot{\mathbf{R}}. \tag{11}$$

$\mathbf{R}$ is the radius vector from the electron to the reference point, $\dot{\mathbf{R}}$ its derivative with respect to t for fixed reference point:

$$\dot{\mathbf{R}} = -\left(\frac{d\xi_1}{dt}, \frac{d\xi_2}{dt}, \frac{d\xi_3}{dt}, ic\right) = -(\mathbf{v}, ic). \tag{11a}$$

We then obtain from (6), carrying out the first three integrations,

$$4\pi^2\Omega/\mu_0 = -e \oint \frac{\dot{\mathbf{R}}}{R^2}\, d\xi_4. \tag{12}$$

[1] See Göttinger Nachr., 1904.

As in (9), the integration is to be carried out about the light point in Fig. 41. However, the locus of the electron ξ_1, ξ_2, ξ_3 is not an independent point of integration, as up to now, but itself depends on the integration variable ξ_4. We must therefore consider the *world line of the electron in the neighborhood of the light point* L and expand R^2 as follows:

$$R^2 = R_L^2 + (\xi_4 - \xi_{4L})\frac{dR^2}{d\xi_4} + \cdots , \tag{13}$$

so as to be able to apply the method of residues.

Here we have

$$R_L^2 = 0 \quad \text{and} \quad \frac{dR^2}{d\xi_4} = \frac{1}{ic}\frac{d(\mathbf{R}\cdot\mathbf{R})}{dt} = \frac{2}{ic}\dot{\mathbf{R}}\cdot\mathbf{R}.$$

The expansion of R^2 becomes hence

$$R^2 = (\xi_4 - \xi_{4L})\frac{2}{ic}\mathbf{R}\cdot\dot{\mathbf{R}} + \cdots \tag{13a}$$

and Eq. (12) passes over into

$$4\pi^2\Omega/\mu_0 = -\frac{e}{2}\frac{ic\dot{\mathbf{R}}}{\mathbf{R}\cdot\dot{\mathbf{R}}}\oint\frac{d\xi_4}{\xi_4 - \xi_{4L}}. \tag{13b}$$

In view of the sign of ξ_4 in the denominator, which is opposite to that in (9), the integral is now equal to $-2\pi i$. We thus obtain

$$4\pi\Omega/\mu_0 = -\frac{ec\dot{\mathbf{R}}}{\mathbf{R}\cdot\dot{\mathbf{R}}}. \tag{14}$$

According to (8a) the vector $\mathbf{R}$ has the time component ir, whereas its space component (light point to reference point) is $\mathbf{r}$; by (11a) the time and space components of $\dot{\mathbf{R}}$ are $-ic$ and $-\mathbf{v}$. Hence

$$\mathbf{R}\cdot\dot{\mathbf{R}} = rc - \mathbf{v}\cdot\mathbf{r} = rc\left(1 - \frac{\mathbf{v}\cdot\mathbf{r}}{rc}\right) = rc\left(1 - \frac{v_r}{c}\right), \tag{14a}$$

where v_r denotes the projection of $\mathbf{v}$ on the direction of $\mathbf{r}$. Substitution in (14) and separation into real and imaginary parts yields the remarkably simple formulas of Lienard (1898) and Wiechert (1900):

$$4\pi\mathbf{A}/\mu_0 = \frac{e}{r}\frac{\mathbf{v}}{1 - v_r/c}, \qquad 4\pi\varepsilon_0\Psi = \frac{e}{r}\frac{1}{1 - v_r/c}. \tag{15}$$

Our derivation shows that, like r, $\mathbf{v}$ and v_r must be taken for the earlier time of the light point. It is interesting to note that the denominator $1 - v_r/c$ will recur in Vol. IV in connection with the Doppler effect.

Actually the original integral form (6) of the four-potential will prove more useful for what follows than the formulas (15) or (10a), where the integration has been carried out.

§30. The Field of the Accelerated Electron

The advantage of Eq. (29.6) rests in the fact that the variables $x_1 \cdots x_4$ of the reference point occur here only in the denominator R^2. We have to differentiate only the latter if we wish to calculate the field of an *electron in any state of motion*. In this manner we obtain from (29.6) first:

$$' \; 4\pi^2 \, \mathrm{Curl}_{nm} \, \Omega \; = \; \mu_0 \int \left(\Gamma_m \frac{\partial}{\partial x_n} \frac{1}{R^2} - \Gamma_n \frac{\partial}{\partial x_m} \frac{1}{R^2} \right) d\xi_1 \cdots d\xi_4. \qquad (1)$$

Since

$$\frac{\partial}{\partial x_n} \frac{1}{R^2} = -2 \frac{x_n - \xi_n}{R^4} = -\frac{2R_n}{R^4},$$

the parenthesis in (1) becomes

$$- \frac{2}{R^4} \left(\Gamma_m R_n - \Gamma_n R_m \right) = + \frac{2}{R^4} \left(\mathbf{\Gamma} \times \mathbf{R} \right)_{nm}.$$

Here we have transferred the usual symbol ($\times$) of the three-dimensional vector product to the product of our two four-vectors, which evidently is a quantity with six components. The same applies for the left side of Eq. (1), where by (26.12) $\mathrm{Curl}_{nm} \, \Omega$ is the nm-component of the six-vector $\mu_0 f$. We thus obtain from (1)

$$2\pi^2 f_{nm} = \int \frac{1}{R^4} \left(\mathbf{\Gamma} \times \mathbf{R} \right)_{nm} d\xi_1 \cdots d\xi_4. \qquad (1a)$$

We now carry out the integration with respect to ξ_1, ξ_2, and ξ_3, in which process, by Eq. (29.11), $\mathbf{\Gamma}$ transforms itself into $-e\dot{\mathbf{R}}$, and ξ_4 refers, from this point on, to the point electron. We find

$$2\pi^2 f = -e \oint \frac{(\dot{\mathbf{R}} \times \mathbf{R})}{R^4} d\xi_4. \qquad (2)$$

The integration is here to be carried out over a circuit about the light point, as in Fig. 41. The difference from the previous calculations consists only in the fact that the denominator now vanishes to the second order, so that we have to carry the expansion in denominator and numerator one term further. If we abbreviate

$$\frac{\xi_4 - \xi_{4L}}{ic} = u,$$

we write in place of (29.13a)

$$R^2 = 2u\,\dot{\mathbf{R}}\cdot\mathbf{R} + u^2\{\ddot{\mathbf{R}}\cdot\mathbf{R} + \dot{\mathbf{R}}\cdot\dot{\mathbf{R}}\} + \cdots$$

$$R^4 = 4u^2(\dot{\mathbf{R}}\cdot\mathbf{R})^2\left(1 + u\,\frac{\ddot{\mathbf{R}}\cdot\mathbf{R} + \dot{\mathbf{R}}\cdot\dot{\mathbf{R}}}{\dot{\mathbf{R}}\cdot\mathbf{R}} + \cdots\right).$$

and, since $\dot{\mathbf{R}} \times \dot{\mathbf{R}} = 0$,

$$\mathbf{R} \times \dot{\mathbf{R}} = (\mathbf{R} \times \dot{\mathbf{R}})_L + u(\mathbf{R} \times \ddot{\mathbf{R}})_L + \cdots$$

Thus, if at this point we transfer the denominator in part to the numerator and suppress the subscript L (2) becomes:

$$2\pi^2 f = \frac{eic}{4(\dot{\mathbf{R}}\cdot\mathbf{R})^2} \oint \frac{du}{u^2}(\mathbf{R} \times \dot{\mathbf{R}} + u(\mathbf{R} \times \ddot{\mathbf{R}}))\left(1 - u\,\frac{\ddot{\mathbf{R}}\cdot\mathbf{R} + \dot{\mathbf{R}}\cdot\dot{\mathbf{R}}}{\dot{\mathbf{R}}\cdot\mathbf{R}}\right).$$

Here we need write out only the term multiplied with u^{-1}, since only this is involved in determining the residue, and can omit the terms with u^{-2}, u^0, $u^1 \cdot$. We thus obtain

$$2\pi^2 f = \frac{eic}{4(\dot{\mathbf{R}}\cdot\mathbf{R})^2} \oint \frac{du}{u}\left(\mathbf{R} \times \ddot{\mathbf{R}} - \mathbf{R} \times \dot{\mathbf{R}}\,\frac{\ddot{\mathbf{R}}\cdot\mathbf{R} + \dot{\mathbf{R}}\cdot\dot{\mathbf{R}}}{\dot{\mathbf{R}}\cdot\mathbf{R}}\right).$$

Since the integration indicated in Fig. 41 amounts simply to the addition of the factor $-2\pi i$, we find finally

$$\frac{4\pi f}{ec} = \frac{\mathbf{R} \times \ddot{\mathbf{R}}}{(\dot{\mathbf{R}}\cdot\mathbf{R})^2} - \mathbf{R} \times \dot{\mathbf{R}}\,\frac{\ddot{\mathbf{R}}\cdot\mathbf{R} + \dot{\mathbf{R}}\cdot\dot{\mathbf{R}}}{(\dot{\mathbf{R}}\cdot\mathbf{R})^3}. \tag{3}$$

According to (29.8a) and (29.11a) the expressions on the right must be formulated specifically for

$$\mathbf{R} = (\mathbf{r}, ir), \qquad \dot{\mathbf{R}} = -(\mathbf{v}, ic), \qquad \ddot{\mathbf{R}} = (-\dot{\mathbf{v}}, 0). \tag{3a}$$

We examine this general representation first for the special case of the

A. Electron in Uniform Motion

Here, since $\ddot{\mathbf{R}} = 0$,

$$\frac{4\pi f}{ec} = -\mathbf{R} \times \dot{\mathbf{R}}\,\frac{\dot{\mathbf{R}}\cdot\dot{\mathbf{R}}}{(\dot{\mathbf{R}}\cdot\mathbf{R})^3} = \mathbf{R} \times \dot{\mathbf{R}}\,\frac{c^2 - v^2}{c^3 r^3(1 - v_r/c)^3}. \tag{4}$$

According to (3a) the six-vector $\mathbf{R} \times \dot{\mathbf{R}}$ is given, in matrix notation, by

$$\mathbf{R} \times \dot{\mathbf{R}} = -\begin{bmatrix} r_x & r_y & r_z & ir \\ v_x & v_y & v_z & ic \end{bmatrix}. \tag{5}$$

We calculate its space-space and space-time components as subdeterminants of the matrix. In the notation of ordinary three-dimensional vector calculus we obtain

$$R \times \dot{R} = \begin{cases} \mathbf{v} \times \mathbf{r} & \text{for the space-space components,} \\ i(r\mathbf{v} - c\mathbf{r}) & \text{for the space-time components.} \end{cases} \tag{5a}$$

If this is substituted on the right side of (4) and f is separated into its space-space portion $\mathbf{H}$ and its space-time portion $-ic\mathbf{D}$ on the left as well, we find

$$\frac{4\pi\mathbf{H}}{e} = \mathbf{v} \times \mathbf{r} \frac{1 - v^2/c^2}{r^3(1 - v_r/c)^3},$$

$$-\frac{4\pi\mathbf{D}}{e} = \left(r\frac{\mathbf{v}}{c} - \mathbf{r}\right)\frac{1 - v^2/c^2}{r^3(1 - v_r/c)^3}. \tag{6}$$

These expressions appear basically different from the expressions (28.14) and (28.14a), with which we represented previously the field of the electron in uniform motion (there designated by $\mathbf{H}'$, $\mathbf{E}'$), but can actually be

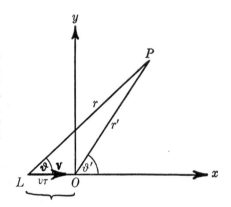

FIG. 42. The field of an electron in uniform motion. The electron moves along the x-axis with the velocity $\mathbf{v}$; O is the location of the electron which is simultaneous with the observation at P, L the light point, so that $LO = v\tau$, where τ is the retarded time of the light signal emitted from L to P.

transformed into each other by elementary geometrical considerations. We will show this in Problem III.3. We will then make use of Fig. 42, which pictures at the same time the different viewpoint of the present and the earlier formulas: In the present formulas $\mathbf{r}$ and r refer to the light point L, in which the electron was at the time $t - r/c$, t being the time coordinate of the reference point P. The earlier formulas, on the other hand, concerned the position of the electron simultaneous with t, which is designated in the figure by O; the coordinates of the reference point with respect to O are given by x', y', z' as in (28.14).

B. The Accelerated Electron

If in (3) we omit the part (4) we obtain the pure "acceleration field"

$$\frac{4\pi f}{ec} = \frac{\mathbf{R} \times \ddot{\mathbf{R}}}{(\dot{\mathbf{R}} \cdot \mathbf{R})^2} - \frac{(\mathbf{R} \times \dot{\mathbf{R}})(\mathbf{R} \cdot \dddot{\mathbf{R}})}{(\dot{\mathbf{R}} \cdot \mathbf{R})^3}. \tag{7}$$

To analyze this, we calculate from (3a)

$$\mathbf{R} \times \ddot{\mathbf{R}} = \begin{cases} -\mathbf{r} \times \dot{\mathbf{v}}, \text{ the space-space portion,} \\ ir\dot{\mathbf{v}}, \quad \text{ the space-time portion} \end{cases} \tag{7a}$$

as well as

$$\mathbf{R} \cdot \ddot{\mathbf{R}} = -\mathbf{r} \cdot \dot{\mathbf{v}}. \tag{7b}$$

We then obtain from (7) with due regard to (5a) and (29.14a)

$$\frac{4\pi \mathbf{H}}{ec} = -\frac{\mathbf{r} \times \dot{\mathbf{v}}}{c^2 r^2 (1 - v_r/c)^2} - \frac{(\mathbf{r} \times \mathbf{v})(\mathbf{r} \cdot \dot{\mathbf{v}})}{c^3 r^3 (1 - v_r/c)^3},$$

$$\frac{4\pi \mathbf{D}}{e} = -\frac{r\dot{\mathbf{v}}}{c^2 r^2 (1 - v_r/c)^2} + \frac{(cr - rv)(\mathbf{r} \cdot \dot{\mathbf{v}})}{c^3 r^3 (1 - v_r/c)^3}. \tag{8}$$

From this we conclude directly

$$r\mathbf{H}/c = \mathbf{r} \times \mathbf{D} \quad \text{and} \quad \mathbf{r} \cdot \mathbf{D} = 0. \tag{8a}$$

H, D, and **r** or, as we might also say, **H, E,** and **r** are mutually perpendicular to each other. Furthermore, taking the absolute value in the first Eq. (8a) in view of the second Eq. (8a), leads to

$$\frac{1}{c} \mid \mathbf{H} \mid = \mid \mathbf{D} \mid \tag{8b}$$

or, expressed in other terms, to

$$\mid \mathbf{H} \mid = \sqrt{\frac{\varepsilon_0}{\mu_0}} \mid \mathbf{E} \mid.$$

We thus have a typical transversal field, as for the plane light wave in Eqs. (6.11) and (6.13). Its strength decreases with increasing r as $1/r$; for the denominators in (8a) have each *one* factor r more than the numerators, not *two* as for the electron in uniform motion (Eq. (6)). Hence at great distances (6) may be neglected as compared to (7), and (7) represents the entire field of the accelerated electron.

C. The Longitudinally Accelerated Electron

Let us assume specifically that **v** and $\dot{\mathbf{v}}$ have the same direction (rectilinear motion, longitudinally accelerated electron); we then see readily that

$$\dot{\mathbf{v}}(\mathbf{r} \cdot \mathbf{v}) = \mathbf{v}(\mathbf{r} \cdot \dot{\mathbf{v}})$$

and hence also

$$(\mathbf{r} \times \dot{\mathbf{v}})(\mathbf{r} \cdot \mathbf{v}) = (\mathbf{r} \times \mathbf{v})(\mathbf{r} \cdot \dot{\mathbf{v}}).$$

If, now, the denominators of the right sides of Eqs. (8) are made the same, two terms cancel each other in each case. These Eqs. (8) then reduce to

$$\frac{4\pi \mathbf{H}}{ec} = -\frac{\mathbf{r} \times \dot{\mathbf{v}}}{c^2 r^2 (1 - v_r/c)^3}, \qquad \frac{4\pi \mathbf{D}}{e} = \frac{-r\dot{\mathbf{v}} + \mathbf{r}\dot{v}_r}{c^2 r^2 (1 - v_r/c)^3}. \qquad (9)$$

If we make the common direction of $\mathbf{v}$ and $\dot{\mathbf{v}}$ the axis $\vartheta = 0$ of a spherical polar coordinate system r, ϑ, φ, we have

$$v_r = v \cos \vartheta, \qquad v_\vartheta = -v \sin \vartheta, \qquad v_\varphi = 0,$$

$$\dot{v}_r = \dot{v} \cos \vartheta, \qquad \dot{v}_\vartheta = -\dot{v} \sin \vartheta, \qquad \dot{v}_\varphi = 0,$$

$$\mathbf{r} \times \dot{\mathbf{v}} = (\mathbf{r} \times \dot{\mathbf{v}})_\varphi, \qquad H = H_\varphi, \qquad D = D_\vartheta$$

and we obtain from (9)

$$4\pi H_\varphi = \frac{e\dot{v}}{cr} \frac{\sin \vartheta}{(1 - \beta \cos \vartheta)^3}, \qquad 4\pi c D_\vartheta = \frac{e\dot{v}}{cr} \frac{\sin \vartheta}{(1 - \beta \cos \vartheta)^3} \qquad (10)$$

These are the same expressions as (19.20), with the addition of the *relativistic denominator* $(1 - \beta \cos \vartheta)^3$, which of course was lacking in the nonrelativistic calculation $(\beta \to 0)$. In fact our earlier factor $\ddot{\mathbf{p}}(t - r/c)$ is the same as our present factor $e\dot{v}$, computed for the light point. Correspondingly we find in place of the radiation S in (19.22)

$$S = \frac{e^2 \dot{v}^2}{16\pi^2 \varepsilon_0 c^3 r^2} \frac{\sin^2 \vartheta}{(1 - \beta \cos \vartheta)^6}. \qquad (11)$$

Accordingly the maximum of the radiation no longer lies at $\vartheta = \pi/2$, but advances, as β approaches 1, from $\vartheta = \pi/2$ toward $\vartheta = 0$.[1] We already referred on p. 155 to this phenomenon, which is characteristic for x-ray theory.

§31. The Maxwell Stresses and the Stress-Energy Tensor

So far we have only dealt with the kinematics of the electron, prescribing its motion and inquiring regarding the accompanying field. We now turn to the *statics* and then to the *dynamics* of the electron. With respect to the

---

[1] By differentiation of (11) with respect to ϑ we obtain as condition for S_{max} a quadratic equation for $\cos \vartheta$, which, for small β, yields

$$\cos \vartheta = 3\beta, \qquad \vartheta = \frac{\pi}{2} - 3\beta,$$

and for β nearly equal to 1,

$$\cos \vartheta = 1 - \frac{1 - \beta^2}{15}, \qquad \vartheta = \sqrt{\frac{1 - \beta^2}{5}}.$$

statics of the electron we have familiarized ourselves till now only with the Lorentz force, acting at the locus of the electron. A field concept cannot be satisfied herewith, however, but must follow up the transfer of force actions in vacuum, where there are no charges. This was Faraday's intimation when he spoke of lines of force as of elastic bands which transmit tension and compression. Maxwell was also here able to place Faraday's notions into clear mathematical focus. This was the origin of Maxwell's *stress tensor*, which may be expanded relativistically into a *stress-energy tensor*.

We proceed from the Lorentz force density in Eq. (28.17),

$$\mathbf{k} = \frac{1}{c} \, \Gamma \cdot F, \qquad ck_n = \sum_{r=1}^{4} \Gamma_r F_{nr}, \tag{1}$$

and replace Γ, in accord with Maxwell's equations (26.16), by the six-vector of the excitation f. We then obtain from (1)

$$ck_n = \sum_{r=1}^{4} \mathbf{Div}_r \, f \cdot F_{nr} = \sum_{r=1}^{4} \sum_{m=1}^{4} \frac{\partial f_{rm}}{\partial x_m} \, F_{nr}. \tag{2}$$

We will show that $\mathbf{k}$ may be expressed as the four-dimensional divergence of a tensor T, i.e. that

$$k_n = \sum_{m=1}^{4} \frac{\partial}{\partial x_m} \, T_{nm} \tag{3}$$

and

$$T_{nm} = -\frac{1}{c} \sum_{r=1}^{4} F_{nr} f_{mr} + \delta_{nm} \Lambda, \tag{4}$$

where Λ denotes the Lagrange density in (26.24).

Since we here enter the domain of tensor quantities the following rather abstract computations with double indices cannot be avoided.

To transform the right side of (2) we utilize the identity

$$\frac{\partial f_{rm}}{\partial x_m} \, F_{nr} = \frac{\partial}{\partial x_m} \, (f_{rm} F_{nr}) - f_{rm} \frac{\partial F_{nr}}{\partial x_m}. \tag{5}$$

Change of the order of summation yields for the first term on the right of (5), summed as indicated in (2),

$$\sum_m \frac{\partial}{\partial x_m} \sum_r f_{rm} F_{nr} = -\sum_m \frac{\partial}{\partial x_m} \sum_r f_{mr} F_{nr}. \tag{6}$$

The second term on the right side of (5) becomes, after carrying out the summation over r and m (including the negative sign)

$$\sum_r \sum_m f_{mr} \frac{\partial F_{nr}}{\partial x_m}.$$

We write this expression once more, reversing both the symbols for the summation subscripts r, m and the sequence of the subscripts of f and F:

$$\sum_r \sum_m f_{mr} \frac{\partial F_{mn}}{\partial x_r}$$

and form half the sum of these equal expressions:

$$\frac{1}{2} \sum_r \sum_m f_{mr} \left(\frac{\partial F_{mn}}{\partial x_r} + \frac{\partial F_{nr}}{\partial x_m} \right). \tag{7}$$

Now we make use of Maxwell's Eq. (26.18). According to it (the three terms are formed by the cyclic interchange of the subscripts m, n, r)

$$\frac{\partial F_{mn}}{\partial x_r} + \frac{\partial F_{nr}}{\partial x_m} + \frac{\partial F_{rm}}{\partial x_n} = 0.$$

Taking care of the negative sign by changing the sequence of subscripts of F, we then can write instead of (7)

$$\frac{1}{2} \sum_r \sum_m f_{mr} \frac{\partial F_{mr}}{\partial x_n}. \tag{7a}$$

This is the result of the summation of the second term on the right side of (5), whereas that for the first term was given by (6). Hence we obtain, finally, from (2), (6), and (7a),

$$ck_n = - \sum_m \frac{\partial}{\partial x_m} \sum_r f_{mr} F_{nr} + \frac{1}{2} \sum_r \sum_m f_{mr} \frac{\partial F_{mr}}{\partial x_n}. \tag{8}$$

Here the first term is already identical with the first half of the representation of the tensor T in (3) and (4). To prove fully the correctness of the representation we must still demonstrate that the second term on the right of (8) is equal to

$$c \sum_{m=1}^{4} \delta_{nm} \frac{\partial \Lambda}{\partial x_m} = c \frac{\partial \Lambda}{\partial x_n}.$$

By (26.24) this is actually the case. The statement (3) is thus proved.

Our expression (4) for T represents a *symmetric tensor of the second rank*. Its symmetry follows directly from the proportionality of f and F and from the meaning of δ_{nm} ; the tensor character in the sense of p. 244 follows from the behavior of the six-vectors f and F in a Lorentz transformation. This calculation, overloaded with indices and formal as it may seem, leads to far-reaching physical consequences.

From (4) we compute the components of T individually, beginning with the diagonal terms of the matrix, all of which have the term with Λ in common. We find

$$T_{11} = -\frac{1}{c}\{f_{12}F_{12} + f_{13}F_{13} + f_{14}F_{14}\} + \Lambda$$

$$= -H_z B_z - H_y B_y + D_x E_x + \tfrac{1}{2}\mathbf{H}\cdot\mathbf{B} - \tfrac{1}{2}\mathbf{D}\cdot\mathbf{E}$$

$$= -\mathbf{H}\cdot\mathbf{B} + H_x B_x + D_x E_x + \tfrac{1}{2}\mathbf{H}\cdot\mathbf{B} - \tfrac{1}{2}\mathbf{D}\cdot\mathbf{E}$$

$$= H_x B_x + D_x E_x - W,$$

where W denotes the energy density. Similarly,

$$T_{22} = H_y B_y + D_y E_y - W,$$

$$T_{33} = H_z B_z + D_z E_z - W.$$

On the other hand

$$T_{44} = -\frac{1}{c}\{f_{41}F_{41} + f_{42}F_{42} + f_{43}F_{43}\} + \Lambda$$

$$= D_x E_x + D_y E_y + D_z E_z + \tfrac{1}{2}\mathbf{H}\cdot\mathbf{B} - \tfrac{1}{2}\mathbf{D}\cdot\mathbf{E}$$

$$= \tfrac{1}{2}\mathbf{D}\cdot\mathbf{E} + \tfrac{1}{2}\mathbf{H}\cdot\mathbf{B} = W.$$

We now turn to the nondiagonal elements, beginning with those having the subscript 4, such as

$$T_{14} = T_{41} = -\frac{1}{c}(f_{42}F_{12} + f_{43}F_{13})$$

$$= ic(D_y B_z - D_z B_y) = -ic\varepsilon_0\mu_0(\mathbf{E}\times\mathbf{H})_x = -\frac{i}{c}S_x.$$

Similarly:

$$T_{24} = T_{42} = -\frac{i}{c}S_y, \qquad T_{34} = T_{43} = -\frac{i}{c}S_z.$$

The remaining nondiagonal elements are

$$T_{12} = T_{21} = -\frac{1}{c}(f_{13}F_{23} + f_{14}F_{24}) = \begin{cases} H_y B_x + D_x E_y = H_y B_x + D_y E_x, \\ H_x B_y + D_z E_y = H_z B_y + D_y E_x, \end{cases}$$

where the equality of the last four expressions again follows from the proportionality of $\mathbf{D}$ and $\mathbf{E}$, and similarly

$$T_{13} = T_{31} = \begin{cases} H_x B_z + D_x E_z, \\ H_z B_x + D_z E_x, \end{cases} \qquad T_{23} = T_{32} = \begin{cases} H_y B_z + D_y E_z, \\ H_z B_y + D_z E_y. \end{cases}$$

Thus the complete T matrix becomes, in abbreviated notation

$$T = \left(\begin{array}{c|c} \sigma & -\dfrac{i}{c}\mathbf{S} \\ \hline -\dfrac{i}{c}\mathbf{S} & W \end{array} \right) \tag{9}$$

Here σ is the *three-dimensional* matrix of the socalled Maxwell stresses:

$$H_x B_x + D_x E_x - W, \quad H_y B_x + D_y E_x, \quad\quad H_z B_x + D_z E_x,$$

$$H_x B_y + D_x E_y, \quad\quad H_y B_y + D_y E_y - W, \quad H_z B_y + D_z E_y, \quad\quad (10)$$

$$H_x B_z + D_x E_z, \quad\quad H_y B_z + D_y E_z, \quad\quad H_z B_z + D_z E_z - W.$$

The electrical portion of σ represents a tension of the magnitude W in the direction of the lines of force and a compression of the same magnitude in the directions perpendicular thereto. This is seen immediately if the x-axis is placed in the direction of $\mathbf{E}$ and $\mathbf{B}$ is put equal to 0. Then

$$\sigma_{xx} = \tfrac{1}{2}\mathbf{D}\cdot\mathbf{E}, \quad \sigma_{yy} = \sigma_{zz} = -\tfrac{1}{2}\mathbf{D}\cdot\mathbf{E}, \quad \sigma_{ik} = 0.$$

The same may be shown for the magnetic lines of force if the x-axis is placed in their direction and $\mathbf{E}$ is put equal to 0. We have thus returned to the model which Faraday had constructed purely on the basis of intuition.

This stress tensor σ, by itself, is however not a legitimate physical quantity in the sense of the theory of relativity. It becomes one only by its extension with the energy quantities $\mathbf{S}$ and W, forming the "stress-energy tensor" T. This has the characteristic property that its "trace" (sum of all the four terms on the principal diagonal) vanishes. We have in fact:

$$T_{11} + T_{22} + T_{33} + T_{44} = \mathbf{H}\cdot\mathbf{B} + \mathbf{D}\cdot\mathbf{E} - 3W + W = 0.$$

We now return to the relationship between T and the Lorentz force density $\mathbf{k}$ as given by (3), and consider first the fourth line of (3). In view of (9) it is

$$k_4 = -\frac{i}{c}\operatorname{div}\mathbf{S} + \frac{\partial W}{\partial x_4}.$$

If we replace k_4 by its value $i\rho L/c$ from (28.17c), we find, since $x_4 = ict$,

$$\frac{\partial W}{\partial t} + \operatorname{div}\mathbf{S} + \rho L = 0, \quad L = \mathbf{v}\cdot\mathbf{E}. \quad\quad (11)$$

This is Poynting's theorem, Eq. (5.7), where the former energy loss by Joule heat is replaced by the work done on the moving charge ρ.

Consider now one of the space components of Eq. (3), e.g. the first line:

$$k_1 = k_x = \operatorname{div}_x \sigma - \frac{i}{c}\frac{\partial S_x}{\partial x_4}$$

which, written out in detail, becomes

$$\frac{\partial \sigma_{xx}}{\partial x} + \frac{\partial \sigma_{yx}}{\partial y} + \frac{\partial \sigma_{zx}}{\partial z} - \frac{1}{c^2}\frac{\partial S_x}{\partial t} = k_x. \quad\quad (12)$$

If we omit the last term on the left, i.e. confine ourselves to a stationary state, we obtain the characteristic equation (8.11) of elastic equilibrium in Vol. II. Just as there the volume force F_x is absorbed and balanced by the stresses σ_{ik}, insofar as they point in the x-direction. The Lorentz force density may be completely replaced by these stresses in our case. They are defined throughout the field by the tensor array (10), even where, in view of the absence of charge density, the Lorentz force is nonexistent. We have thus attained the goal set at the beginning of this section, of following up the transmission of the force through vacuum (without the use of a test body).

However, what do we know of the nonstationary state and the term with $\partial \mathbf{S}/\partial t$ which is then added in (12)? The answer is given by Eq. 14.1 in Vol. II, where the corresponding term, there designated by $-\rho \partial^2 \mathbf{s}/\partial t^2$, represented the *inertial resistance* of unit volume of the elastic body or, with positive sign, its *change in momentum*. We learn from this that there exists a momentum per unit volume $\mathbf{G}$ also in the electromagnetic field, and that it is to be defined, in direction and magnitude, by

$$\mathbf{G} = \frac{1}{c^2}\,\mathbf{S}. \tag{13}$$

We already know that the electromagnetic field possesses *energy* and how this is to be localized in space. We now see that we must also attribute to the field *momentum*, continuously distributed through space wherever there is an energy flux $\mathbf{S}$ and of the same direction with the latter.

Correspondingly a *light wave* carries momentum and exerts a pressure on a nonreflecting (black) body on which it is incident—the *light pressure* discovered by Maxwell. Similarly, if a light wave is emitted by a body, it imparts to it a recoil which is equal and opposite to the momentum carried by it. We call the latter body the "transmitter," the former the "receiver," and assume that both were at rest for $t < 0$. At $t = 0$, when the wave is emitted by the transmitter, the latter receives a recoil. The center of gravity of the transmitter and receiver is then set into motion and remains in motion for the duration $0 < t < T$. At the time $t = T$ the light wave is absorbed by the receiver (without reflection, as we shall assume for the sake of brevity). The wave then imparts to the receiver an equal impulse forward. From this point on the center of gravity of the two bodies is once more at rest, though it has been displaced a certain distance during the interim T, corresponding to the backward motion of the transmitter. This contradiction with the law governing the center of gravity vanishes only if we assign a momentum to the light wave itself during its lifetime T. Then momentum is neither created nor destroyed, both in emission and in absorption, and the center of gravity remains *permanently* at rest.

As is well known, it is difficult to demonstrate the pressure of light in the laboratory. The radiometers constructed for this purpose indicate generally convection currents of residual gases, caused by the thermal effect of the radiation. The proof of the pressure of light in the heavens is much grander. The tails of the comets, pointed away from the sun, show it (Lebedew), also the solar corona, where luminous particles are balanced by the pressure of light or radiation at a height equal to as much as the radius of the sun. The inner constitution of the sun and the bright fixed stars generally is also controlled by the common action of the pressure of radiation and the thermodynamic gas pressure (Karl Schwarzschild, †1916, for the surface of the sun; quite generally, A. S. Eddington, †1944). We must here content ourselves with pointing out some general relationships between momentum, energy, and light pressure.

From the definition of $\mathbf{S}$ we obtain for a transversal plane wave $\left(\mathbf{H} \perp \mathbf{E}, |\mathbf{H}| = \sqrt{\dfrac{\varepsilon_0}{\mu_0}} |\mathbf{E}| \right)$:

$$|\mathbf{S}| = |\mathbf{E} \times \mathbf{H}| = |\mathbf{E}||\mathbf{H}| = \left\{ \begin{array}{l} \sqrt{\dfrac{\varepsilon_0}{\mu_0}}\, \mathbf{E}^2 = \varepsilon_0\, c \mathbf{E}^2 \\[2ex] \sqrt{\dfrac{\mu_0}{\varepsilon_0}}\, \mathbf{H}^2 = \mu_0\, c \mathbf{H}^2 \end{array} \right\} = cW.$$

From this follows, in view of the definition (13) of $\mathbf{G}$

$$|\mathbf{G}| = \frac{W}{c}. \tag{14}$$

The momentum incident on a screen is hence equal in absolute magnitude, but for the factor $1/c$, to the *energy density in front of the screen* (this applies not only for vacuum, but for any non-absorbing medium).

We consider a bundle of parallel rays, a "wave packet", of length l and cross section q. Let W be the energy contained in it, G the momentum contained in it:

$$W = qlW, \qquad G = qlG = gl\,\frac{W}{c} = \frac{W}{c}\,; \tag{15}$$

the last follows from (14). We speak of a "photon" or a "light quantum" if the energy W of the bundle is equal to $h\nu$ (h = Planck's constant, ν = number of vibrations per second). By (15) the momentum of this bundle is

$$G = \frac{h\nu}{c}. \tag{15a}$$

In the theory of light quanta the light pressure is thus identified with a "hail of photons", to which every photon contributes the quantity $h\nu/c$.

We test this statement once again with the aid of the representation (10) of the stress tensor σ. Let the light wave, assumed plane, be incident perpendicularly, in the positive x-direction, on a plate. Because of the transversal nature of light E_x, D_x, B_x, and H_x are zero and the first row of (10) reduces to

$$\sigma_{xx} = -W, \qquad \sigma_{xy} = \sigma_{xz} = 0.$$

If we are dealing with the light bundle described by (15), the force

$$-\sigma_{xx}q = Wq$$

acts on the plate during the time $T = l/c$; its time integral yields the impulse imparted to the plate. We calculate:

$$G = \int_0^T Wq \, dt = \frac{1}{c} \int_0^l Wq \, dl = \frac{Wql}{c} = \frac{W}{c},$$

which agrees with (15).

If the light wave is not incident perpendicularly on the plate, but at an angle α with respect to the normal, the quadratic character of the coefficients in the tensor transformation formula (28.20) leads to

$$\sigma_{xx} = W \cos^2 \alpha.$$

This dependence on angle is reasonable since the area bombarded by the light pencil is now $q/\cos \alpha$ and only the component in the x-direction of the momentum of the light rays is effective as light pressure.

§32. Relativistic Mechanics

Unlike electrodynamics, which fits the requirements of the theory of relativity from the very start so that we could actually base this theory on it, classical mechanics must undergo fundamental revision to harmonize it with the theory of relativity. This revision even affects, for the individual particle, the definition of its *momentum* (its "quantitas motus") as a *four-vector*. In agreement with Vol. I, §2 we assume it to be proportional to the four-vector $\mathbf{V}$ of the velocity in Eq. (27.18) and call it again $\mathbf{G}$:

$$\mathbf{G} = m_0 \mathbf{V}, \qquad \mathbf{V} = \frac{dx_1}{d\tau}, \quad \frac{dx_2}{d\tau}, \quad \frac{dx_3}{d\tau}, \quad \frac{dx_4}{d\tau}. \tag{1}$$

The coefficient m_0 is the *rest mass* of the particle, $d\tau = \sqrt{1 - \beta^2}\, dt$ is the differential of the intrinsic time. We can also write in place of (1)

$$\mathbf{G} = \frac{m_0}{\sqrt{1 - \beta^2}} (\mathbf{v}, ic), \qquad \mathbf{v} = \frac{dx}{dt}, \quad \frac{dy}{dt}, \quad \frac{dz}{dt}. \tag{1a}$$

The quantity

$$m = \frac{m_0}{\sqrt{1 - \beta^2}} \tag{2}$$

is called the *mass in motion*. It is not constant, as in classical mechanics, but increases for $\beta \to 1$, $v \to c$ toward infinity; accordingly it depends on the frame of reference and is hence not a legitimate world entity. This applies not only for the electron, but for every mass—though a large mass cannot be accelerated to velocities close to c in the same manner as an electron. (However, in cosmic rays with their tremendous energies the variation of mass finds expression also for the heavy and semi-heavy particles, the protons and mesons.)

The law of inertia, Newton's *first law*, now becomes in relativistic formulation

$$\mathbf{G} = \text{const.} \tag{3}$$

Correspondingly the *second law* may be written as four-dimensional vector equation as follows:

$$\frac{d\mathbf{G}}{d\tau} = \mathbf{F}. \tag{4}$$

Here $\mathbf{F}$ is the external force, extended to a four-vector. We know that for the individual electron the Lorentz force density $\mathbf{k}$ is such a four-vector, but not the Lorentz force $\mathbf{K}$ itself. The latter becomes one only after it has been divided by $\sqrt{1 - \beta^2}$ and has thus been placed into the invariant relationship with $\mathbf{k}$ which is expressed by Eqs. (28.19) and (28.19a). This leads to the following definition of the four-vector $\mathbf{F}$ for the individual electron in terms of $\mathbf{K}$:

$$F_{1,2,3} = \frac{\mathbf{K}}{\sqrt{1 - \beta^2}} = \frac{e(\mathbf{E} + \mathbf{v} \times \mathbf{B})}{\sqrt{1 - \beta^2}}, \qquad F_4 = \frac{ei}{c\sqrt{1 - \beta^2}} \, \mathbf{v} \cdot \mathbf{E}. \tag{4a}$$

When this is substituted in (4) the factor $\sqrt{1 - \beta^2}$ on the right cancels the factor $\sqrt{1 - \beta^2}$ contained in $d\tau$. We thus obtain instead of the first three components of (4)

$$\frac{d}{dt}(m\mathbf{v}) = \frac{d}{dt} \frac{m_0 \mathbf{v}}{\sqrt{1 - \beta^2}} = \mathbf{K}. \tag{5}$$

This is the equation of motion (4.6) in Vol. I. As was first pointed out by Planck,[1] the *Lorentz force* $\mathbf{K}$ here takes the place of the classical *Newtonian force*. For the sake of distinction the four-force $\mathbf{F}$ in (4) is designated as *Minkowski force*.

As was already shown in Vol. I, §4, (5) becomes, for *longitudinal* and *transversal* direction of the force ($\mathbf{K} \parallel \mathbf{v}$ and $\mathbf{K} \perp \mathbf{v}$),

$$\frac{m_0}{(1 - \beta^2)^{3/2}} \frac{d\mathbf{v}}{dt} = \mathbf{K} \quad \text{and} \quad \frac{m_0}{(1 - \beta^2)^{1/2}} \frac{d\mathbf{v}}{dt} = \mathbf{K}, \quad \text{respectively.}$$

[1] Verhandl. d. deutsch. phys. Ges. *4*, p. 136, 1906.

The designations *longitudinal* and *transversal* mass for $m_0(1 - \beta^2)^{-3/2}$ and $m_0(1 - \beta^2)^{-1/2}$ were discussed and criticized at the same place in Vol. I.

We supplement Eq. (5) by the fourth energetic component, which follows from (4) and (4a):

$$\frac{d}{dt} \frac{m_0 c^2}{\sqrt{1 - \beta^2}} = e\mathbf{v}\cdot\mathbf{E} = \mathbf{v}\cdot\mathbf{K}. \tag{6}$$

The equality of $e\mathbf{v}\cdot\mathbf{E}$ and $\mathbf{v}\cdot\mathbf{K}$ postulated here follows in the electrodynamic case simply from the fact that $\mathbf{v}\cdot(\mathbf{v}\times\mathbf{B}) = 0$. Applied to an arbitrary force law it signifies that the four-force $\mathbf{F}$ must be *perpendicular* to the world line of the particle (see p. 243).

$\mathbf{v}\cdot\mathbf{K}$ is the work done on the moving particle by the force $\mathbf{K}$ in unit time, i.e. it is equal to dA/dt. Accordingly the left side of Eq. (6) is simply the change in the kinetic energy T effected by the force $\mathbf{K}$. We hence have

$$T = \frac{m_0 c^2}{\sqrt{1 - \beta^2}} + \text{const.} \tag{6a}$$

In Exercise III.4 we shall convince ourselves of the fact that Eq. (6) may be derived from the equation of motion (5) also by the formalism customary in the derivation of the energy theorem in elementary mechanics, namely scalar multiplication with $\mathbf{v}$. Since, by definition, T must vanish for $v \to 0$, the constant in (6a) must be put equal to $-m_0 c^2$. Hence, in view of (2),

$$T = m_0 c^2 \left(\frac{1}{\sqrt{1 - \beta^2}} - 1\right) = (m - m_0)c^2. \tag{7}$$

The classical expression $T = mv^2/2$ follows from this by passing to the limit $c \to \infty$, as was already noted at the end of §4 in Vol. I.

A. The Equivalence of Energy and Mass

Just as we considered the rest mass m_0 apart from the mass in motion m, we introduce apart from the *energy in motion* E the *rest energy* E_0, where then $T = E - E_0$. We can hence write in place of (7)

$$E - E_0 = (m - m_0)c^2. \tag{7a}$$

We render this equation more specific by the statement

$$E = mc^2 \tag{8}$$

and the consequent relation

$$E_0 = m_0 c^2. \tag{8a}$$

This is the theorem of the *inertia of energy*, which according to Einstein is the most important result of the (special) theory of relativity. We quote Einstein literally: "The mass of a body is a measure of its energy content; if the energy changes by ΔE, the mass changes in the same direction by $\Delta E/c^2$. It is not out of question that for bodies whose energy content is variable in a high degree (e.g. for radium salts) a test of the theory may be successful."[1]

This test has since been carried out on a huge scale: The atomic transformations which have been discovered in the meantime and been studied in detail for most of the light elements have led, by the use of the equivalence theorem, to an undreamed of increase *in the precision of the chemical atomic weights*[2] and the fission of the heaviest element, uranium—more precisely, the uranium isotope of atomic weight 235—which was discovered by Otto Hahn only toward the end of 1938 has, in accord with the loss of mass occurring in it, had a terrifying effect in the *destruction caused by the uranium bomb*. We shall concern ourselves here only with the second example and this only briefly and superficially.

The uranium 235 atom, after capture of a neutron (atomic weight 1) has assumed the atomic weight $M = 236$, but retained the atomic number $Z = 92$ of the original uranium atom. It may, e.g., split into krypton, $Z = 36$, and barium, $Z = 56$, or into xenon, $Z = 54$, and strontium, $Z = 38$. Both fission possibilities are observed. The *conservation of the nuclear charge eZ* is here assured, since

$$92 = 36 + 56 = 54 + 38.$$

However, the mass is not conserved. Instead, the mass excess of the atomic weight over the integer 235 (the so-called "packing fraction") is set free, i.e. transformed into energy. If we assume that it amounts to one unit in the first decimal (for the heavier isotopes the atomic weights are not yet precisely known), we obtain for the energy available from one gram-atom

$$0.1\ c^2 = 9\cdot10^{19}\ \text{g}\cdot\text{cm}^2\cdot\text{sec}^{-2} = 9\cdot10^{12}\ \text{joules}.$$

Computed for a kilogram of fissioned uranium 235 it is $1000/235$ times as much, or $38\cdot10^{12}$ joules. We transform this into heat units (one large calorie $\cong 4.2\cdot10^3$ joules) and obtain

$$\frac{38}{4.2\cdot10^3}\cdot10^{12}\ \text{cal} \cong 10^{10}\ \text{cal}.$$

[1] A. Einstein, "Does the inertia of a body depend on its energy content?", Ann. Physik, Vol. 17, 1905. Einstein here explains the specialization of Eq. (7a) to the equivalence theorem (8) by an imaginary experiment: a moving body emits radiation and is observed from a system at rest.

[2] H. Bethe, Phys. Rev. *47*, 633, 1935; Oliphant, Kempton, and Rutherford, Proc. Roy. Soc. London *149*, 406, 1935. The almost simultaneous publication of these two papers on the two sides of the Atlantic shows once more the inevitable course of development of the understanding of physics as prescribed by the experimental material available at the time.

If we note that the energy transfers of ordinary molecular processes lie in the range from 100 to 1000 calories, we see that our uranium process supplies many million times as much energy. On this basis we may understand both the terrible effect of the *uranium bomb* and the beneficial effect of the *uranium engine*, i.e. a controllable, continuously operating uranium process, which could remove all economic ills of the times. The fact that the practical realization of the uranium process differs from that here considered, i.e. that it is carried out by way of a transuranium element (plutonium), does not require mention. The validity of the proof indicated by our simplified process is not affected thereby.

B. Relationship between Momentum and Energy

In classical mechanics the components of momentum are *derivatives of the kinetic energy* with respect to the velocity components, e.g. for an individual particle in Cartesian coordinates:

$$G_k = m\dot{x}_k = \frac{\partial T}{\partial \dot{x}_k}, \qquad T = \frac{m}{2}(\dot{x}_1^2 + \dot{x}_2^2 + \dot{x}_3^2) \quad \text{with } m = \text{const.} \quad (9)$$

This no longer applies in relativistic mechanics. It may readily be verified however that the relativistic momentum components (1a) are *derivatives of the following quantity*:

$$K = -m_0 c^2 \sqrt{1 - \beta^2} + \text{const.} \qquad (9a)$$

with respect to the $\dot{x}_k$. Following Helmholtz,[1] a function which accomplishes this is called a "kinetic potential". If, again, K is normalized so that it vanishes for $\beta = 0$, the constant must be chosen equal to $m_0 c^2$, yielding

$$K = m_0 c^2 (1 - \sqrt{1 - \beta^2}). \qquad (9b)$$

Hence the definition of the momentum of an individual point mass replacing (9) becomes

$$G_k = \frac{\partial K}{\partial \dot{x}_k} = \frac{m_0 \dot{x}_k}{\sqrt{1 - \beta^2}} \qquad (10)$$

in agreement with the definition (1a). For $c \to \infty$ K evidently passes over into T and (10) into (9).

C. The Principles of D'Alembert and Hamilton

What are the consequences of this changed meaning (10) of the momentum coordinates for the general principles of mechanics? We shall first discuss *D'Alembert's principle*. The inertial reaction forces introduced by D'Alembert (see Vol. I, Eq. (10.1)) are also now given by $-\dot{G}_k$. (The

---

[1] In his general studies on the principle of least action.

usual definition as mass $\times$ acceleration is of course now invalid). The statements of D'Alembert's principle in Vol. I, §10 then continue to apply literally: "The inertial reaction forces balance themselves against the physically impressed forces" (Vol. I, p. 57). "The sumtotal of the lost forces is in equilibrium on the system." (Vol. I, p. 58). The condition on p. 49 of Vol. I serves as definition of the word "mechanical system:" "The virtual work of the reactions within the system is equal to zero."

Hamilton's principle is derived from D'Alembert's principle in the manner of §33 of Vol. I. Here T is to be replaced by K and, for forces possessing a potential, (33.12) of Vol. I is replaced by

$$\delta \int_{t_0}^{t_1} (\mathrm{K} - V)\, dt = 0. \tag{11}$$

Here the variation is to be carried out as in the earlier example: The space coordinates are varied, whereas the endpoints of the path and the time for its transversal remain fixed.

Application to a single point mass yields

$$\delta \int_{t_0}^{t_1} \mathrm{K}\, dt = \delta \int_{t_0}^{t_1} m_0 c^2 (1 - \sqrt{1 - \beta^2})\, dt = -\delta \int_{t_0}^{t_1} m_0 c^2 \sqrt{1 - \beta^2}\, dt. \tag{11a}$$

Here we have already taken account of the fact that the times t_0 and t_1 are not to be varied, i.e. that $\delta(t_1 - t_0) = 0$. Hence (11) becomes

$$\delta \int_{t_0}^{t_1} \left(m_0 c^2 \sqrt{1 - \beta^2} + V \right) dt = 0. \tag{12}$$

We can readily convince ourselves that this variational prescription agrees with our equation of motion (5), and this not only for the Lorentz force **K**, to which (5) was limited, but for any given potential energy and an arbitrary force $\mathbf{K} = -\mathrm{grad}\, V$ derived from it.

In the variation we must replace x, y, z by $x + \delta x, y + \delta y, z + \delta z$, obtaining

$$\delta \frac{dx}{dt} = \frac{d\delta x}{dt}, \cdots, \qquad \delta V = \frac{\partial V}{\partial x} \delta x + \cdots = -K_x \delta x - \cdots,$$

where the $\cdots$ represent corresponding expressions in y and z. Similarly we must form

$$\delta \sqrt{1 - \beta^2} = \delta \sqrt{1 - \frac{1}{c^2}\left\{ \left[\frac{dx}{dt}\right]^2 + \left[\frac{dy}{dt}\right]^2 + \left[\frac{dz}{dt}\right]^2 \right\}}$$

$$= \frac{1}{2\sqrt{1 - \beta^2}} \left[-\frac{2}{c^2} \right]\left[\frac{dx}{dt}\frac{d\delta x}{dt} + \cdots \right]$$

and hence

$$\delta \int_{t_0}^{t_1} m_0 c^2 \sqrt{1 - \beta^2}\, dt = \int_{t_0}^{t_1} \left\{ -\frac{m_0}{\sqrt{1 - \beta^2}} \frac{dx}{dt}\frac{d\delta x}{dt} + \cdots \right\} dt. \tag{12a}$$

An integration by parts, in which the terms without integral sign vanish (because $\delta x = 0$ for $t = t_0$ and $t = t_1$) according to our original assumption, transforms this into

$$\int_{t_0}^{t_1} \left\{ \frac{d}{dt} \left[\frac{m_0}{\sqrt{1 - \beta^2}} \frac{dx}{dt} \right] \delta x + \cdots \right\} dt. \tag{12b}$$

Thus (12) yields altogether

$$\int_{t_0}^{t_1} \left\{ \left[\frac{d}{dt} \left(\frac{m_0}{\sqrt{1 - \beta^2}} \frac{dx}{dt} \right) - K_x \right] \delta x + \cdots \right\} dt = 0. \tag{12c}$$

Since δx, δy, and δz are independent of each other the factor of δx must vanish, as well as those of δy and δz. We thus obtain in fact our earlier equation of motion (5) for arbitrary $\mathbf{K}$; it is valid, incidentally, even if $\mathbf{K}$ cannot be derived from a potential energy.

If there are no external forces ($V = $ const) (12) may be abbreviated to

$$\delta \int_{t_0}^{t_1} \sqrt{1 - \beta^2} \, dt = \delta \int_{\tau_0}^{\tau_1} d\tau = 0. \tag{13}$$

This is *Fermat's principle of least time*, which now however does not relate to the conventional time t, but to the Lorentz-invariant intrinsic time. Since $d\tau$ corresponds to the four-dimensional line element ds but for the factor ic, we can also write in place of (13)

$$\delta \int_A^E ds = 0. \tag{13a}$$

This is the *principle of the shortest path* for given starting point A and end point E, or, as we called it in Vol. I, Eq. (37.14), the *principle of the geodetic line*, extended to four dimensions and made Lorentz-invariant. We shall therefore call it more precisely the *principle of the shortest world line*.

D. Lagrange Function and Lagrange Equations

In our formulation (11) of Hamilton's principle the relativistic Lagrange function

$$L_{\mathbf{rel}} = \mathbf{K} - V \tag{14}$$

replaces the classical Lagrange function $L_{kl} = T - V$ in Vol. I, Eq. (33.13). The *general Lagrange equations* for arbitrary position and velocity coordinates are derived from L_{rel} by carrying out the variation prescribed in (11), just as they are derived from L_{kl} in §34 of Vol. I:

$$\frac{d}{dt} \frac{\partial L_{\mathrm{rel}}}{\partial \dot{q}_k} - \frac{\partial L_{\mathrm{rel}}}{\partial q_k} = 0. \tag{14a}$$

In spite of their similarity with those of classical mechanics, these equations, when applied to specific cases, yield results which differ decidedly from those of the latter. For example, in the Kepler problem of the hydrogen atom they lead to an ellipse with precessing perihelion instead of to a closed ellipse as a consequence of the relativistic variation of mass; see also §38 regarding the perihelion of mercury.

E. Schwarzschild's Principle of Least Action

In his fundamental papers "On Electrodynamics" Schwarzschild[1] introduced with the designation "electrokinetic potential" the quantity

$$L = \Psi - \mathbf{v} \cdot \mathbf{A}. \tag{15}$$

We shall show that when multiplied with the charge density ρ this is a relativistic invariant. To this end we form the scalar product of the four-vector $\mathbf{\Gamma}$ of current density (Eq. (28.16)) and the four-potential $\mathbf{\Omega}$ (Eq. (26.4)). We obtain

$$\mathbf{\Gamma} \cdot \mathbf{\Omega} = \rho(\mathbf{v} \cdot \mathbf{A} - \Psi).$$

We call $-\mathbf{\Gamma} \cdot \mathbf{\Omega}$ the *Schwarzschild invariant*. In view of (15)

$$\mathbf{\Gamma} \cdot \mathbf{\Omega} = -\rho L. \tag{16}$$

Schwarzschild adds to this invariant the Lagrange density Λ from (26.24) which, we know, is also Lorentz-invariant, and forms, with $T =$ kinetic energy,

$$T - \Lambda - \rho L. \tag{17}$$

We shall replace (17) by

$$K' = T' - 2\Lambda - \rho L = T' - 2\Lambda + \mathbf{\Gamma} \cdot \mathbf{\Omega}. \tag{17a}$$

Here T' represents the relativistic value of the kinetic energy from (7), where, however, the rest mass m_0 is to be replaced by the rest-mass density μ_0. (Also the remaining terms in (17a) are densities, referring to unit volume.) Hence

$$T' = \mu_0 c^2((1 - \beta^2)^{-\frac{1}{2}} - 1). \tag{17b}$$

The factor 2 of Λ in (17a), on the other hand, derives from our basic distinction between the entities of quantity f and of intensity F; Schwarzschild, who puts $\mathbf{D} = \mathbf{E}$ and $\mathbf{B} = \mathbf{H}$ and hence writes in our Eq. (26.24)

---

[1] K. Schwarzschild, Göttinger Nachr. 1903. See in particular the first of the three papers. The notation L is the same as Schwarzschild's; Schwarzschild uses φ in place of our Ψ. Note the date of publication 1903! Thus Schwarzschild arrived intuitively at the correct postulate of the theory of invariants six years ahead of Minkowski.

$\mathbf{H}^2 - \mathbf{E}^2$ instead of $\mathbf{H} \cdot \mathbf{B} - \mathbf{D} \cdot \mathbf{E}$, gains a factor 2 in the variation, which we must supply in (17a). In detail our formula (17a) becomes

$$K' = \mu_0 c^2((1 - \beta^2)^{-\frac{1}{2}} - 1) - 2\Lambda + \mathbf{\Gamma} \cdot \mathbf{\Omega}. \tag{18}$$

From this point on we follow Schwarzschild's procedure. He integrates (17) over an arbitrary region of space-time and constructs in this manner an action function W, which he subjects to the requirement $\delta W = 0$. We form correspondingly

$$W = \iiiint K' \, dx \, dy \, dz \, dt \tag{19}$$

and also set

$$\delta W = 0. \tag{19a}$$

According to Schwarzschild this variation is to be carried out in following fashion:

a. The components Ω_1, Ω_2, Ω_3, and Ω_4 of the potential and

b. the coordinates x_1, x_2, x_3, and x_4 of the electrons are subjected to arbitrary small variations; these variations are to vanish on the boundaries of the region. The variations a and b are independent of each other[1] and may be carried out individually, e.g. also for each component of $\mathbf{\Omega}$. If several electrons are present we can limit ourselves to one of them since the effects of the rest on it are contained in the potential $\mathbf{\Omega}$. The fact that we use the four-potential $\mathbf{\Omega}$, originally introduced for convenience of calculation, rather than the six-vector F as fundamental field quantity represents a new departure, to which we shall return in §37.

a. Since the first term on the right of Eq. (18) is independent of $\mathbf{\Omega}$ we are only concerned with Λ and $\mathbf{\Gamma} \cdot \mathbf{\Omega}$. If the variation is limited to $\delta\Omega_1$ we find

$$\delta(\mathbf{\Gamma} \cdot \mathbf{\Omega}) = \Gamma_1 \delta\Omega_1. \tag{20}$$

In the expression (26.24) for Λ we must imagine F as expressed (by (26.11)) by c Curl $\mathbf{\Omega}$, whereas f is to be regarded as an unknown. Hence, for the specific variation mentioned above, (-2Λ) reduces to the following three terms (the remaining terms of the Curl, to be formed with Ω_2, Ω_3, and Ω_4, drop out):

$$\delta(-2\Lambda) = f_{12} \frac{\partial \delta\Omega_1}{\partial x_2} + f_{13} \frac{\partial \delta\Omega_1}{\partial x_3} + f_{14} \frac{\partial \delta\Omega_1}{\partial x_4} = -\left[\frac{\partial f_{12}}{\partial x_2} + \frac{\partial f_{13}}{\partial x_3} + \frac{\partial f_{14}}{\partial x_4}\right] \delta\Omega_1 + \cdots.$$

[1] Schwarzschild does not take account of the auxiliary condition Div $\mathbf{\Omega} = 0$ which was satisfied automatically in our method of integration in §29. We adhere to Schwarzschild's prescription also in this respect.

The dots refer to partial derivatives with respect to the coordinates which vanish in the later integration over our world region (since $\delta\Omega = 0$ on its boundary). Together with (20) we thus find for the factor of $\delta\Omega_1$ in the integrand of (19)

$$-\left[\frac{\partial f_{12}}{\partial x_2} + \frac{\partial f_{13}}{\partial x_3} + \frac{\partial f_{14}}{\partial x_4}\right] + \Gamma_1.$$

It must vanish since $\delta W = 0$. If we substitute for f_{ik} the values in the array (26.14) and for Γ its value from (28.16) we find

$$-\frac{\partial H_z}{\partial y} + \frac{\partial H_y}{\partial z} + \frac{\partial D_x}{\partial t} + \rho v_x = 0. \tag{21}$$

This is exactly the first component of the three Maxwell equations $\dot{\mathbf{D}} + \mathbf{J} = \operatorname{curl} \mathbf{H}$, where here, for vacuum, the convection current density $\rho\mathbf{v}$ represents $\mathbf{J}$. The second and third component are evidently obtained similarly by the variation of Ω_2 and Ω_3 and the condition div $\mathbf{D} = \rho$ from that of Ω_4. We cannot, of course, expect to derive the other set of Maxwell equations and the condition div $\mathbf{B} = 0$ in the same manner, since these are already implicit in the existence of the potential.

This clarifies also the basis for our earlier name "Lagrange density" for Λ (see p. 220): In our electrodynamical variation principle Λ takes the place of the earlier Lagrange function L or L_{rel} (Eq. (32.14)).

b. The potential Ω is not varied; hence $\delta\Lambda = 0$.

On the other hand, the world line of the electron is to be compared with neighboring world lines, so that the first term on the right of (18) and the Schwarzschild invariant $\Gamma \cdot \Omega$ are to be varied. We shall first deal with the term $\Gamma \cdot \Omega$. It is here convenient to replace the world volume element $dx\,dy\,dz\,dt = dV\,dt$ in (19) by $dV_n\,d\tau$, where dV_n represents its three-dimensional cross section perpendicular to the world line. Since the charge density ρ occurring in Γ is concentrated on the world line of the electron we obtain in the integration over dV_n (not in that over $dV!$) the electron charge e.

At the same time, in the expressions for Γ and Ω, we pass from the coordinates $x_1 \cdots x_4$ used so far to the coordinates $\xi_1 \cdots \xi_4$ of the world line element considered at the moment $(d\xi_4 = ic\,d\tau,\ d\tau =$ element of the intrinsic time). Then $d\xi_j/d\tau$ replaces dx_j/dt and we obtain

$$\iiiint \Gamma \cdot \Omega\, dV_n\, d\tau = e \int \sum_j \frac{d\xi_j}{d\tau} \Omega_j\, d\tau. \tag{22}$$

We must note that in the variation not only ξ_j is changed by $\delta\xi_j$, but also Ω_j is changed by

$$\delta\Omega_j = \Omega_j(\xi_1 + \delta\xi_1, \cdots, \xi_4 + \delta\xi_4) - \Omega_j(\xi_1, \cdots, \xi_4) = \sum_i \frac{\partial\Omega_j}{\partial\xi_i} \delta\xi_i$$

(the charge e is of course conserved). Hence (22) leads to

$$\delta \iiiint \mathbf{\Gamma} \cdot \mathbf{\Omega} \, dV_n \, d\tau = e \int \left(\sum_j \frac{d\delta\xi_j}{d\tau} \Omega_j + \sum_j \sum_i \frac{d\xi_j}{d\tau} \frac{\partial\Omega_j}{\partial\xi_i} \delta\xi_i \right) d\tau. \quad (22a)$$

The first of the two terms on the right is transformed by integration by parts and yields (since $\delta\xi_j = 0$ on the boundary of the world region)

$$-\sum_j \frac{d\Omega_j}{d\tau} \delta\xi_j = -\sum_j \sum_i \frac{\partial\Omega_j}{\partial\xi_i} \frac{d\xi_i}{d\tau} \delta\xi_j.$$

Thus (22a), after interchange of the subscripts i, j in the double sum, becomes

$$e \int \left(\sum_j \sum_i \frac{d\xi_i}{d\tau} \left[\frac{\partial\Omega_i}{\partial\xi_j} - \frac{\partial\Omega_j}{\partial\xi_i} \right] \delta\xi_j \right) d\tau. \quad (23)$$

Here we have in the parenthesis $\text{Curl}_{ji}\, \mathbf{\Omega}$, i.e. except for the factor c the component F_{ji} of the field (see (26.11)). On the other hand, we have, by (28.17)

$$\mathbf{\Gamma} \cdot F = c\mathbf{k} \quad (\mathbf{k} = \text{force density}) \quad (24)$$

and by (28.19)

$$\int \mathbf{k} \, dV_n = \frac{\mathbf{K}}{\sqrt{1 - \beta^2}} \quad (\mathbf{K} = \text{Lorentz force}).$$

Hence our expression (23), which was obtained by carrying out the integration over V_n, signifies simply

$$\int \sum_j K_j \delta\xi_j \frac{d\tau}{\sqrt{1 - \beta^2}} = \int \sum_j K_j \, d\xi_j \, dt. \quad (25)$$

We must add to this from the first term of the right side of (18), if we again put $dV \, dt = dV_n \, d\tau$ and integrate over the world line cross section dV_n :

$$m_0 c^2 \delta \int \left((1 - \beta^2)^{-\frac{1}{2}} - 1 \right) d\tau = m_0 c^2 \delta \int (1 - \sqrt{1 - \beta^2}) \, dt = \delta \int K \, dt.$$

We have already carried out the variation of this integral over K in Eq. (11a) and the succeeding equations. We found there, translated into the present notation ξ_j of the world line coordinates (see (12b)):

$$-\int \sum_j \frac{d}{dt} \left(\frac{m_0}{\sqrt{1 - \beta^2}} \frac{d\xi_j}{dt} \right) \delta\xi_j \, dt. \quad (26)$$

Together with (25) we obtain as variation of the action integral

$$\delta W = -\int \left\{ \sum_j \frac{d}{dt} \left[\frac{m_0}{\sqrt{1 - \beta^2}} \frac{d\xi_j}{dt} \right] - K_j \right\} \delta\xi_j \, dt. \quad (27)$$

We require that this integral should vanish for arbitrary displacements. This is only possible if the $\{\ \}$ vanishes for $j = 1, 2, 3, 4$. In this manner we have derived our earlier Eq. (5) including the corresponding fourth component, and this in more explicit form: Our present derivation yields not only this equation of motion, but also the Lorentz force impressed on the electron by the field. *Schwarzschild's principle of least action thus combines Maxwell's electrodynamics and the Lorentz electron theory in a single four-dimensionally invariant formulation.*

From a historical point of view it may be noted that Schwarzschild, starting with the kinetic potential (17), also obtains the Maxwell equations and the equation of motion of the electron including the expression for the Lorentz force. Only the variation of mass of the electron escapes him, since, in (17), he employs the classical value of the kinetic energy. It is true that his derivation of the Maxwell equations is not quite correct from our point of view because of the missing factor 2 in Λ, which is compensated in Schwarzschild's treatment by putting f and F ($\mathbf{D} = \mathbf{E}, \mathbf{H} = \mathbf{B}$) equal. In our representation the proportionality of f and F is also contained in the Schwarzschild principle. It is only necessary to eliminate Γ from the Eqs. (21) and (26.5), which we can write

$$\operatorname{Div} f = \Gamma \qquad \text{and} \qquad \operatorname{Div} F = c\mu_0\Gamma.$$

Schwarzschild's action principle is very suggestive. It could be made the starting point of the theory and the Maxwell equations be regarded as its consequences. There would be at the same time the inviting possibility of refining the Maxwell equations by extending the kinetic potential (18) (addition of other field invariants, taking account of interactions between the electrons, their magnetic moment and spin). We will enter upon such questions in §37.

§33. Electromagnetic Theory of the Electron

At the turn of the century interest was focused on the variable mass of the electron. The assumption of the *rigid electron*, which appeared appropriate in the theory of the absolute ether, led to a different, much more complicated law of transformation (Max Abraham) than Lorentz's assumption of the *deformable electron*, which soon afterwards attained an assured basis in the theory of relativity. The experiments of Kaufmann, Bucherer, Neumann and many others were concerned with this law of transformation.

The theoretical treatment of the problem (also for the rigid electron) rested on the definition (31.13) of the electromagnetic momentum. Without detaining ourselves with the rigid electron we shall show that the same starting point, with the relativistic treatment of the momentum, leads to the same law (32.2) of the variation of the mass as the theory of relativity, which however extends it immediately to any arbitrary mass m.

We shall obtain as a by-product an interesting formula for the rest mass m_0 of the electron.

Below we understand by $\mathbf{G}$ the *total momentum* of the field in infinite space; we shall call the momentum per unit volume, designated by $\mathbf{G}$ in (31.13), $\mathbf{g}$. With dV as three-dimensional volume element we then have

$$\mathbf{G} = \int \mathbf{g} \, dV = \frac{1}{c^2} \int \mathbf{S} \, dV. \tag{1}$$

To be able to carry out the indicated integration we utilize the ideas and symbols of Eqs. (28.12). Let x, y, z be the frame of reference moving with the electron, x', y', z' a coordinate system at rest, with respect to which the electron has the instantaneous velocity v in the positive x'-direction. In the x, y, z system we then have of course $\mathbf{G} = 0$; the field is electrostatic so that $\mathbf{H} = 0$ and $\mathbf{S} = 0$. We are interested in the momentum $\mathbf{G}'$ and more particularly in its x-component:

$$G'_x = \frac{1}{c^2} \int S'_x \, dV' = \frac{1}{c^2} \int (E'_y H'_z - E'_z H'_y) \, dV'. \tag{2}$$

We express the primed quantities in terms of the unprimed ones in accord with Eq. (28.12a), in which however, in view of the opposite direction of motion, the sign of v must be changed:

$$E'_x = E_x, \qquad E'_y = \frac{1}{\sqrt{1 - \beta^2}} E_y, \qquad E'_z = \frac{1}{\sqrt{1 - \beta^2}} E_z,$$

$$H'_x = 0, \qquad H'_y = \frac{-v}{\mu_0 c^2 \sqrt{1 - \beta^2}} E_z, \qquad H'_z = \frac{B'_z}{\mu_0} = \frac{v}{\mu_0 c^2 \sqrt{1 - \beta^2}} E_y, \tag{2a}$$

$$dV' = dV\sqrt{1 - \beta^2} \qquad \text{(Lorentz contraction).} \tag{2b}$$

Thus we obtain from (2)

$$G'_x = \frac{v}{\mu_0 c^4 \sqrt{1 - \beta^2}} \int (E_y^2 + E_z^2) \, dV. \tag{3}$$

In the xyz-system the E-field is spherically symmetrical, so that

$$\int E_x^2 \, dV = \int E_y^2 \, dV = \int E_z^2 \, dV = \frac{1}{3} \int E^2 \, dV. \tag{3a}$$

The same applies for the charge distribution. It seems most natural to spread the charge e uniformly over a sphere of radius a ("radius of the electron"). Then, as follows e.g. from (7.6a),

$$E = E_r = \begin{cases} 0 & \text{for} \quad r < a \\[2mm] \dfrac{e}{4\pi\varepsilon_0 r^2} & \text{for} \quad r \geqq a \end{cases} \tag{3b}$$

and hence

$$\int E^2 \, dV = 4\pi \int E_r^2 r^2 \, dr = \frac{e^2}{4\pi\varepsilon_0^2} \int_a^\infty \frac{dr}{r^2} = \frac{e^2}{4\pi\varepsilon_0^2 a}. \tag{3c}$$

(3) and (3a) then lead to

$$G_x' = \frac{v}{\mu_0 c^4 \sqrt{1 - \beta^2}} \cdot \frac{2}{3} \frac{e^2}{4\pi\varepsilon_0^2 a} = \frac{e^2}{6\pi\varepsilon_0 c^2 a} \frac{v}{\sqrt{1 - \beta^2}}. \tag{4}$$

We may also readily convince ourselves that,

$$G_y' = G_z' = 0 \tag{4a}$$

as must be expected for the spherical symmetry in the xyz-system. For if we form

$$G_y' = \frac{1}{c^2} \int S_y' \, dV' = \frac{1}{c^2} \int (E_z' H_x' - E_x' H_z') \, dV'$$

in analogy to (2) and again make use of Eqs. (2a, b),

$$G_y' = \frac{v}{\mu_0 c^2} \int E_x E_y \, dV = 0$$

since for a spherically symmetrical field E_x and E_y are proportional to x and y, and xy integrated over the sphere vanishes.

Eqs. (4) and (4a) can be combined to

$$\mathbf{G}' = m\mathbf{v}, \qquad m = \frac{m_0}{\sqrt{1 - \beta^2}}. \tag{5}$$

The mass factor m here introduced has thus the dependence on velocity familiar to us from §32. For the rest mass m_0 we find from (4):

$$m_0 = \frac{e^2}{6\pi c^2 \varepsilon_0 a}. \tag{6}$$

The reader may check the dimensional correctness of this formula, i.e. the independence of the choice of the unit of charge Q and the unit of length M. The factor ε_0, which in the Gaussian system is set equal to 1, is from our point of view indispensable. If it is suppressed the formula becomes dimensionally meaningless.

With the value of the rest mass computed by (6) Eq. (5) states: *The mechanical momentum of the electron is equal to the momentum contained in the electromagnetic field as defined by Eq. (2):*

$$\mathbf{G}_{\text{electron}} = \mathbf{G}_{\text{field}}. \tag{7}$$

We read directly in Eq. (6) that the transition to the limit $a \to 0$ is unfortunately impossible; it would lead to $m_0 \to \infty$ and $e/m_0 \to 0$. To determine the numerical value of a, we must know the experimental values of e and e/m_0. In MKSQ units, with Q = 1 coulomb, these are:

$$e = 1.60 \cdot 10^{-19}Q, \qquad e/m_0 = 1.76 \cdot 10^{11}Q/K. \tag{8}$$

From this we compute

$$m_0 = 0.9 \cdot 10^{-30}K. \tag{8a}$$

Eq. (6) then yields, with (7.18a),

$$a = \frac{2}{3}\frac{e^2}{4\pi\varepsilon_0 c^2 m_0} = \frac{2}{3}\frac{e^2}{10^7 m_0}\frac{MK}{Q^2} = \frac{2}{3}\frac{(1.60 \cdot 10^{-19})^2}{0.9 \cdot 10^{-23}}M \cong 2 \cdot 10^{-13} \text{ cm.} \tag{9}$$

This is a subatomic dimension, of the same order of magnitude as nuclear dimensions.

It is of course quite arbitrary that we have here assumed a surface charge. We might equally well have distributed the electron charge e uniformly over the electron volume.

If we then call its radius once more a, we find[1] instead of (3b)

$$E = E_r = \begin{cases} \dfrac{er}{4\pi\varepsilon_0 a^3} & \text{for} \quad r \lessgtr a \\[2ex] \dfrac{e}{4\pi\varepsilon_0 r^2} & \text{for} \quad r \gtrless a \end{cases} \tag{10}$$

and instead of (3c)

$$\int E^2\, dV = \frac{e^2}{4\pi\varepsilon_0^2}\left\{\int_0^a \frac{r^4\, dr}{a^6} + \int_a^\infty \frac{dr}{r^2}\right\} = \frac{e^2}{4\pi\varepsilon_0^2 a}\left\{\frac{1}{5} + 1\right\}. \tag{10a}$$

The factor 6/5 is thus to be added to the formula for $\mathbf{G}'$ in (4) so that we obtain in place of (6)

$$m_0 = \frac{e^2}{5\pi c^2\,\varepsilon_0\, a}. \tag{10b}$$

The order of magnitude of the value of a found in (9) is not affected.

The following remark is of greater importance: Who can guarantee that the Maxwell equations can be extrapolated right up to the surface or into the interior of the electron? May not their simplicity and linearity be a consequence of the fact that they are exactly valid only for weak fields and that they must be corrected, in the immediate neighborhood of concentrated charges, by higher terms, in some such manner as the theory of

---

[1] The first line of (10) evidently follows from the fact that elements of charge whose distance from the center is less than r may be thought of as concentrated at the center of the sphere, while those distant by more than r from the center do not contribute to the field strength.

dilute solutions in thermochemistry? We shall return to this question in §37. It will here merely be emphasized that the derivation of the law governing the variation of mass with velocity is not subject to this criticism, since, in §32, it could be derived from the general principles of relativistic mechanics, whereas our present computation of m_0 is affected; the latter is anyhow beyond experimental verification, in view of the hypothetical character of the electron radius. The derivation of the mass-velocity law in §32 is, like all considerations of the special theory of relativity, only tied to the condition that the occurring relative motions should be nearly uniform. We express this here by the demand that the electron motion be *quasistationary*. We mean hereby that its velocity change in the time taken by a light wave to sweep over the electron (i.e. the time $2a/c$) be small compared to v. We thus demand only:

$$\dot{\mathbf{v}} \frac{2a}{c} \ll v. \tag{11}$$

All processes in vacuum tubes satisfy this requirement.

With respect to formula (6) for the rest mass we note furthermore that it may be derived in the following very elementary manner: We consider a *slowly* moving electron. Its mass is equal to the rest mass m_0 and its kinetic energy

$$T = \frac{m_0}{2} v^2. \tag{12}$$

If this is of electromagnetic origin we must set it equal to the *magnetic* energy of the field since the electric energy is constant for small fields, i.e. not proportional to v^2. We hence put

$$T = \frac{\mu_0}{2} \int \mathbf{H}^2 \, dV. \tag{12a}$$

Here we can substitute for $\mathbf{H}$ the value (15.12) from the law of Biot-Savart

$$\mathbf{H} = \frac{ev}{4\pi} \frac{\sin \vartheta}{r^2}.$$

We then obtain for surface charge

$$T = \frac{\mu_0}{2} \left(\frac{ev}{4\pi}\right)^2 \int_a^\infty \frac{dr}{r^2} \int_0^\pi \sin^3 \vartheta \, d\vartheta \int_0^{2\pi} d\varphi.$$

The three integrals are, in sequence,

$$\frac{1}{a}, \frac{4}{3}, 2\pi.$$

Hence

$$T = \frac{\mu_0}{2} \frac{e^2 v^2}{6\pi} \frac{1}{a}. \tag{13}$$

Comparison with (12) yields

$$m_0 = \frac{\mu_0 e^2}{6\pi} \frac{1}{a}$$

which is identical with (6) since $\varepsilon_0 \mu_0 c^2 = 1$.

Since the kinetic energy of the electron computed with (6) proved to be equal to the magnetic energy of the surrounding field we may suspect that its *rest energy* will correspond to the *electrostatic energy* of the Coulomb field. In the simple case of surface charge we find that this is equal to

$$E_{\text{stat}} = \frac{\varepsilon_0}{2} \int \mathbf{E}^2 \, dV = \frac{\varepsilon_0}{2} 4\pi \int_{r=a}^{\infty} E_r^2 \, r^2 \, dr$$

and obtain in view of (3c)

$$E_{\text{stat}} = \frac{e^2}{8\pi\varepsilon_0 a}. \tag{14}$$

In contrast to this Eq. (6) yields for the rest energy of our electron by Einstein's law of equivalence of mass and energy:

$$E_0 = m_0 c^2 = \frac{e^2}{6\pi\varepsilon_0 a}. \tag{15}$$

Thus only $\frac{3}{4}$ of this rest energy is explained electromagnetically by our preceding (admittedly primitive) *considerations.* The program indicated by the title of this section is as yet incapable of realization.

As was already said on p. 236, the electron is a stranger in electrodynamics. The forces which, opposing the Coulomb forces, prevent its explosion are unknown to us, just like the theory of the elementary particles in general. Poincaré introduced (as early as 1906, in the Rendiconti di Palermo) a cohesion pressure of unknown origin which was supposed to envelop the electron at rest like a membrane under uniform tension; the missing quarter of the rest energy was supposed to be hidden herein. The hypothesis of rigidity of the absolute theory could transfer this cohesion pressure to the electron in motion. It did not suffice, however, for a purely electromagnetic description of the electron. Even the assumption of rigidity contradicts the group-theoretical nature of Maxwell's electrodynamics which, as we know, demands the deformable electron of Lorentz.

Altogether, we should face the fact that our electrodynamic theory of the electron is as yet very incomplete. We have known for 20 years that the electron possesses in addition to its charge a quite definite spin and a quite definite magnetic moment. Both can only be defined on the basis of the quantum theory and are inaccessible to Maxwell's electrodynamics. The secret of the spin was first discovered in the more precise analysis of the Zeeman effect; the secret of the magnetic moment was actually, as we know now, clearly and tangibly demonstrated in ferromagnetism. It is strange that practical electronics remained untouched by these fundamental facts and could get along with the notion of the charged point mass or the minute charged sphere.

Our Problems III.5 to III.10 deal with this application of electron theory. The varied electron trajectories which occur in vacuum tubes and which, in the e/m experiments, first served to clarify the nature of the electron are at the same time in a way the simplest and best defined examples of the mechanics of an isolated point mass.

PART IV

MAXWELL'S THEORY FOR MOVING BODIES AND OTHER ADDENDA

§34. Minkowski's Equations for Moving Media

The extension of Maxwell's theory from media at rest to those in motion was a favorite problem of the older electrodynamics. Heinrich Hertz had failed in this effort (see his paper cited in footnote 2 on p. 2) because he adhered consistently to classical theory (the "Galilei transformation"). His friend Emil Cohn[1] came closer to the goal, but was not yet (in 1902!) in possession of the necessary tools, the Lorentz transformation. Even H. A. Lorentz did not quite attain the final form in his papers in the Enzyklopädie (1903), particularly not for magnetizable bodies. Einstein called his paper of 1905 "On the electrodynamics of moving bodies" and indicated in this manner a principal goal of his theory of relativity; however he does not enter upon the general structure of the equations for ponderable bodies but confines himself instead to the questions arising for the isolated electron. Minkowski, in 1908, at long last in full possession of the principle of relativity, was the first to solve the problem completely.[2]

Minkowski's logic was simple: The Maxwell equations for a state of rest apply within the laboratory. Consider a point of space-time P of a body moving[3] with respect to the laboratory at the laboratory time t; let it have the velocity $\mathbf{v}$. Let P be transformed to rest by the introduction of the coordinates x', y', z', t' for the description of the processes in the neighborhood of P, t. In this system Maxwell's equations for a state of rest apply to the quantities $\mathbf{E}'$, $\mathbf{B}'$, $\mathbf{D}'$, $\mathbf{H}'$, $\mathbf{J}'$, ρ':

$$\frac{\partial \mathbf{B}'}{\partial t'} = -\operatorname{curl} \mathbf{E}', \qquad \frac{\partial \mathbf{D}'}{\partial t'} + \mathbf{J}' = \operatorname{curl} \mathbf{H}',$$

$$\operatorname{div} \mathbf{D}' = \rho', \qquad \operatorname{div} \mathbf{B}' = 0, \tag{1}$$

with material constants differing from those for vacuum:

$$\mathbf{D}' = \varepsilon \mathbf{E}', \qquad \mathbf{B}' = \mu \mathbf{H}', \qquad \mathbf{J}' = \sigma \mathbf{E}'. \tag{2}$$

[1] Göttinger Nachr. 1901, p. 74; Ann. Physik 7, 29, 1902.

[2] Göttinger Nachr. 1908, p. 53; Gesammelte Werke II, p. 352.

[3] The motion may be variable in space and time and must merely be capable of quasistationary treatment in the sense of Eq. (33.11). Thus $\mathbf{v}$ need not be a pure translation and the body need not be rigid. Only the fixed value of $\mathbf{v}$ in the space-time point P, t enters in the following Lorentz transformations.

These constants have the same values as if the body were at rest with respect to the laboratory, since it knows nothing of its motion. The operations curl and div in (1) refer of course, just like the time t', to the primed system. Now the inverse Lorentz transformation is to be carried out, which transforms the primed system back into the original one of the laboratory. In the latter Eqs. (1) apply once more if all primes are omitted, in view of the basic property of covariance of the Maxwell equations with respect to the Lorentz transformations. However, Eqs. (2), transformed to the unprimed system, take on a new form.

We know the relationship of the $\mathbf{E}'$, $\mathbf{B}'$ and the $\mathbf{E}$, $\mathbf{B}$ from Eqs. (28.8a) and (28.11):

$$E'_{||} = (\mathbf{E} + \mathbf{v} \times \mathbf{B})_{||}, \qquad E'_{\perp} = \left[\frac{\mathbf{E} + \mathbf{v} \times \mathbf{B}}{\sqrt{1 - \beta^2}} \right]_{\perp}$$

$$B'_{||} = \left(\mathbf{B} - \frac{1}{c^2} \mathbf{v} \times \mathbf{E} \right)_{||}, \qquad B'_{\perp} = \left\{ \frac{\mathbf{B} - \dfrac{1}{c^2} \mathbf{v} \times \mathbf{E}}{\sqrt{1 - \beta^2}} \right\}_{\perp} \qquad (3)$$

$||$ and $\perp$ signify as before "parallel" and "perpendicular to the velocity $\mathbf{v}$". We shall supplement this by the corresponding relations between $\mathbf{D}'$, $\mathbf{H}'$ and $\mathbf{D}$, $\mathbf{H}$. In view of the definition of the six-vectors

$$f = (\mathbf{H}, -ic\mathbf{D}), \qquad F = (c\mathbf{B}, -i\mathbf{E})$$

they are obtained from (3) by replacing $\mathbf{E}$ by $c\mathbf{D}$ and $\mathbf{B}$ by $\mathbf{H}/c$. We thus obtain

$$D'_{||} = \left(\mathbf{D} + \frac{1}{c^2} \mathbf{v} \times \mathbf{H} \right)_{||}, \qquad D'_{\perp} = \left[\frac{\mathbf{D} + \dfrac{1}{c^2} \mathbf{v} \times \mathbf{H}}{\sqrt{1 - \beta^2}} \right]_{\perp},$$

$$H'_{||} = (\mathbf{H} - \mathbf{v} \times \mathbf{D})_{||}, \qquad H'_{\perp} = \left[\frac{\mathbf{H} - \mathbf{v} \times \mathbf{D}}{\sqrt{1 - \beta^2}} \right]_{\perp}, \qquad (4)$$

Substitution of (3) and (4) in (2) yields, for both the parallel and the perpendicular components for which the denominator cancels on the two sides,

$$\mathbf{D} + \frac{1}{c^2} \mathbf{v} \times \mathbf{H} = \varepsilon (\mathbf{E} + \mathbf{v} \times \mathbf{B})$$

$$\mathbf{B} - \frac{1}{c^2} \mathbf{v} \times \mathbf{E} = \mu (\mathbf{H} - \mathbf{v} \times \mathbf{D}). \qquad (5)$$

Here $\mathbf{B}$ may, for example, be eliminated in the first equation by means of the second, so that $\mathbf{D}$ is expressed only in terms of $\mathbf{E}$ and $\mathbf{H}$;[1] similarly

---

[1] Here we make use of the transformation $\mathbf{A} \times (\mathbf{D} \times \mathbf{C}) = \mathbf{B}(\mathbf{A} \cdot \mathbf{C}) - \mathbf{C}(\mathbf{A} \cdot \mathbf{B})$ and of the relation $\varepsilon_0 \mu_0 c^2 = 1$. It should be noted that according to (5a, b) the identity in direction of $\mathbf{D}$ and $\mathbf{E}$ as well as that of $\mathbf{B}$ and $\mathbf{H}$ has ceased to apply even for the isotropic medium.

elimination of **D** leads to an expression of **B** in terms of **E** and **H**. The resulting equations become simpler if they are written separately for the components $\parallel$ and $\perp$. They then become

$$D_{||} = \varepsilon E_{||}, \qquad B_{||} = \mu H_{||} \tag{5a}$$

$$\left[1 - \frac{\varepsilon\mu}{\varepsilon_0\mu_0} \beta^2 \right] \begin{cases} D_\perp = \varepsilon(1 - \beta^2)E_\perp + (\varepsilon\mu - \varepsilon_0\mu_0)\mathbf{v} \times \mathbf{H}, \\ B_\perp = \mu(1 - \beta^2)H_\perp + (\varepsilon_0\mu_0 - \varepsilon\mu)\mathbf{v} \times \mathbf{E}. \end{cases} \tag{5b}$$

Having taken care of the first two Eqs. (2) we now turn to the third Eq. (2), "Ohm's law for moving conductors". What is the relationship of **J'** and **J**? We know from Eq. (26.6) that **J** is the space component of a four-vector **Γ**, whose time component is $ic\rho$. We also know that every four-vector transforms itself like the coordinate vector x_1, x_2, x_3, x_4. We therefore have for the specialized Lorentz transformation (**v** $\parallel$ x):

$$J'_x = \frac{J_x - v\rho}{\sqrt{1 - \beta^2}}, \qquad J'_y = J_y, \qquad J'_z = J_z, \qquad \rho' = \frac{\rho - \dfrac{v}{c^2}J_x}{\sqrt{1 - \beta^2}}.$$

For an arbitrary direction of **v** this becomes

$$J'_{||} = \left[\frac{\mathbf{J} - \rho\mathbf{v}}{\sqrt{1 - \beta^2}} \right]_{||}, \qquad J'_\perp = J_\perp, \qquad \rho' = \frac{\rho - \dfrac{1}{c^2}\mathbf{v}\cdot\mathbf{J}}{\sqrt{1 - \beta^2}}, \tag{6}$$

where we may also write the middle equation in the form

$$J'_\perp = (\mathbf{J} - \rho\mathbf{v})_\perp, \tag{6a}$$

since by definition $v_\perp = 0$.

With this meaning of **J'** and the meaning (3) of **E'** our Ohm's law becomes

$$\left(\frac{\mathbf{J} - \rho\mathbf{v}}{\sqrt{1 - \beta^2}} \right)_{||} = \sigma(\mathbf{E} + \mathbf{v} \times \mathbf{B})_{||},$$

$$(\mathbf{J} - \rho\mathbf{v})_\perp = \sigma\left(\frac{\mathbf{E} + \mathbf{v} \times \mathbf{B}}{\sqrt{1 - \beta^2}} \right)_\perp. \tag{7}$$

These two equations can also be combined into a single one, though only in a somewhat artificial manner. We here make use of the following notation, which is customary also elsewhere in the literature and will be useful later on:[1]

$$\mathbf{E}^* = \mathbf{E} + \mathbf{v} \times \mathbf{B}, \qquad \mathbf{H}^* = \mathbf{H} - \mathbf{v} \times \mathbf{D}. \tag{8}$$

---

[1] The * here employed of course bears no relation to the earlier * of the dual six-vector.

We can then write in place of (7)

$$J - \rho v = \sigma \frac{E^* - \frac{v}{c}\left(\frac{v}{c} \cdot E^*\right)}{\sqrt{1 - \beta^2}}. \tag{9}$$

Since $v_\perp = 0$ the perpendicular component of this is identical with the second Eq. (7). Furthermore, since $v \cdot E^* = v E_{||}$ the parallel component of (9) is

$$(J - \rho v)_{||} = \frac{\sigma}{\sqrt{1 - \beta^2}}\left[E_{||}^* - \frac{v}{c} \cdot \frac{v}{c} E_{||}^*\right] = \frac{\sigma}{\sqrt{1 - \beta^2}} E^*(1 - \beta^2),$$

which agrees with the first Eq. (7). We call

$$J_l = J - \rho v \tag{9a}$$

the "conduction current".

Eq. (9) expresses the fact that the *convection current* ρv and the *conduction current* J_l are superposed and that their differentiation depends on the *frame of reference of the observer*. The reason for this evidently rests in the four-dimensional combination of J and ρv in the four-vector Γ. Just as for the six-vector F the distinction between its electric and magnetic aspect depended on the reference frame of the observer (see p. 241), a change in the reference frame now adds the time component $ic\rho$ of the Γ vector and the corresponding convection current ρv to the conduction current J_l. The former, like the latter, produces a *magnetic field*.

This conclusion was contained already in the *Rowland effect* discovered in 1878. Since we are here dealing exclusively with charge in motion and since therefore the conduction term in (9) is lacking, the convection current ρv alone is magnetically active and takes the place of J in the appropriate Maxwell equation.

The question naturally arises whether also the so-called "free charge",[1] which occurs at the surface of a homogeneous dielectric in an electric field, is magnetically active when the dielectric is set into motion. This led Roentgen to his fundamental experiment:[2] A dielectric plate is placed in a

---

[1] We have avoided this notation elsewhere (like Röntgen, who expressly designated his dielectric plate as uncharged) since the "free charge" is not a charge dimensionally, but a divergence of field strength (see p. 40), in our case a surface divergence of the electric field strength.

[2] W. C. Roentgen, Ann. Physik Vol. 35, p. 264, 1888. In a supplement to this paper Roentgen reports the negative result of an experiment with a rotatably suspended condenser so oriented with respect to the motion of the earth that the "ether wind" passed through the condenser plates. Does this ether wind generate a magnetic field and, as a result, a deflection of the condenser? From our present relativistic point of view the negative result of the experiment is a foregone conclusion. A similar, refined, arrangement became famous at a later date in the Trouton-Noble experiment.

plate condenser, parallel to the plate electrodes, and is moved perpendicularly to the lines of force in the condenser (it was rotated about an axis normal to the condenser plates in the experiment). Does this motion produce a magnetic field? Roentgen could answer this affirmatively and Lorentz, as a result, named the current equivalent to the motion the *Roentgen current*. In agreement with later experiments and considerations of Eichenwald[1] the magnitude of this current for the experimental arrangement in question is:

$$\mathbf{R} = \mathbf{v}(\varepsilon - \varepsilon_0) \mid \mathbf{E}_0 \mid = \mathbf{v} \mid \mathbf{P}_0 \mid . \tag{10}$$

The plate is here assumed to be unmagnetic ($\mu = \mu_0$), its motion a parallel displacement $\mathbf{v}$; $\mathbf{E}_0$ is the field strength in the charged condenser and $\mathbf{P}_0$ the corresponding polarization of the plate, both referring to the plate at rest, as indicated by the subscript 0. Since the "free charge" is concen-

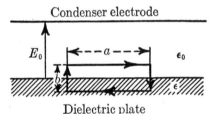

Condenser electrode

Dielectric plate

FIG. 43. Explanation of the Röntgen current. Section perpendicular to the direction of motion of the dielectric plate and the condenser electrode. The location of the Röntgen current is the surface of the plate. The portion linked by a rectangular loop a, b, a, b is indicated by a heavy line.

trated on the surface of the dielectric plate, the Roentgen current is also a pure surface current: It is absent both from the air gap and from the interior of the plate and occurs only at their interface; its direction is that of $\mathbf{v}$, just as for the Rowland current.

We shall show that (10) follows from (5a, b) if $\beta^2 = (\mid \mathbf{v} \mid / c)^2$ is neglected (which is of course fully justified under the conditions of the experiment) and if furthermore $\mu = \mu_0$, and on the right the values of $\mathbf{E}$ and $\mathbf{H}$ for the plate at rest are substituted, namely $E_\perp = E_0$, $E_{\parallel} = 0$, $H = H_0 = 0$. We then find

$$D_{\parallel} = B_{\parallel} = 0, \qquad D = D_\perp = \varepsilon E_0 , \qquad B = B_\perp = \mu_0(\varepsilon_0 - \varepsilon)\mathbf{v} \times \mathbf{E}_0 .$$

Fig. 43 represents a section normal to $\mathbf{v}$ ($\mathbf{v}$ is directed into the plane of the paper) in which the shaded portion below indicates the dielectric plate, the upper portion, the air gap of the condenser. We compute the line integral of $\mathbf{H} = \mathbf{B}/\mu_0$ about the rectangular loop which has been drawn, the direction of the integration being related to the direction of $\mathbf{v}$ by a right-handed screw motion. In view of the direction of $\mathbf{v} \times \mathbf{E}_0$, $\mathbf{H}$ has the direction of the arrow on the upper side of the rectangle a, but

[1] A. Eichenwald, Ann. Physik, Vol. 11, pp. 1 and 241, 1903.

vanishes on it since $\varepsilon = \varepsilon_0$; the same applies for the sides b of the rectangle. Thus there remains only the lower side a of the rectangle, which is traversed in a direction opposite to a. It yields

$$\oint \mathbf{H} \cdot d\mathbf{s} = -a(\varepsilon_0 - \varepsilon)\mathbf{v} \times \mathbf{E}_0 .$$

This magnetic circuit is equal to the surface current flowing through its interior, which in the figure is indicated by the heavy line through the middle of the rectangle. It is $a \cdot \mathbf{R}$ if we call the surface current per unit length $\mathbf{R}$. We thus obtain

$$\oint \mathbf{H} \cdot d\mathbf{s} = -a\,\mathbf{R}, \qquad \mathbf{R} = (\varepsilon - \varepsilon_0)\mathbf{v}\,|\,\mathbf{E}_0\,| = \mathbf{v}\,|\,\mathbf{P}_0\,|, \qquad (10a)$$

where by the vectorial symbol $\mathbf{v}$ we also indicate the positive direction of $\mathbf{R}$ (pointing into the plane of the paper like $\mathbf{v}$). Thus Eq. (10) is verified·

The experiments of Eichenwald in which the dielectric plate and the two condenser plates were rotated about their common normal as a unit, so that the convection currents $\mathbf{v}\,|\,\mathbf{D}\,|$ of the condenser plates (surface density $\omega = |\,\mathbf{D}\,|$) are added to the Roentgen currents $\mathbf{v}\,|\,\mathbf{P}_0\,|$ on the dielectric plate, are of special interest. Since $\mathbf{D}$ and $\mathbf{P}$ differ, a residual magnetic field arises here also, contrary to Hertz's earlier theory and in spite of the opposite signs of the Rowland and Roentgen currents. (The sign of the condenser charge is opposite to that of the charge on the plate induced by it.) Eichenwald (on p. 331) states expressly regarding this residual field: "The magnetic effect is independent of the material of the dielectric." In fact, $\mathbf{D} - \mathbf{P} = \varepsilon_0 \mathbf{E}$ is the vacuum component of $\mathbf{D}$, for which the term "dielectric displacement" is not particularly appropriate, but which is very characteristic for Maxwell's theory and its optical application.

We finally want to mention a kind of inversion of Roentgen's experiment, the experiment of H. A. Wilson[1]: A hollow dielectric cylinder is placed between the electrodes of an uncharged cylindrical condenser in a uniform magnetic field parallel to the cylinder axis. *If the cylinder is rotated the condenser is charged.*

We have followed Minkowski closely so far and believe to have thus even improved on the clarity of the otherwise insurpassable representation in W. Pauli's article in the Enzyklopädie. We shall now establish contact with H. A. Lorentz's article in the Enzyklopädie which, in its mathematical formulation, follows the paper of H. Hertz (1891) and older papers of Helmholtz. To this end we introduce for the quantities referred to the

---

[1] Phil. Trans. Vol. 204, p. 121, 1904; see also H. A. Wilson and M. Wilson, Proc. Roy. Soc., Vol. 89, p. 99, 1913.

laboratory (i.e. the unprimed quantities) on the right of Eq. (1) in place of $\mathbf{E}$ and $\mathbf{H}$ the quantities $\mathbf{E}^*$ and $\mathbf{H}^*$ from (8):

$$\frac{\partial \mathbf{B}}{\partial t} = -\operatorname{curl} \mathbf{E}^* + \operatorname{curl} (\mathbf{v} \times \mathbf{B}),$$

$$\frac{\partial \mathbf{D}}{\partial t} + \mathbf{J} = \operatorname{curl} \mathbf{H}^* + \operatorname{curl} (\mathbf{v} \times \mathbf{D}). \tag{11}$$

We shift the last terms on the right over onto the left and take account of the auxiliary conditions in (1):

$$\operatorname{div} \mathbf{B} = 0, \qquad \operatorname{div} \mathbf{D} = \rho. \tag{11a}$$

We then can write instead of (11)

$$\frac{\partial \mathbf{B}}{\partial t} + \mathbf{v} \operatorname{div} \mathbf{B} - \operatorname{curl} (\mathbf{v} \times \mathbf{B}) = -\operatorname{curl} \mathbf{E}^*,$$

$$\frac{\partial \mathbf{D}}{\partial t} + \mathbf{v} \operatorname{div} \mathbf{D} - \operatorname{curl} (\mathbf{v} \times \mathbf{D}) + \mathbf{J} - \rho\mathbf{v} = \operatorname{curl} \mathbf{H}^*. \tag{11b}$$

We have encountered the aggregates on the left already in Vol. II, (18.7c). There we computed for an arbitrary vector $\mathbf{A}$ and a surface element $d\sigma$ which moves with the velocity $\mathbf{v}$, varying from point to point, and in the process changes size and shape itself, the "$\mathbf{A}$-flux through $d\sigma$"

$$\frac{d}{dt} (A_n \, d\sigma) = \left[\frac{\partial \mathbf{A}}{\partial t} + \mathbf{v} \operatorname{div} \mathbf{A} - \operatorname{curl} (\mathbf{v} \times \mathbf{A}) \right]_n d\sigma.$$

Here we employ the abbreviation introduced by Lorentz[1]

$$\dot{\underline{\mathbf{A}}} = \frac{\partial \mathbf{A}}{\partial t} + \mathbf{v} \operatorname{div} \mathbf{A} - \operatorname{curl} (\mathbf{v} \times \mathbf{A}). \tag{12}$$

The preceding equation then passes into

$$\frac{d}{dt} (A_n \, d\sigma) = \dot{\underline{A}}_n \, d\sigma \tag{12a}$$

or, for a finite surface σ,

$$\frac{d}{dt} \int A_n \, d\sigma = \int \dot{\underline{A}}_n \, d\sigma. \tag{12b}$$

We then obtain in place of (11b) the basic form of the *Maxwell-Minkowski equations in moving bodies (viewed from our laboratory)* given by Lorentz and Pauli:

$$\dot{\underline{B}}_n = -\operatorname{curl} \mathbf{E}^*, \tag{13}$$

$$\dot{\underline{\mathbf{D}}} + \mathbf{J} - \rho\mathbf{v} = \operatorname{curl} \mathbf{H}^*.$$

[1] See Enzyklopaedie, Vol. V, part 2, p. 75, Eq. (5).

Their advantage rests in the fact that they lead directly to the *integral form*:

$$\int \dot{\underline{B}}_n \, d\sigma = - \oint \mathbf{E}^* \cdot ds,$$

$$\int \underline{C}_n \, d\sigma = \oint \mathbf{H}^* \cdot ds, \qquad \underline{C} = \dot{\underline{D}} + \mathbf{J} - \rho\mathbf{v}. \tag{14}$$

At the left we integrate over a surface moving with the velocity $\mathbf{v}$, on the right, over its boundary s, the direction of traversal of s and the normal n of σ being correlated by the right-screw rule. We here recall footnote 3 on p. 280, according to which $\mathbf{v}$ can be arbitrary, i.e. σ and s be attached to an arbitrarily moving and deformed body. In this manner we have arrived at a formulation which is closely related to our original axioms, Eqs. (3) and (4) of §3, and generalizes them greatly. Only in the present generalized form do they do justice to the facts, already mentioned in §3, of the induction for moving conductors and moving magnets.

The *boundary conditions* for moving bodies also follow from Eqs. (14), in the same manner as those for media at rest in §3. *They require the continuity of the tangential components of* $\mathbf{E}^*$ *and* $\mathbf{H}^*$ *as well as of the normal component of* $\mathbf{B}$. Here we must note that the velocity $\mathbf{v}$ occurring in $\mathbf{E}^*$ and $\mathbf{H}^*$ is to be regarded as a constant of the Lorentz transformation, by which the point P of the moving body is transformed to rest. Thus $\mathbf{v}$ has the same value on the two sides of the boundary surface, namely that in the point P, or is at least continuous in passing through the boundary.

The situation is different if $\mathbf{v}$ jumps discontinuously from the value 0 (laboratory) to the value $\mathbf{v}$ (moving solid body). We consider in particular the case, important for unipolar induction, that the field is stationary ($\partial/\partial t = 0$) and the surface of the body is displaced with the velocity $\mathbf{v}$ ($v = v_{\text{tang}}$). We shall show that then *not the tangential components of* $\mathbf{E}^*$, $\mathbf{H}^*$, but *the tangential components* (as seen from the laboratory) *of* $\mathbf{E}$, $\mathbf{H}$ *must be continuous along the boundary layer.*

We note in preparation that the two conditions which are here compared signify the same for a tangential direction of $\mathbf{E}$ or $\mathbf{H}$ *parallel* to $\mathbf{v}$ (because of the meaning of the vector products in (8)), but that they are actually contradictory for every other tangential direction, in particular that *at right angles* to $\mathbf{v}$.

Let us consider now, just as in Fig. 3, a rectangular loop $\Delta s \, \Delta h$, which is initially placed normal to the boundary surface; this is now distorted, since the side Δs parallel to the boundary within the body is displaced, whereas the opposite side, in vacuum, remains fixed. Then, since $\partial \mathbf{B}/\partial t = 0$ and div $\mathbf{B} = 0$, $\dot{\underline{B}}$ is, by (12), equal to $-\text{curl } (\mathbf{v} \times \mathbf{B})$; the integral on the left of the first Eq. (14) becomes, making use of Stokes' theorem,

$$-\int \text{curl}_n \, (\mathbf{v} \times \mathbf{B}) \, d\sigma = -\oint (\mathbf{v} \times \mathbf{B}) \cdot d\mathbf{s},$$

i.e. in general *not equal to* 0 as for constant or continuously varying $\mathbf{v}$.

On the other hand, the integral on the right of the same equation, carried out over the same distorted loop, becomes in view of the meaning of $\mathbf{E}^*$

$$-\oint \mathbf{E} \cdot d\mathbf{s} - \oint (\mathbf{v} \times \mathbf{B}) \cdot d\mathbf{s}.$$

Equating the two expressions leads to the requirement

$$\oint \mathbf{E} \cdot d\mathbf{s} = 0, \quad \text{i.e. continuity of } E_{\text{tang}},$$

in accord with our earlier conclusion (3.9).

The same consideration applied to the second Eq. (14) yields (since $\partial \mathbf{D}/\partial t = 0$ and $\text{div} \, \mathbf{D} = \rho$)

$$\underline{\mathbf{C}} = \mathbf{v}\rho - \text{curl} \, (\mathbf{v} \times \mathbf{D}) + \mathbf{J} - \rho\mathbf{v} = -\text{curl} \, (\mathbf{v} \times \mathbf{D}) + \mathbf{J}$$

and for the integrals on the left and right sides of the second Eq. (14)

$$-\oint (\mathbf{v} \times \mathbf{D}) \cdot d\mathbf{s} + \int J_n \, d\sigma \quad \text{and} \quad \oint \mathbf{H} \cdot d\mathbf{s} - \oint (\mathbf{v} \times \mathbf{D}) \cdot d\mathbf{s}, \text{respectively.}$$

Equating of the two leads to

$$\int J_n \, d\sigma = \oint \mathbf{H} \cdot d\mathbf{s}.$$

If in the limit $\Delta h \to 0$ the surface integral over $\mathbf{J}$ is put equal to zero as in Eq. (3.8) (see also footnote 1 at that point) we obtain

$$\oint \mathbf{H} \cdot d\mathbf{s} = 0, \quad \text{i.e. continuity of } H_{\text{tang}}. \tag{15a}$$

This closes our consideration of the boundary conditions in the special case of a moving interface between two different media.

The existence of the *Rowland* and *Roentgen currents* attests the fact that the preceding theory is not only of importance for the large velocities of the theory of relativity. The same follows from the problem of *unipolar induction* which has been famous since the days of Arago and Faraday. The literature on this subject is voluminous and by no means free of contradictions, since this problem is concerned with the exact laws of the electrodynamics of moving bodies. We shall discuss this problem only qualitatively here and defer quantitative considerations to Problem IV.1. Furthermore, we are primarily interested in the fields which occur here; hence we pass over the phenomena of motion, which are realized in apparatus of many types and have claimed most attention in experimental work.

If, for example a bar magnet, suitably supported, is rotated about its axis, induction currents arise in a wire of which one end glides, for example, on the middle of the magnet while the other is connected to the bearing at one of the ends of the axis of the magnet. Since here only the magnet pole adjoining this end is effective, we speak of "unipolar induction". This arrangement has been employed not only in laboratory experiments, but at times also on a large scale in electric generators.

We simplify the statement of the problem if we separate the conductor from the body generating the magnetic field. Let us consider for example a copper disk between the pole pieces of an electromagnet. It is known that such a "Faraday disk" is raised to incandescence if maintained in rotation and that alternatively an initial rotational momentum of the disk is rapidly damped by the magnetic field.

This occurs, however, only for an inhomogeneous field, such as is normally realized experimentally, where the disk extends beyond the innermost homogeneous portion of the field of the pole pieces. In order to deal with a well-defined and easily solvable problem we assume that the magnetic field is uniform throughout and introduce into it a metal rod with its axis perpendicular to the magnetic field $\mathbf{B}$, which we set into uniform translatory motion along its axis. Its surface is charged hereby. No Joule heat arises in the interior since the conduction current is everywhere zero; the total charge is of course also zero. Interior and exterior field join continuously, but with discontinuous normal gradient, corresponding to the presence of surface charge. The *interior field* is perpendicular both to the axis of the rod and to $\mathbf{B}$ and can be given immediately for any form of the cross section.

In contrast with the interior field, the *exterior field* cannot be given immediately but requires the solution of a *boundary-value problem*: the continuous fitting of the potential in the exterior to the surface values of the potential known from the interior field. In the special case of the circular cross section this boundary-value problem is readily solved; see Problem IV.1.

Along with the interior field the potential difference between two surface points is determined. If, by means of sliding contacts on a connecting wire, it is to be used for the generation of current, the interior is no longer free of current. The present description of the interior field then becomes invalid.

A static magnetic field, arising from the Rowland currents at the surface of the rod, occurs along with the electric field. It is, however, evidently very small compared to the original inducing field and can hence be neglected.

Our description applies throughout for an observer at rest in the laboratory; for an observer moving with the rod the electric field within the rod is zero.

In the actual realization of the experiment the rod is of course replaced by a metallic body of revolution and the translation by a rotation about the axis of symmetry of the latter. A mathematical difficulty which arises here is also indicated in Problem IV.1.

§35. The Ponderomotive Forces and the Stress-Energy Tensor

We return to §31 and generalize the concepts introduced there from vacuum to a body of arbitrary ε, μ, which, however, we shall assume to be both *homogeneous* and *isotropic*, although the anisotropic body would be of special interest in connection with electro- and magnetostriction. We shall, furthermore, regard the body as at rest, since, for the questions at issue, we can place our frame of reference on this body.

Our earlier definitions of §26

$$F = (c\mathbf{B}, -i\mathbf{E}), \qquad f = (\mathbf{H}, -ic\mathbf{D}), \qquad \mathbf{\Gamma} = (\mathbf{J}, ic\rho) \qquad (1)$$

as well as the Maxwell equations in the differential form given there

$$\mathbf{Div}\ F^* = 0, \qquad \mathbf{Div}\ f = \mathbf{\Gamma} \qquad (1a)$$

retain their validity; it can be readily demonstrated that the factors c arising in (1) are not derived from the vacuum constants ε_0, μ_0, but from the time measurement $x_4 = ict$, which, in the special theory of relativity, applies quite generally for all ponderable media. However, the relation

$$f = \sqrt{\frac{\varepsilon_0}{\mu_0}}\ F \qquad (2)$$

between excitation and field is to be changed to

$$f = \left\{ \begin{matrix} \mu_0/\mu \\ \varepsilon/\varepsilon_0 \end{matrix} \right\} \sqrt{\frac{\varepsilon_0}{\mu_0}}\ F, \qquad (2a)$$

where the upper line refers to the space-space, the lower to the space-time components of f and F.

A comparison of (1) and (2a) shows that this change leads in fact to the required relations between excitation and field:

$$\mathbf{H} = \frac{\mu_0}{\mu} \sqrt{\frac{\varepsilon_0}{\mu_0}}\ c\mathbf{B}, \quad \text{i.e.} \quad \mathbf{H} = \mathbf{B}/\mu,$$

$$-ic\mathbf{D} = \frac{\varepsilon}{\varepsilon_0} \sqrt{\frac{\varepsilon_0}{\mu_0}}\ (-i\mathbf{E}), \quad \text{i.e.} \quad \mathbf{D} = \varepsilon\mathbf{E}.$$

We see from (2a) that the simple proportionality (2) between f and F which was characteristic for vacuum passes over, for the ponderable body, into a type of linear vector function with two different proportionality constants for the space-space and the space-time components. In the anisotropic body this is replaced by a much more general vector function (see p. 28) with in general 12 different material constants.

Starting from the universally valid representation of the Lorentz force density in (31.1) we convince ourselves, by a critical consideration of the individual steps, that the transformations up to Eq. (31.8) remain unaltered, and are not influenced by the different proportionality factors for the electric and magnetic quantities in (2a). The same statement applies also for the diagonal elements of the tensor T, so that the diagonal sum of the latter retains its earlier value

$$\sum_{n=1}^{4} T_{nn} = 0. \tag{3}$$

The same holds for the nondiagonal elements T_{nm}, provided only that n and m differ from 4. On the other hand we compute from (31.4) e.g.

$$T_{14} = -\frac{1}{c} (F_{12} f_{42} + F_{13} f_{43}) \qquad T_{41} = -\frac{1}{c} (F_{42} f_{12} + F_{43} f_{13})$$

$$= -ic(B_z D_y - B_y D_z) \qquad = -\frac{i}{c} (E_y H_z - E_z H_y) \tag{4}$$

$$= -ic\varepsilon\mu(\mathbf{E} \times \mathbf{H})_x = -\frac{i}{c} \frac{\varepsilon\mu}{\varepsilon_0 \mu_0} S_x, \qquad = -\frac{i}{c} (\mathbf{E} \times \mathbf{H})_x = -\frac{i}{c} S_x.$$

If the subscript 1 is here replaced by 2 or 3 the result remains the same except that the subscript x of $\mathbf{S}$ is replaced by y and z, respectively.

This different behavior of the two groups of the T_{nm} (n and $m \neq 4$ as against n or $m = 4$) has the result that whereas the three-dimensional stress tensor can still be written in the form (31.10), the complete four-dimensional tensor T takes on the *asymmetric* form

$$T = \left(\begin{array}{c|c} \sigma & -\dfrac{i}{c} \dfrac{\varepsilon\mu}{\varepsilon_0 \mu_0} \mathbf{S} \\ \hline -\dfrac{i}{c} \mathbf{S} & W \end{array} \right). \tag{5}$$

This asymmetry has questionable consequences. We know from hydrodynamics and the theory of elasticity that an asymmetric stress tensor leads to torques which do not correspond to observation (see e.g. Vol. II, §10 and §8). Also in electrodynamics torques may be deduced from the asymmetric character of our tensor T with respect to its principal diagonal; these torques are very small and scarcely observable, but are even so improbable. M. Abraham has hence proposed a *symmetric* form of the tensor T, differing from Minkowski's, and M. von Laue has followed Abraham's suggestion.[1] The two points of view are compared with respect to their physical consequences by W. Pauli in his oft-quoted article in the Enzyklopädie, p. 665.

Following once again Minkowski, we deduce from the matrix (5) that the fourth component of our earlier Eq. (31.3) remains unchanged and corresponds to *Poynting's theorem* also in a ponderable body. The first three components of the same equation, which are affected by the changed upper portion of our matrix (5) on the other hand, lead to a definition of the electromagnetic momentum density differing from (31.13). Whereas we found for vacuum

$$\mathbf{G} = \frac{1}{c^2} \mathbf{S} = \frac{1}{c^2} \mathbf{E} \times \mathbf{H} \tag{6}$$

we now obtain the different (though, of course, dimensionally equal) expression

$$\mathbf{G} = \frac{1}{c^2} \frac{\varepsilon \mu}{\varepsilon_0 \mu_0} \mathbf{S} = \mathbf{D} \times \mathbf{B}. \tag{6a}$$

This conclusion of Minkowski's theory is also not universally accepted.

As noted initially, we have been able to confine ourselves to bodies at rest in this section. In view of the behavior of the world tensor T in a Lorentz transformation, known to us from (28.20), our formulas can be transferred directly to bodies in motion. The problem of the ponderomotive forces would be solved for them also as soon as the ultimate form of our tensor T for bodies at rest had been determined. The fact that this has not been accomplished in a unique fashion signifies physically really only an esthetic defect and is certainly no serious objection to the theory of relativity. In fact from our present electron-theoretical standpoint all processes take place in vacuum, for which the question of the stress-energy tensor has received a satisfying and generally recognized solution in §31. From this point of view the ponderable bodies with their continuous material constants ε, μ are simply convenient abstractions and are not physical realities.

---

[1] In his excellent textbook "Die Relativitätstheorie," Vol. I: "Das Relativitätsprinzip der Lorentztransformation," and Vol. II: "Die allgemeine Relativitätstheorie und Einsteins Lehre von der Schwerkraft," which have been published as Nrs. 38 and 58 of the series "Wissenschaft" by Vieweg.

36. The Energy Loss of an Accelerated Electron by Radiation and Its Reaction on the Motion

We know that, unlike the electron in uniform motion, the accelerated electron radiates. According to (19.24) the energy radiated per unit time is, for a velocity small compared with c,

$$S = \frac{e^2 \, \dot{v}^2}{6\pi\varepsilon_0 \, c^3}.$$

(1)

This energy loss must of course find expression in the equation of motion of the electron. To take account of it we will replace it by an equivalent force. Consider the effect of a short acceleration interval from t_1 to t_2. Before and after the interval, as well as at its limits, the motion is to be regarded as uniform, i.e. $\dot{v}$ as equal to zero. In view of the briefness of the interval the velocity is changed but little, so that we may put $\beta_1 \cong \beta_2 \cong \beta$. We call the desired force the "reaction force of the radiation" and denote it by $\mathbf{R}$. (Please excuse the use of the same symbol $\mathbf{R}$ as for the Roentgen current and, before that, for the impedance operator!) It must satisfy the condition that the work done by it on the electron in the interval from t_1 to t_2 be equal to the negative radiation loss of the electron, i.e.

$$\int \mathbf{R} \cdot d\mathbf{s} = -\int_{t_1}^{t_2} S \, dt.$$

(2)

From the identity

$$\dot{v}^2 = \frac{d}{dt}(v\dot{v}) - v\ddot{v}$$

and our assumption $\dot{v}(t_1) = \dot{v}(t_2) = 0$ we find

$$\int \dot{v}^2 \, dt = v\dot{v} \Big|_{t_1}^{t_2} - \int \ddot{v} v \, dt = -\int \ddot{v} \, ds,$$

so that, by (1) and (2),

$$\int \mathbf{R} \cdot d\mathbf{s} = \frac{e^2}{6\pi\varepsilon_0 \, c^3} \int \ddot{v} \, ds.$$

(3)

We thus obtain as the simplest formulation

$$|R| = \frac{e^2 \, \ddot{v}}{6\pi\varepsilon_0 \, c^3};$$

(4)

furthermore, it may be shown that other expressions consistent with (3) deviate from (4) only by terms of a smaller order of magnitude (see the discussion after Eq. (27)). In the magnetic cgs-system this becomes, by (16.30),

$$| R | = \frac{2}{3} \frac{e^2 \ddot{v}}{c}; \tag{4a}$$

this may be compared with Larmor's formula (19.24b).

We shall study the effect of the reaction force for the very simple case of an electron vibrating about its position of rest, which, as in §19c, may serve as an idealized model of a light source. Let the vibration be recti-linear; in view of the reaction force it is damped. We call the distance of the electron from its position of rest ξ and set

$$\xi = \xi_0 \, e^{-i\omega t}, \qquad \omega = \omega_0(1 + \alpha). \tag{5}$$

$\omega_0 = 2\pi/\tau$ is the angular frequency in the absence of damping, τ the corre-sponding period, α a complex number; it is very small in absolute value and, for our expression for ξ, must have a negative imaginary part. We shall demonstrate both facts.

The equation of motion of the electron is

$$m_0 \ddot{\xi} + f\xi = R. \tag{6}$$

The restoring force, which may arise in some fashion from the atomic binding, has been set equal to $-f\xi$ and been transferred to the left. We divide (6) by m_0 and put

$$\frac{f}{m_0} = \omega_0^2, \tag{7}$$

$$\frac{R}{m_0} = \frac{a}{c} \dddot{\xi}. \tag{7a}$$

(7) follows from Eq. (5), according to which ω_0 is the characteristic fre-quency of the oscillation for $R = 0$; in (7a) a signifies, by (4), a length of the same order of magnitude as the electron radius in (33.6). Eq. (6) then becomes

$$\ddot{\xi} + \omega_0^2 \xi = \frac{a}{c} \dddot{\xi}. \tag{8}$$

Substitution from (5) yields, after cancellation of $\omega_0^2\xi$,

$$-(1 + \alpha)^2 + 1 = i \frac{a}{c} \omega_0(1 + \alpha)^3 = 2\pi i \frac{a}{\lambda} (1 + \alpha)^3. \tag{9}$$

λ is the wave-length of the emitted light; even in the x-ray region it is very large compared to the radius a of the electron. We may hence neglect α as compared with 1 on the right side of the equation and α^2 as compared with 2α on the left side. We thus find from (9)

$$\alpha = -\pi i \frac{a}{\lambda}. \tag{10}$$

The sign of our result agrees with the expectation expressed at (5). The fact that, in our approximate calculation, α has become purely imaginary indicates that the period of the oscillation is not changed appreciably by the reaction force, just as in (18.9d), where the period of the quasistationary current oscillation did not depend materially on the resistance. Substitution of (10) in (5) yields

$$\left|\frac{\xi}{\xi_0}\right| = \exp\left(-\pi \frac{a}{\lambda} \omega_0 t\right) = \exp\left(-2\pi^2 \frac{a}{\lambda} \frac{t}{\tau}\right). \tag{11}$$

The amplitude is hence reduced by a factor $1/e$ in a time, measured in periods of the oscillation,

$$\frac{t}{\tau} = \frac{\lambda}{2\pi^2 a}. \tag{12}$$

The corresponding light path measured in wave-lengths, x/λ, has the same value. For $\lambda = 4\cdot 10^{-5}$ cm, $a = 2\cdot 10^{-13}$ cm (12) yields

$$\frac{x}{\lambda} = 10^7, \qquad x = 400 \text{ cm} = 4\text{M}. \tag{12a}$$

The "distance of coherence" of light waves, measured for particularly sharp (i.e. particularly monochromatic) spectral lines, is of the same order of magnitude. There are no absolutely sharp spectral lines. Every broken or damped wave train, when subjected to Fourier analysis (see Vol. VI, Exercise I.4), yields a finite spectral width (more precisely, half-value width). The Doppler effect, which is the result of the thermal motion of the emitting particles and is hence temperature-dependent, has the same consequence. The reciprocal of the damping time t given by (12) is defined as the *natural classical line width*. With the notation $D = 1/t$ and with $\tau = \lambda/c$ we obtain

$$D = 2\pi^2 \frac{ac}{\lambda^2}$$

or, with the value of a given by (33.6),

$$D = \frac{\pi}{3} \frac{e^2}{\varepsilon_0 m_0 c} \frac{1}{\lambda^2}. \tag{12b}$$

This classical line width is the lower limit of the observable line width, at the lowest possible temperatures (elimination of the Doppler effect) and

the lowest possible pressures (elimination of so-called collision damping)
We specialize the expression for the cgs-units customary in spectroscopy
($e^2/\varepsilon_0 = 4\pi c^2 e_{\mathrm{magn}}^2$ by Eq. (16.30)) and find

$$D = \frac{4\pi^2}{3} \frac{c e_{\mathrm{magn}}^2/m_0}{\lambda^2},$$

$$e_{\mathrm{magn}} = 1.60 \cdot 10^{-20}, \quad c = 3.10^{10},$$

$$e_{\mathrm{magn}}/m_0 = 1.76 \cdot 10^7.$$

(12c)

If in particular we set $\lambda = 4 \cdot 10^{-5}$ cm as before, we obtain

$$D = 7 \cdot 10^7 \ \mathrm{sec}^{-1}.$$

(12d)

This is very small, even compared with the minute frequency separation
of the hydrogen doublet of the first Balmer line:

$$\Delta\nu_H = \frac{Rc\alpha^2}{2^4} \cong 10^{10} \ \mathrm{sec}^{-1}$$

$Rc = $ Rydberg frequency

$\alpha = $ fine-structure constant

The reaction force **R** plays an important role in Planck's theory of
black-body radiation (see Vol. V). It determines the amplitude to which
a linear oscillator is raised in equilibrium with thermal radiation, from
which it then accomplishes its emission of quanta.

So far we have considered only the slowly moving electron insofar as we
have determined the force of reaction only from the standpoint of an
observer moving with the electron. The theory of relativity makes it
possible, however, to change the frame of reference and to determine then
the reaction force for an electron moving with arbitrary velocity. The
fact that a result differing materially from (4) will be obtained follows from
the fact that Eq. (1) then ceases to be valid and, instead, (30.11) yields
for the energy radiated per unit area and per unit time at an angle ϑ with
respect to the direction of motion

$$S = \frac{e^2 \dot{v}^2}{16\pi^2 \varepsilon_0 c^3 r^2} \frac{\sin^2 \vartheta}{(1 - \beta \cos \vartheta)^6}.$$

(13)

The acceleration is here assumed to be longitudinal, i.e. in the direction
of motion. It should furthermore be noted that (13) refers to the time
scale of the observer at rest, whereas we must know for the determination
of **R** the radiation per unit time in the system of the moving electron. The
retardation relation (29.10b) exists between the two time scales (t, time of
the observer, τ, time of the electron as well as "intrinsic time" of the
latter):

$$\tau = t - \frac{r}{c}, \qquad d\tau = dt + \frac{v_r}{c} d\tau, \qquad \frac{dt}{d\tau} = 1 - \frac{v_r}{c} = 1 - \beta \cos \vartheta.$$

Thus to refer S to the time scale of the moving electron we must multiply 13) by

$$\frac{dt}{d\tau} = 1 - \beta \cos \vartheta. \tag{14}$$

Furthermore, to pass from S to the total radiation S of the electron we must integrate over the sphere of radius r and with the element of angle $l_\omega = 2\pi \sin \vartheta \, d\vartheta$. With the abbreviation $u = 1 - \beta \cos \vartheta$ we thus obtain

$$2\pi \int_0^\pi \frac{\sin^2 \vartheta \sin \vartheta \, d\vartheta}{(1 - \beta \cos \vartheta)^5} = \frac{2\pi}{\beta} \int_{1-\beta}^{1+\beta} \left\{ 1 - \left(\frac{1 - u}{\beta} \right)^2 \right\} \frac{du}{u^5}.$$

This integral may be evaluated in an elementary manner and yields

$$\frac{2\pi}{3} \frac{4}{(1 - \beta^2)^3}.$$

Hence

$$S = \frac{e^2 \dot{v}^2}{6\pi \varepsilon_0 c^3} \frac{1}{(1 - \beta^2)^3}. \tag{15}$$

The value of **R** which is now sought—we shall call it **R′**—must (for longitudinal acceleration) be implicitly related to (15) by (2). However, it is not determined uniquely hereby; for this it would be necessary, following Abraham, to add conservation of momentum to conservation of energy. We hence prefer to proceed from the relativistic equation of motion of the electron, which will yield an explicit value for **R′**; at the same time we need no longer limit ourselves to the longitudinal case.

In accord with Eq. (32.4) etc. and adding the desired reaction force **R′** we write the equation of motion in the form:

$$m_0 \mathbf{W} = \mathbf{F} + \mathbf{R}'. \tag{16}$$

W is the four-vector of the acceleration introduced in (27.18b), **F**, the four-force acting at any moment, which in the electrodynamic case is related to the Lorentz force **K** by Eq. (28.19b):

$$\mathbf{F} = \eta \mathbf{K}, \qquad \eta = (1 - \beta^2)^{-1/2}. \tag{17}$$

W and **F** are perpendicular to the world line of the electron in the four-dimensional meaning of the term, i.e. (see (27.18c) and (28.19c))

$$\mathbf{V} \cdot \mathbf{W} = 0 \quad \text{and} \quad \mathbf{V} \cdot \mathbf{F} = 0. \tag{18}$$

Accordingly (16) leads to the requirement that **R′** also be perpendicular to the world line:

$$\mathbf{V} \cdot \mathbf{R}' = 0. \tag{18a}$$

It would seem most reasonable, following (4), to define $\mathbf{R}'$ by

$$\mathbf{R}' = b\,\dot{\mathbf{W}}, \qquad b = \frac{e^2}{6\pi\varepsilon_0\,c^3}, \qquad \dot{\mathbf{W}} = \frac{d\mathbf{W}}{d\tau} \tag{19}$$

(dots will indicate differentiation with respect to the intrinsic time also in what follows). This would, in fact, be a Lorentz-invariant definition of $\mathbf{R}$ which, specialized to the frame of reference of the electron, would agree directly in its space components with (4). However, this definition would contradict the requirement (18a). We hence modify it to

$$\mathbf{R}' = b(\dot{\mathbf{W}} + \alpha\mathbf{V}) \tag{19a}$$

and determine the constant α here introduced from the condition

$$\mathbf{V}\cdot\dot{\mathbf{W}} + \alpha\mathbf{V}\cdot\mathbf{V} = 0, \qquad \alpha = -\frac{\mathbf{V}\cdot\dot{\mathbf{W}}}{\mathbf{V}\cdot\mathbf{V}}. \tag{20}$$

Like (19) this definition of $\mathbf{R}'$ satisfies the requirement of being identical with (4) in the frame of reference of the electron, since the first three components of $\mathbf{V}$ vanish here; furthermore if, as in (4), correction terms of a smaller order of magnitude are neglected the definition is also unique.

The value (20) of α can be further simplified: First, by (27.18a) $\mathbf{V}\cdot\mathbf{V} = -c^2$; second, we may deduce from $\mathbf{V}\cdot\mathbf{W} = 0$ by differentiation with respect to τ:

$$\mathbf{V}\cdot\dot{\mathbf{W}} + \dot{\mathbf{V}}\cdot\mathbf{W} = 0, \qquad \mathbf{V}\cdot\dot{\mathbf{W}} = -\dot{\mathbf{V}}\cdot\mathbf{W} = -\mathbf{W}\cdot\mathbf{W}.$$

Hence we may write

$$\alpha = \frac{1}{c^2}\,\mathbf{W}\cdot\mathbf{W} \tag{20a}$$

and

$$\mathbf{R}' = b\left(\dot{\mathbf{W}} - \frac{\mathbf{W}\cdot\mathbf{W}}{c^2}\,\mathbf{V}\right). \tag{21}$$

This is the very concise formulation of the reaction force, valid in every frame of reference. The conciseness is lost when we pass over to three dimensions, to permit comparison with the formulations known from the literature. We proceed from Eqs. (27.18) and (27.18b):

$$\mathbf{V} = (\eta\mathbf{v}, ic\eta), \qquad \mathbf{W} = \dot{\mathbf{V}} = (\dot{\eta}\mathbf{v} + \eta\dot{\mathbf{v}}, ic\dot{\eta}) \tag{22}$$

and compute with the aid of the definition (17) of η

$$\eta^2 = \frac{1}{1-\beta^2}, \qquad \eta\dot{\eta} = \frac{1}{(1-\beta^2)^2}\frac{\mathbf{v}\cdot\dot{\mathbf{v}}}{c^2} = \frac{\eta^4}{c^2}\,\mathbf{v}\cdot\dot{\mathbf{v}}, \qquad \frac{\dot{\eta}}{\eta} = \frac{\eta^2}{c^2}\,\mathbf{v}\cdot\dot{\mathbf{v}}. \tag{22a}$$

We then obtain from (22)

$$W = \eta\left(\frac{\eta^2}{c^2}(\mathbf{v}\cdot\dot{\mathbf{v}})\mathbf{v} + \dot{\mathbf{v}}, \frac{i}{c}\eta^2\mathbf{v}\cdot\dot{\mathbf{v}}\right),$$

$$\mathbf{W}\cdot\mathbf{W} = \eta^2\left\{\frac{\eta^4}{c^4}(\mathbf{v}\cdot\dot{\mathbf{v}})^2 v^2 + \frac{2\eta^2}{c^2}(\mathbf{v}\cdot\dot{\mathbf{v}})^2 + \dot{\mathbf{v}}^2 - \frac{\eta^4}{c^2}(\mathbf{v}\cdot\dot{\mathbf{v}})^2\right\}. \tag{22b}$$

The first and last term of the { } can be combined to yield

$$\frac{\eta^4}{c^2}(\mathbf{v}\cdot\dot{\mathbf{v}})^2\left(\frac{v^2}{c^2} - 1\right) = -\frac{\eta^2}{c^2}(\mathbf{v}\cdot\dot{\mathbf{v}})^2,$$

which combines with the second term. We thus obtain

$$\mathbf{W}\cdot\mathbf{W} = \eta^2\left\{\frac{\eta^2}{c^2}(\mathbf{v}\cdot\dot{\mathbf{v}})^2 + \dot{\mathbf{v}}^2\right\} \tag{23}$$

$$\frac{\mathbf{W}\cdot\mathbf{W}}{c^2}\mathbf{V} = \frac{\eta^3}{c^2}\left\{\frac{\eta^2}{c^2}(\mathbf{v}\cdot\dot{\mathbf{v}})^2 + \dot{\mathbf{v}}^2\right\}\mathbf{v}, \cdots. \tag{23a}$$

In the last equation we have written down only the three space components of the four-dimensional vector. We do the same in the computation of $\dot{\mathbf{W}}$, i.e. the differentiation of (22b) with respect to τ:

$$\dot{\mathbf{W}} = \frac{3\eta^2\dot{\eta}}{c^2}(\mathbf{v}\cdot\dot{\mathbf{v}})\mathbf{v} + \frac{\eta^3}{c^2}\{\dot{v}^2\mathbf{v} + (\mathbf{v}\cdot\ddot{\mathbf{v}})\mathbf{v} + (\mathbf{v}\cdot\dot{\mathbf{v}})\dot{\mathbf{v}}\} + \dot{\eta}\dot{\mathbf{v}} + \eta\ddot{\mathbf{v}}.$$

If we take account of (22a) we find

$$\dot{\mathbf{W}} = \frac{3\eta^5}{c^4}(\mathbf{v}\cdot\dot{\mathbf{v}})^2\mathbf{v} + \frac{\eta^3}{c^2}\{\dot{v}^2\mathbf{v} + (\mathbf{v}\cdot\ddot{\mathbf{v}})\mathbf{v} + 2(\mathbf{v}\cdot\dot{\mathbf{v}})\dot{\mathbf{v}}\} + \eta\ddot{\mathbf{v}}. \tag{24}$$

We obtain for the difference of (24) and (23a)

$$\frac{\mathbf{R}'}{b} = \frac{2\eta^5}{c^4}(\mathbf{v}\cdot\dot{\mathbf{v}})^2\mathbf{v} + \frac{\eta^3}{c^2}\{(\mathbf{v}\cdot\ddot{\mathbf{v}})\mathbf{v} + 2(\mathbf{v}\cdot\dot{\mathbf{v}})\dot{\mathbf{v}}\} + \eta\ddot{\mathbf{v}}. \tag{25}$$

Finally we pass from the intrinsic time of the electron, to which $\dot{\mathbf{v}}$ and $\ddot{\mathbf{v}}$ are referred, to the time scale t of the observer, in which we shall denote the corresponding quantities by $\mathbf{v}'$ and $\mathbf{v}''$. We set

$$\dot{\mathbf{v}} = \frac{d\mathbf{v}}{d\tau} = \eta\frac{d\mathbf{v}}{dt} = \eta\mathbf{v}', \qquad \dot{\eta} = \frac{\eta^4}{c^2}\mathbf{v}\cdot\mathbf{v}'$$

$$\ddot{\mathbf{v}} = \frac{d}{d\tau}(\eta\mathbf{v}') = \dot{\eta}\mathbf{v}' + \eta^2\mathbf{v}'' = \frac{\eta^4}{c^2}(\mathbf{v}\cdot\mathbf{v}')\mathbf{v}' + \eta^2\mathbf{v}''.$$

Thus we obtain from (25)

$$\frac{\mathbf{R}'}{b} = \frac{3\eta^7}{c^4}(\mathbf{v}\cdot\mathbf{v}')^2\mathbf{v} + \frac{\eta^5}{c^2}(\mathbf{v}\cdot\mathbf{v}'')\mathbf{v} + \frac{3\eta^5}{c^2}(\mathbf{v}\cdot\mathbf{v}')\mathbf{v}' + \eta^3\mathbf{v}''. \tag{25a}$$

This value of $\mathbf{R}'$ is to be employed in the equation of motion (16). If w
transform the latter into the customary form of the momentum equatio
(32.5), i.e. replace $m_0\mathbf{W}$ by $d\mathbf{G}/dt = m_0\mathbf{W}/\eta$ and $\mathbf{F}$ by $\mathbf{K} = \mathbf{F}/\eta$, $\mathbf{R}'$ mus
also be replaced by $\mathbf{R}^* = \mathbf{R}'/\eta$. We thus obtain for $\mathbf{R}^*$ from (25a) afte
substituting the values of b and η:

$$\mathbf{R}^* = \frac{e^2}{6\pi\varepsilon_0 c^3} \frac{1}{1 - \beta^2}$$

$$\left(\mathbf{v}'' + \mathbf{v}' \frac{3\mathbf{v}\cdot\mathbf{v}'}{c^2(1 - \beta^2)} + \frac{\mathbf{v}}{c^2(1 - \beta^2)}\left[\mathbf{v}\cdot\mathbf{v}'' + \frac{3(\mathbf{v}\cdot\mathbf{v}')^2}{c^2(1 - \beta^2)}\right]\right). \tag{26}$$

For small velocities ($\beta \to 0$) we obtain, of course, $\mathbf{R}^* = \mathbf{R}$, i.e. the valu
from Eq. (4). Eq. (26) was derived first by Abraham from electrodynamic
and by v. Laue from the theory of relativity. Our procedure follows th
suggestions of Pauli.[1]

We must still establish the limits of validity of the formulas derive
here. By using the relativity transformation for uniform motion we hav
assumed the acceleration to be "small" without otherwise restricting th
magnitude of the velocity. We are hence dealing with a process of approxi
mation or a series expansion which is broken off. H. A. Lorentz[2] carrie
out this process, representing the retarded potentials as power series o
the relaxation time $t - r/c$ and computing the mechanical force of th
intrinsic field. The first term of this expansion is the *inertial reaction* o
the electron, which, for small velocities,[3] is given by

$$-m_0\dot{\mathbf{v}} = -\frac{e^2\dot{\mathbf{v}}}{6\pi\varepsilon_0 c^2 a}. \tag{27}$$

The reaction force $\mathbf{R}$ of Eq. (4) appears as second term:

$$\mathbf{R} = \frac{e^2\dddot{\mathbf{v}}}{6\pi\varepsilon_0 c^3}. \tag{27a}$$

Lorentz emphasizes that $\mathbf{R}$ is the only term of the expansion which doe
not depend on a ("on the shape of the electron"). The higher terms, whic
are not computed, are of the form

$$m_0\left(\frac{a}{c}\right)^2 \dddot{\mathbf{v}}, \qquad m_0\left(\frac{a}{c}\right)^3 \ddddot{\mathbf{v}}, \cdots \tag{27b}$$

Here a/c is the time required by the light to traverse the "radius of th
electron". The term (27a) can be brought into the same form; in view o

[1] Enzykl. d. Math. Wiss. Vol. V₂, p. 654.

[2] "The Theory of Electrons," Teubner, 1909, Note 18, p. 251.

[3] $\mathbf{v}'$, $\mathbf{v}'' \cdots$ are then identical with, $\dot{\mathbf{v}}$ $\ddot{\mathbf{v}} \cdots$.

the meaning of m_0 it can be written

$$m_0 \, \frac{a}{c} \, \ddot{\mathbf{v}}.$$

The terms of the series must decrease in order that the series may converge, i.e. be practically useful. Hence we must have

$$|\ddot{\mathbf{v}}| < \frac{c}{a} |\dot{\mathbf{v}}|, \qquad |\dddot{\mathbf{v}}| < \frac{c}{a} |\ddot{\mathbf{v}}|, \cdots \tag{28}$$

It is clear however that in processes involving very large energies, such as the flight of an electron close to an atomic nucleus, not only very great accelerations, but also very great changes in acceleration can occur. The termination of the series with the term **R** would then be no longer permissible. The same applies for the acceleration process in the betatron (see Problem III.10) and synchrotron. The formulation of the radiation resistance in such extreme cases constitutes an as yet unsolved problem which has led to many discussions (Wessel, Dirac, Bopp, Stückelberg).[1]

§37. *Approaches to the Generalization of Maxwell's Equations and to the Theory of the Elementary Particles*

Gustav .Mie took the first step in this direction in 1912 in his famous papers[2] "Foundations of a Theory of Matter." Their goal is no less than the generalization of the Maxwell equations so that they include the *existence of the electron.* In order that the generalization may not lose itself in limitless possibilities it is subjected from the start to the principle of relativity and derived from a "world function" which may depend only on Lorentz-invariant quantities. Here a distinction is made—possibly for the first time in a consistent fashion—between entities of intensity and entities of quantity, i.e. written in our notation and units, between

$$F = (c\mathbf{B}, \, -i\mathbf{E}), \qquad \Omega = \left(\mathbf{A}, \frac{i}{c} \, \Psi \right)$$

on the one hand and

$$f = (\mathbf{H}, \, -ic\mathbf{D}), \qquad \Gamma = (\rho\mathbf{v}, \, i\rho c)$$

---

[1] The most recent contributions to this question are given by the papers of W. Heitler and H. W. Peng, Proc. Cambridge Phil. Soc. *38*, 296 (1942) and, from the standpoint of Einstein's latest methods, N. Hu, Proc. R. Irish Academy, *57*, 87 (1947).

[2] Ann. d. Phys: First communication, Vol. 37, p. 511; second communication, Vol. 39, p. 1. The third communication (Vol. 40, p. 1, 1913) deals with the theory of gravitation and is of course outdated, having been originated before the *general* theory of relativity.

on the other. Mie tests the invariants which may be set up with the entities of intensity and the entities of quantity respectively. We record three[1] of the former (see (26.24), (26.21), and (26.4)), suppressing constant factors insofar as they are dimensionally superfluous:

$$\Lambda = \frac{\varepsilon_0}{2} \, (c^2 \mathbf{B}^2 - \mathbf{E}^2), \tag{1}$$

$$M = c\mathbf{E} \cdot \mathbf{B}, \tag{2}$$

$$|\, \mathbf{\Omega} \,|^{\,2} = \mathbf{A}^2 - \Psi^2/c^2. \tag{3}$$

He finds that only the invariants (1) and (3) need be considered for the description of quasistationary processes and constructs a world function[2] W such that at large distance from the electron the ordinary Maxwell equations apply, whereas the equations are modified at the electron and in its immediate neighborhood. Like Schwarzschild's function, the world function is to be integrated over an arbitrary region of the four-dimensional world and to be varied in suitable manner.

What must be the form of the world function if it is to yield the ordinary Maxwell equations at an adequate distance from the electron? According to our experience with Schwarzschild's principle of action we must then have $W = \Lambda$. In fact, in pure vacuum ($\mathbf{\Gamma} = 0$ and no kinetic energy of matter) the kinetic potential K' in Eq. (32.18) reduces to the middle term, proportional to Λ and when subjected to the variation, yields the Maxwell equations of vacuum. At the same time the change in the world function in the neighborhood of the electron is to be such that from it a definite value e of the electron charge (or, at least, of the specific charge e/m of the electron) may be computed. This is certainly not so if the two invariants Λ and $|\, \mathbf{\Omega} \,|^{2}$ are superposed linearly, since then also the resulting differential equations would be linear in the field and potential components respectively and their integrals would consequently involve coefficients which could be chosen arbitrarily. On the other hand, both requirements might be satisfied by the formula

$$W = \Lambda + a \,|\, \mathbf{\Omega} \,|^{\,n} \tag{4}$$

where n is a sufficiently large number. Actually, the second term may then be neglected at sufficiently large distance from the electron since it

---

[1] Weyl, in §28 of his book to be quoted on p. 321, points out a fourth field invariant, constructed from F and Ω.

[2] Mie himself calls the function constructed with entities of intensity "Hamiltonian function H" and designates as "world function" that constructed with entities of quantity. We have taken the liberty of reversing the nomenclature so as to establish correspondence with Schwarzschild's action function. The entities of intensity are then obtained from Mie's world function by differentiation with respect to the entities of quantity.

vanishes as r^{-n}, whereas a singularity of high order occurs at the location of the electron. For mathematical reasons Mie puts specifically $n = 6$; in this manner he obtains a spatially highly concentrated charge distribution, which, however, is not stable in the field of another electron.

It would, after all, have been indeed surprising if the fundamental problem of the elementary particles could have been solved by clever guessing. Today we are convinced that much experimental preparation will be required instead. Nevertheless, blazing the path to the problem was an act of great merit, as is evident from the fact that all later workers in the field have followed in Mie's tracks.

Pauli, in Nr. 64 of his paper in the Enzyklopädie, had already emphasized that dependence on the absolute values of the electrodynamic potentials led to serious difficulties in Mie's theory. Hence we shall avoid use of invariant (3) in the formulations to be discussed below. Born and Infeld[1] in particular utilize in their theory a world function which depends only on invariants (1) and (2). The nonlinearity of the electromagnetic field, which is required here also, follows from the choice of W as nonlinear function of Λ and M. The conjugate four-current Γ drops out along with the potential Ω and, just as in Lorentz's electron theory, must be brought in as a foreign element.

W is chosen so that an infinity of the field at the location of the electron is avoided. In this manner a finite value for the self-energy of the electron is obtained and a difficulty of classical theory, which yields infinite energy for the point electron $(a = 0)$, is circumvented. For quasistationary problems, in which the above invariant (2) does not enter, the formulation of Born and Infeld is

$$W = \varepsilon_0 b^2 \left\{ \sqrt{1 + \frac{2\Lambda}{\varepsilon_0 b^2}} - 1 \right\}. \tag{5}$$

The universal constant b here introduced has the dimension of an electric field strength since, by (1), Λ has the dimension $\varepsilon_0 \mathbf{E}^2$.

This formulation follows the pattern of the action function of classical and relativistic mechanics. In the classical mechanics of the point mass not acted upon by forces we have as the integrand of the Hamiltonian principle the kinetic energy

$$T = \frac{m}{2} v^2. \tag{6}$$

In relativistic mechanics this is replaced by the "kinetic potential" in Eq. (32.9b)

$$K = m_0 c^2 \left\{ 1 - \sqrt{1 - \frac{v^2}{c^2}} \right\}. \tag{6a}$$

[1] M. Born, Proc. Roy. Soc. London (A) *143*, 410, 1933/34; M. Born and L. Infeld, loc. cit. *144*, 425, 1934. M. Born, Ann. de l'Inst. Henri Poincaré, Tome VII.

which passes over into (6) for $v \ll c$ and then becomes independent of c. Similarly (5) passes for $\Lambda \ll \varepsilon_0 b^2$ over into the value $W = \Lambda$, which corresponds to Maxwell's theory, and becomes independent of b. Whereas (6a) sets an upper limit c to v, (6) imposes no restriction on v. Similarly (5) sets an upper limit b to the field strength $\mathbf{E}$ in the electrostatic case ($\mathbf{B} = 0$, $\Lambda = -\varepsilon_0 \mathbf{E}^2/2$), whereas the formula $W = \Lambda$ permits an unlimited increase in the field strength.

The Maxwell equations in vacuum for $\mathbf{E}$, $\mathbf{B}$ and $\mathbf{D}$, $\mathbf{H}$ are retained in the theory of Born and Infeld. In the electrostatic case, to which we shall limit ourselves in the following, we obtain for a centrally symmetric field and a point charge e at $r = 0$, just as for the conventional theory

$$D_r = \frac{e}{4\pi} \frac{1}{r^2}. \tag{7}$$

D_r may also be determined from the world function by the general rule of Mie's theory:

$$D_r = -\frac{\partial W}{\partial E_r}. \tag{7a}$$

According to (5) this yields, with $\mathbf{B} = 0$ and $W = \varepsilon_0 b^2 \{\sqrt{1 - \mathbf{E}^2/b^2} - 1\}$

$$D_r = \frac{\varepsilon_0 E_r}{\sqrt{1 - E_r^2/b^2}}. \tag{8}$$

It follows that

$$\varepsilon_0 E_r = \frac{D_r}{\sqrt{1 + D_r^2/(\varepsilon_0^2 b^2)}}. \tag{9}$$

If (7) is substituted in (9) we obtain

$$E_r = \frac{e}{4\pi\varepsilon_0} \frac{1}{\sqrt{r_0^4 + r^4}}, \qquad r_0 = \sqrt{\frac{e}{4\pi\varepsilon_0 b}}. \tag{10}$$

The quantity r_0 may be regarded as the electron radius. E_r is now everywhere finite, since for $r = 0$ we have $E_r = b = e/(4\pi\varepsilon_0 r_0^2)$. $\mathbf{D}$, on the other hand, becomes infinitely large at the same place. For $r > r_0$, E_r differs little from the Coulomb field $e/(4\pi\varepsilon_0 r^2)$. The electrostatic potential is

$$\Psi(r) = \int_r^\infty E_r \, dr = \frac{e}{4\pi\varepsilon_0 r_0} f\left(\frac{r}{r_0}\right) \tag{11}$$

with

$$f(x) = \int_x^\infty \frac{dy}{\sqrt{1 + y^4}}.$$

At the origin we have

$$\Psi(0) = \frac{e}{4\pi\varepsilon_0 r_0} f(0), \qquad f(0) = 1.854.$$

The Hamiltonian function H is related to our world function W by the general formula

$$H = W + \mathbf{E}\cdot\mathbf{D} - \mathbf{B}\cdot\mathbf{H}.$$

Thus we find in the electrostatic case

$$H = \varepsilon_0 b^2(\sqrt{1 - E_r^2/b^2} - 1) + E_r D_r. \tag{12}$$

If E_r is eliminated with the aid of (9) this contracts to

$$H = \varepsilon_0 b^2(\sqrt{1 + D_r^2/(\varepsilon_0^2 b^2)} - 1), \tag{13}$$

We obtain therefore for the total energy

$$W = 4\pi \int_0^\infty H r^2 dr.$$

Substitution of D_r from (7) yields finally

$$W = \frac{e^2}{4\pi\varepsilon_0 r_0} \int_0^\infty (\sqrt{1 + y^4} - y^2)\, dy \tag{14}$$

The numerical value of the integral is 1.236. If (14) is put equal to the selfenergy of the electron $m_0 c^2$ we obtain

$$r_0 = 1.236 \frac{e^2}{4\pi\varepsilon_0 m_0 c^2}, \tag{15}$$

i.e. very nearly the classical radius a of the electron from (33.9). Then (10) yields for b the value

$$b = \frac{e}{4\pi\varepsilon_0 r_0^2} \cong \frac{e}{4\pi\varepsilon_0 a^2},$$

very nearly equal to the classical field strength E at the "edge of the electron".

In this manner an electron radius of the proper order of magnitude and a very high critical field strength b are obtained. The field of several point charges, also, can be determined in unique fashion. Within these limits the theory of Born and Infeld thus leads to sensible results, although its fundamental formula (5) can claim only heuristic validity.

We shall finally discuss the problem of the "*scattering of light by light*". This problem arose from Dirac's theoretical discovery of the *positron* and *pair production*. (Pair production, i.e. the simultaneous generation of an electron and a positron from hard gamma radiation, was realized experimentally soon afterwards by Irene and Frederic Joliot-Curie, whereas the positron was observed in cosmic radiation by Anderson and Blackett and Occhialini.) It is clear that this problem also involves a change in the Maxwell equations for vacuum which is equivalent to a "non-linear theory of the electromagnetic field".[1] The linear Maxwell equations could not account for such scattering, but would imply that the fields of two inter-penetrating light waves are simply to be superposed.

The problem has been treated quantum mechanically by Euler and Kockel under the direction of Heisenberg[2] and hence lies entirely outside of the framework of our presentation. We can merely indicate the procedure for its solution.

The world function is here chosen so that for weak fields it reduces, as before, to the Lagrange density Λ. In the second approximation it is written as a function of the second degree in Λ and M. Only even powers of M may occur here, however, since for a transition from a right-handed to a left-handed coordinate system $\mathbf{B}$ and hence M change their sign (in the terminology of Euler M is not "mirror-invariant"). The next term of the expansion must hence have the form

$$\alpha\Lambda^2 + \beta\varepsilon_0^2 M^2. \tag{17}$$

In order that α and β may be pure numbers and this second term have the same dimension as the first term Λ, we divide (17) by $\varepsilon_0 \times$ the square of a critical field strength, which, as before, we shall call b. It is here defined as the field strength "at the edge of the electron", i.e. at a distance $r = a$ from its "center" (a = classical electron radius), multiplied by the fine-structure constant $1/137$).

We obtain thus the formula

$$W = \Lambda + \frac{1}{\varepsilon_0 b^2} (\alpha\Lambda^2 + \beta\varepsilon_0^2 M^2) + \cdots \tag{18}$$

As in Mie's theory the components of the entities of quantity $\mathbf{D}$ and $\mathbf{H}$ are obtained from this world function as partial derivatives with respect to the corresponding components of $\mathbf{E}$ and $\mathbf{B}$:

$$\mathbf{D} = -\frac{\partial W}{\partial \mathbf{E}}, \qquad \mathbf{H} = \frac{\partial W}{\partial \mathbf{B}}. \tag{19}$$

---

[1] This is the title of the paper of Born mentioned in the last footnote.

[2] H. Euler and B. Kockel, Naturwiss, *23*, 1935; Euler, Leipzig thesis, Ann. d. Phys. *26*, 1936; Heisenberg and Euler, Z. Physik *98*, 1936. See also the simplified representation in the paper of M. Born cited at the end of footnote 1 on p. 303. The problem of pair production was approached simultaneously from a different angle by R. Serber and E. A. Uhling, Phys. Rev. *48*, 1933.

Since, by (1) and (2),

$$\frac{\partial \Lambda}{\partial \mathbf{E}} = -\varepsilon_0 \mathbf{E}, \qquad \frac{\partial M}{\partial \mathbf{E}} = c\mathbf{B}, \qquad \frac{\partial \Lambda}{\partial \mathbf{B}} = \varepsilon_0 c^2 \mathbf{B} = \mathbf{B}/\mu_0,$$

$$\frac{\partial M}{\partial \mathbf{B}} = c\mathbf{E},$$

we obtain from (19)

$$\mathbf{D} = \varepsilon_0 \left\{ \mathbf{E} + \frac{2}{\varepsilon_0 b^2} (\alpha\Lambda\mathbf{E} - \beta\varepsilon_0 M c\mathbf{B}) + \cdots \right\},$$

$$\mathbf{H} = \frac{1}{\mu_0} \left\{ \mathbf{B} + \frac{2}{\varepsilon_0 b^2} (\alpha\Lambda\mathbf{B} + \beta\varepsilon_0 M \mathbf{E}/c) + \cdots \right\}.$$

(20)

The most difficult part is the determination of the numerical coefficients α and β from Dirac's theory of pair production, for which we refer to the original papers.

The final result is the following: Maxwell's equations for vacuum, relating $\mathbf{B}$, $\mathbf{E}$ and $\mathbf{D}$, $\mathbf{H}$ retain their form. However, just as in the theory of Born and Infeld, $\mathbf{D}$ is no longer proportional to $\mathbf{E}$; a correction term occurs in the expressions for $\mathbf{D}$ and $\mathbf{H}$ which depends on $\mathbf{E}$, $\mathbf{B}$, Λ, and M and is negligible compared with the principal term for weak fields. No arbitrary assumptions of any kind are made here; it is merely presumed that an expansion in ascending powers of the field strength is possible, this being indicated in Eqs. (18) and (20) by $\cdots$. In the paper of Heisenberg and Euler the expansion has been extended by an additional term and has even been expressed in closed form. In any case, this work demonstrates the necessity of modifying Maxwell's equations for extremely strong fields even in vacuum.

§38. General Theory of Relativity; Unified Theory of Gravitation and Electrodynamics

In this paragraph also we must limit ourselves to a mere outline. A full presentation of the subject would require a separate textbook; it would be premature to write such a one at this time since many pertinent questions are as yet undecided.

For the present we shall follow the original presentation of Einstein as recorded particularly effectively in his Princeton lectures.[1] He proceeds entirely in the spirit of Klein's Erlangen program: *Classical physics* belongs to the group of elementary geometry (isotropy of space within itself),

---

[1] The four lectures on the theory of relativity, held at Princeton in May 1921, have been reprinted in A. Einstein, The Meaning of Relativity, 3rd Ed., Princeton University Press, 1950.

extended by the displacement of the time axis along itself. The *special theory of relativity* is founded on the group of the linear orthogonal transformations of the four world coordinates x_1, x_2, x_3, and $x_4 = ict$ (isotropy of the four-dimensional world, Lorentz transformations). We are led to the *general theory of relativity* if we start from the broader group of point transformations which, in Vol. I, p. 16 we have characterized by the formulas

$$x'_k = f_k(x_1, x_2, x_3, x_4), \qquad k = 1, 2, 3, 4. \tag{1}$$

In this manner all possible frames of reference become legitimate, not only those which move with a constant velocity $v < c$ relative to each other. "Space and time lose the last vestige of their absolute character postulated by Newton and become merely means for the description of physical phenomena." Such a program had already been set up by Ernst Mach. However, he gave up with a negative point of view (which, strangely enough, he denoted as positivism) and remained an opponent of Einstein's theory of relativity to the end of his days. The latter, on the other hand, assumed a *positive* attitude by inquiring into those space time relations which are conserved in all point transformations. The general theory of relativity signifies the *invariant* or *covariant* theory of this group of trans-formations.

The basis for this had been created in part by Gauss[1] in his theory of surfaces and by Riemann[2] in his initiation lecture.

Gauss studied the *inner* properties of a surface, apart from its external shape and position in three-dimensional space. For this purpose he repre-sents the line element ds, i.e. the separation of two neighboring points of the surface, by the formula

$$ds^2 = E \, dp^2 + 2F dp \, dq + G \, dq^2. \tag{2}$$

p and q are parameters of two families of (in general not orthogonal) curves on the surface, and E, F, and G particular functions of p and q. For a pure bending of the surface (without dilatation or shearing) the totality of line elements and hence also the system of the coefficients E, F, and G is con-served.

Gauss shows that the *measure of curvature*

$$K = \frac{1}{R_1 R_2} \tag{3}$$

---

[1] Disquisitiones generales circa superficies curvas 1827, Ges. Werke, Vol. IV, trans-ted into German in Ostwalds Klassiker Nr. 5.

[2] Über die Hypothesen, welche der Geometrie zugrunde liegen, 1854, Ges. Werke 2nd Edition, p. 272.

introduced by him (R_1 and R_2 are the two "principal radii of curvature" of the surface) may be expressed by the E, F, G and their first and second derivatives with respect to p and q. In this manner he arrives at his "Theorema egregium": If a curved surface is bent into another shape (without dilatation!) the measure of curvature remains invariant in all points. The measure of curvature hence expresses an *inner property* of the surface, whereas the definition $1/(R_1 R_2)$ (just like that by the "spherical image") appears to depend on the external shape of the surface and does not indicate its invariance.

Gauss commends his method of fixing attention on the inner properties of surfaces as "most worthy of being diligently exploited by geometers". We shall see that this challenge was heeded by Riemann and Einstein.

The character of the *geodetic*, or *shortest*, *lines* is, of course, also conserved in the bending since it rest solely on the extremal property of the integrated line element. We also mention the approximation of the surface by one of its tangential planes, although it does not belong to the inner relations of the surface; locations will here be indicated not in the curvilinear coordinates p, q, but in ordinary Cartesian coordinates.

We now consider, with Riemann, an n-dimensional manifold of very general structure. As the generalization of (2) and already written in Einstein's notation, its line element is

$$ds^2 = \sum_\mu \sum_\nu g_{\mu\nu}\, dx_\mu\, dx_\nu, \qquad g_{\mu\nu} = g_{\nu\mu}(\mu,\nu = 1, 2 \cdots n). \tag{4}$$

The $g_{\mu\nu}$ are given functions of the quite arbitrarily chosen parameters x_1, $x_2 \cdots x_n$. Riemann studies the inner, invariant (in more general terms, covariant or contravariant) properties of such a manifold.

To begin with, however, we shall answer the simple question: What must be the dimension N of an Euclidean space in order that the n-fold manifold may be embedded in it? We shall employ Cartesian coordinates X_1, $\cdots X_N$ in this Euclidean space. On the n-fold manifold they may be represented as functions of the n parameters x_1, $\cdots x_n$:

$$X_1 = F_1(x_1, \cdots x_n); \qquad \cdots X_N = F_N(x_1, \cdots x_n). \tag{5}$$

If we form the Euclidean line element

$$dX_1^2 + dX_2^2 + \cdots + dX_N^2$$

this contains the first derivatives of the functions F_1, $\cdots F_N$. In order that it may assume the form (4) on the embedded n-fold manifold with arbitrarily prescribed $g_{\mu\nu}$, the number N of the arbitrarily prescribable F must suffice for the determination of the also arbitrarily prescribable $g_{\mu\nu}$

and hence be equal to the number of the $g_{\mu\nu}$, which is $n(n + 1)/2$. We thus have[1]

$$N = \frac{n(n + 1)}{2}.$$ (6)

In the case of Einstein's four-dimensional world we have

$$N = \frac{4 \cdot 5}{2} = 10;$$ (6a)

in the Gaussian case of the two-parameter surface we have of course

$$N = \frac{2 \cdot 3}{2} = 3.$$ (6b)

For every point of the n-fold manifold a "plane" (Euclidean) manifold may be constructed which, departing from the n-fold manifold, plays the same role as the tangential plane in three-dimensional Euclidean space.

One of the inner properties of the n-fold manifold is, in particular, the minimal property of the geodetic (shortest or straightest) lines. As on the two-parameter surface, they are at the same time the paths of a point mass not subjected to forces. We demonstrate this with the aid of an example: Let a plane table top be covered with a cloth under which there may lie a stone. The geodetic paths are in general straight, but curved close to the stone. The mass point, which is assumed not to be acted upon by forces, including gravity, will here be deflected out of its straight path, in accord with the prevailing surface curvature. This is the simplest example of Einstein's theory of gravitation (stone = sun, point mass = planet).

Riemann investigates the generalization of the Gaussian concept of curvature to the n-dimensional manifold. Following Riemann, Christoffel has defined his three-indices symbols and Einstein his $\Gamma_{\mu\nu}^{\sigma}$ which are identical with them (see Appendix I, Eq. (3) in Vol. II of these Lectures). They depend only on the components $g_{\mu\nu}$ of the "fundamental tensor" and its derivatives with respect to the coordinates and are hence inner properties of the manifold. The "Riemannian curvature tensor" is formed from the g, Γ, and their derivatives; its vanishing is the condition for the manifold being "plane" (Euclidean). The Riemannian "symmetric curvature tensor" $R_{\mu\nu}$ is derived from it by "reduction" (summation with respect to one of the pairs of indices). The "Riemannian scalar" R is derived from the curvature tensor $R_{\mu\nu}$ in similar fashion; it is the generalization of the Gaussian measure of curvature K.

[1] See e.g. H. Lang, Ann. Physik, Vol. 61, 1919 (Munich thesis). I am informed that theorem (6) was stated by Schläfli as early as 1871 (Ann. Mat. pura appl. 5, p. 190), and has been proved by E. Cartan and M. Janet (E. Cartan, La géométrie riemannienne et ses généralizations. Encycl. Française, t.1, 1937).

As we saw from our primitive example of the table top, the curvature properties find expression in the paths of mass points subjected to no other forces; they act on them like forces of physical origin. Einstein recognized herein the *origin of gravitation*, giving quantitative content to an idea of Mach. As the most general force action, superseding all other physical agencies, it is attributed by Einstein solely to the curvature conditions of the space-time continuum.

However, how are these curvature conditions determined? *They are determined by the energies distributed in space and time.* Space and time exist only by virtue of the physical processes which occur in them. Their structure is derived from the latter. We are inclined to recall Goethe's grand vision of the "Mothers" in Faust II (corresponding, in a sense, to the Platonic ideas which existed before the creation of the world):

> Göttinen thronen hehr in Einsamkeit,
> Um sie kein Raum, noch wen'ger eine Zeit.
> Von ihnen sprechen ist Verlegenheit.
> Nichts wirst du sehn in ewig leerer Ferne,
> Den Schritt nicht hören, den du tust,
> Nichts Festes finden, wo du ruhst.

The approach to the curvature conditions of Einstein's world which will now be described may seem as disconcerting to the reader as the voyage to the Mothers seemed to Faust; we shall guide the reader along a less forbidding path presently.

The curvature tensor $R_{\mu\nu}$ is to be related to the stress-energy tensor $T_{\mu\nu}$ of material and electromagnetic phenomena by the system of 10 equations ($\mu, \nu = 1, 2, 3, 4$)

$$R_{\mu\nu} - \tfrac{1}{2}g_{\mu\nu}R = -\kappa T_{\mu\nu}, \tag{7}$$

as is shown by Einstein. The factor of proportionality κ here introduced is in essence the constant G of Newton's law of gravitation. Since the $R_{\mu\nu}$ and R may be expressed by the Γ's and g's, and the Γ's, in turn, by the g's and their derivatives, Eqs. (7) are in effect a system of differential equations for the $g_{\mu\nu}$. A general solution of this system would of course be extremely involved. Einstein could show however that they lead in a first approximation to the statements of Newton's theory of gravitation for weak fields or $g_{\mu\nu}$ which differ only little from the Euclidean ones of the special theory of relativity (or, more exactly, the pseudo-Euclidean ones, in view of the negative sign of $dx_4{}^2$).[1]

---

[1] Einstein wrote the author November 28, 1915:

"Last month I passed through one of the most exciting, absorbing and, at the same time, most productive periods of my life. I could not think of writing.

I recognized that my former field equations of gravitation were quite without basis. This is indicated by the following factors . . .

Having lost all confidence in the earlier theory, I saw clearly that a satis-

312 MAXWELL'S THEORY FOR MOVING BODIES AND OTHER ADDENDA 38

Even the first approximation, i.e. the mere fact of Newtonian attraction, reveals the non-Euclidean structure of the scale determination. In the second approximation there occur deviations from the Newtonian law, which are of course greatest in the neighborhood of large concentrations of energy. Hence the anomaly in the path of Mercury, the planet closest to the sun, the deflections (observable only during solar eclipses) of light rays passing very close to the edge of the sun, and the red shift of the spectral lines of the white dwarfs resulting from their extraordinarily high densities.

Gravitational and Inertial Mass

These are the abstract mathematical foundations of Einstein's theory of gravitation. At a much earlier date,[1] almost immediately after the discovery of the special theory of relativity, Einstein recognized a concrete physical basis in the *equivalence of gravitation and acceleration*. The phenomena observed in an elevator which is imagined to be freed from the influence of gravitation and is moving upward with the constant acceleration g are exactly the same as in the same system at rest or in uniform motion when it is subject to the influence of gravity. In both cases a thrown body describes a parabola, a body at rest on the floor presses against it with the force mg, and a pendulum of equal length has the same period of oscillation. Conversely, an elevator falling freely in a gravitational field is not subject to the influence of gravitation: The pressure on the floor ceases, the period of oscillation of a pendulum becomes infinite, and a thrown body describes a straight line. Such a freely falling system realizes a world free from gravitation and curvature, in which pseudo-Euclidean measure is valid, and hence corresponds to the tangential plane to the Riemannian space which was discussed before.

The prerequisite for this is the *identical character of gravitational and*

factory solution could be attained only on the basis of general covariant theory, i.e. of Riemann's covariant $R_{\mu\nu}$. Unfortunately I have immortalized the last errors of this conflict in the Academy papers which I shall send you soon. The final result is the following: . . . The Christoffel symbols $\begin{pmatrix} \mu\nu \\ \sigma \end{pmatrix}$ are to be regarded as the natural representation of the "components" of the gravitational field . . .

The splendid thing which I experienced was not only that now Newton's theory was obtained as first approximation, but that in addition, the precession of the perihelion of Mercury (43″ per century) followed as second approximation. The magnitude of the deflection of light at the sun became twice as large as before."

And on February 8 he remarks on a post card:

"You will be convinced by the general theory of relativity when you have studied it. Hence I do not lose a word to defend it to you."

[1] Jahrbuch f. Radioakt. und Elektronik, Vol. 4, 1907, further elaborated in Ann. d. Phys., Vol. 35, 1911.

inertial mass, which was expressed in Vol. I, §3 by the equation

$$m_g = m_i . \tag{8}$$

Only if this is satisfied is the "weight" $m_{grav} \, g$ equal to the "inertial reaction" $m_{inert} \, g$ and only then is the period of oscillation the same for all pendulums of equal length. In detail the formula for this period is

$$\tau = 2\pi \sqrt{\frac{m_{inert} \, l}{m_{grav} \, g}} . \tag{8a}$$

Already Newton saw that a profound physical problem was hidden herein and Bessel pursued the problem by making extremely careful measurements on pendulums of different materials.[1] R. Eötvös increased the precision of such measurements by powers of ten with his torsion balance. However, Einstein was the first to interpret Eq. (8) in the final form

gravitation = inertia (= world curvature).

We shall show that this equivalence principle suffices for the elementary calculation of the $g_{\mu\nu}$ in a specific case,[2] i.e. to solve a problem which was formulated generally in Eq. (7), but was postponed as being too difficult.

Consider a centrally symmetric gravitational field, e.g. that of the sun, of mass M, which may be regarded as at rest. Let a box K_∞ fall in a radial direction toward M. Since it falls freely, K_∞ is not aware of gravitation and therefore carries continuously with itself the Euclidean metric valid at infinity. Let the coordinates measured within it be x_∞ (longitudinal, i.e. in the direction of motion), y_∞, z_∞ (transversal), and t_∞. K_∞ arrives at the distance r from the sun with the velocity v. v and r are to be measured in the system K of the sun, which is subject to gravitation. In it we use as coordinates r, ϑ, φ, and t. Between K_∞ and K there exist the relations of the special Lorentz transformation, where K_∞ plays the role of the system "moving" with the velocity $v = \beta c$, K that of the system "at rest". The relations are

$$dx_\infty = dr/\sqrt{1 - \beta^2} \qquad \text{(Lorentz contraction)},$$

$$dt_\infty = dt \cdot \sqrt{1 - \beta^2} \qquad \text{(Einstein dilatation)},$$

$$dy_\infty = r d\vartheta$$

$$dz_\infty = r \sin \vartheta \; d\vartheta \; d\varphi \qquad \text{(Invariance of the transversal lengths)}$$

[1] F. W. Bessel, "Experiments on the Force, with which the Earth Attracts Different Kinds of Bodies," Abhandlgen d. Preuss. Akad. 1830; "Studies on the Length of the Second Pendulum", loc. cit. 1826—reprinted in Ostwald's Klassiker Nr. 7.

[2] On the basis of an unpublished paper of W. Lenz, which he kindly communicated to the author in 1944. In the planned publication he will render the argument given in the text more rigorous. He also intends, following Schwarzschild (see below), to extend the consideration to the interior of a sphere filled with an incompressible fluid.

Hence the Euclidean world line element

$$ds^2 = dx_\infty^2 + dy_\infty^2 + dz_\infty^2 - c^2\, dt_\infty^2 \tag{9}$$

passes over into

$$ds^2 = \frac{dr^2}{1 - \beta^2} + r^2(d\vartheta^2 + \sin^2 \vartheta d\varphi^2) - c^2 dt^2(1 - \beta^2). \tag{9a}$$

The factor $1 - \beta^2$, which occurs here twice, is meaningful so far only in connection with our specific box experiment. In order to determine its meaning in the system of the sun we write down the energy equation for K_∞, as interpreted by an observer on K. Let m be the mass of K_∞, m_0 its rest mass. The equation then is:

$$(m - m_0)c^2 - \frac{GmM}{r} = 0. \tag{10}$$

At the left we have the sum of the kinetic energy in accord with Eq. (32.7) and of the (negative) potential energy of gravitation. The energy constant on the right was to be put equal to zero since at infinity $m = m_0$ and $r = \infty$. We have computed the potential energy from the Newtonian law, which we shall consider as a first approximation. We divide (10) by mc^2 and obtain then, since $m = m_0/\sqrt{1 - \beta^2}$,

$$1 - \sqrt{1 - \beta^2} = \frac{\alpha}{r}, \qquad \alpha = \frac{GM}{c^2} = \frac{\kappa M}{8\pi} \ \text{(see Eq. (7a))}. \tag{10a}$$

With $M = 3.3 \cdot 10^5 \, M_{\text{earth}}$ and $g = G\, M_{\text{earth}}/R^2$, $R = $ radius of earth $= (2/\pi)\ 10^7$ meter we obtain

$$\alpha = 3.3 \cdot 10^5\, g \left(\frac{2}{\pi}\frac{10^7}{3 \cdot 10^8}\right)^2 = 14.6 \cdot 10^2 \text{ meter} \cong 1 \text{ km}.$$

It follows from (10a) that

$$\sqrt{1 - \beta^2} = 1 - \frac{\alpha}{r}, \qquad 1 - \beta^2 \cong 1 - \frac{2\alpha}{r}, \tag{11}$$

and hence, from (9a),

$$ds^2 = \frac{dr^2}{1 - 2\alpha/r} + r^2(d\vartheta^2 + \sin^2 \vartheta d\varphi^2) - c^2(1 - 2\alpha/r)\, dt^2. \tag{12}$$

This is the line element derived by K. Schwarzschild[1] from Einstein's Eqs. (7). In Eddington's presentation[2] the 40 components $\Gamma_{\mu\nu}^{\sigma}$ of the gravi-

[1] Preuss. Akad., Sitzungsber. 1916, p. 189.

[2] See his excellent book: "The Mathematical Theory of Relativity," Cambridge, 1923.

tational field are computed and (12) is shown to be the exact solution of the ten equations contained in (7). *Our derivation* claims only to yield an approximation, since it utilizes the Newtonian law as first approximation and neglects, in the second Eq. (11), the term $(\alpha/r)^2$; nevertheless, *our result* is, as shown by Schwarzschild and Eddington, exact in the sense of Einstein's theory.

It might be asked at this point: What is the relativistically exact formulation of the Newtonian law? The question is wrongly put if a vector law is meant hereby. The gravitational field is not a vector field, but has a much more complex tensor character. For the single point mass it is completely described by the four coefficients g of the line element (12) and the vanishing of the remaining $g_{\mu\nu}$.

B. Observable Deductions from the General Theory of Relativity

We shall first deduce the anomaly of the perihelion of Mercury from the line element (12); the general formalism of tensor calculus will not be required here.

The law of the geodetic paths demands

$$\delta \int ds = 0. \tag{13}$$

Of the four coordinates r, ϑ, φ, t in (12) we choose φ as "independent variable" and hence write in place of (13)

$$\delta \int v \, d\varphi = 0, \tag{13a}$$

$$v^2 = \frac{\dot{r}^2}{1 - 2\alpha/r} + r^2(\dot{\vartheta}^2 + \sin^2 \vartheta) - c^2 \dot{t}^2(1 - 2\alpha/r), \tag{14}$$

$$\dot{r} = \frac{dr}{d\varphi}, \qquad \dot{\vartheta} = \frac{d\vartheta}{d\varphi}, \qquad \dot{t} = \frac{dt}{d\varphi}. \tag{14a}$$

We designate the "dependent variables" r, ϑ, t collectively by q. The method of the calculus of variation, which we utilized in Vol. I, §34 for the proof of the Lagrange equations, leads to the "Euler equation" (see the first footnote in the section referred to)

$$\frac{d}{d\varphi} \frac{\partial v}{\partial \dot{q}} - \frac{\partial v}{\partial q} = 0. \tag{15}$$

For $q = \vartheta$ (14) yields

$$\frac{\partial v}{\partial \dot{\vartheta}} = \frac{r^2 \dot{\vartheta}}{v}, \qquad \frac{\partial v}{\partial \vartheta} = \frac{r^2 \sin \vartheta \cos \vartheta}{v},$$

and hence, by (15),

$$\frac{d}{d\varphi}\frac{r^2\dot\vartheta}{v} = \frac{r^2\sin\vartheta\cos\vartheta}{v}.$$

The last equation is fulfilled for $\vartheta = \text{const} = \pi/2$; the other possibility $\vartheta = \text{const} = 0$ represents no *planetary orbit*, but a *meteor* falling centrally straight into the sun. Our denoting the plane of the planetary orbit by $\vartheta = \pi/2$ is obviously simply a convenient choice of our system of polar coordinates.

For $q = t$ (14) yields

$$\frac{\partial v}{\partial \dot t} = -\frac{c^2\dot t(1-2\alpha/r)}{v}, \qquad \frac{\partial v}{\partial t} = 0, \qquad \text{so that } \frac{d}{d\varphi}\frac{\dot t(1-2\alpha/r)}{v} = 0.$$

From this we conclude

$$\frac{\dot t(1-2\alpha/r)}{v} = \text{const.} \tag{16}$$

Since $\dot t$ is real and v (as well as ds) is purely imaginary, the constant also is purely imaginary. We put it equal to ik and find from (16)

$$\dot t = \frac{ikv}{1-2\alpha/r}. \tag{17}$$

k is a first integration constant of the path of the planet.

For $q = r$ (15) would yield a differential equation for $\dot r$, whose integration would provide a second integration constant of the problem. It is simpler, however, to employ a general theorem, which we have established in Vol. I for an arbitrary variation problem and which corresponds in mechanics to the law of conservation of energy, i.e. Eq. (41.18a). We replace L in this equation by v in our present problem and the constant on the right side by ih (numerical value of the Hamiltonian function H in Eq. (41.18) of Vol. I and, at the same time, second integration constant of our problem). We thus find

$$\sum_j \frac{\partial v}{\partial \dot q_j}\dot q_j - v = ih \tag{18}$$

or, after multiplication with v,

$$\frac{1}{2}\sum_j \frac{\partial v^2}{\partial \dot q_j}\dot q_j - v^2 = ihv. \tag{18a}$$

We evaluate the left side with the aid of (14), whereby all terms with $\dot r$, $\dot\vartheta$, and $\dot t$ cancel. It then reduces to

$$-r^2\sin^2\vartheta = -r^2 \text{ since } \vartheta = \pi/2.$$

Hence (18a) yields

$$v = -\frac{r^2}{ih}.$$ (19)

(17) thus becomes

$$\dot{t} = -\frac{k}{h}\frac{r^2}{1 - 2/\alpha r}$$ (20)

and (14) yields, with $\vartheta = \pi/2$,

$$-\frac{r^4}{h^2} = \frac{\dot{r}^2}{1 - 2\alpha/r} + r^2 - \frac{c^2 k^2}{h^2}\frac{r^4}{1 - 2\alpha/r}.$$ (21)

This is a differential equation for r which takes the place of the (once integrated Euler) equation for the dependent variable $q = r$ in Eq. (15). It is simplified if $u = 1/r$ is introduced as a new variable and if it is multiplied with $1 - 2\alpha/r = 1 - 2\alpha u$:

$$\dot{u}^2 + u^2(1 - 2\alpha u) + \frac{1 - c^2 k^2 - 2\alpha u}{h^2} = 0,$$

for which we may also write

$$\dot{u}^2 + u^2 = \frac{2\alpha u}{h^2} + 2\alpha u^3 - \frac{1 - c^2 k^2}{h^2}.$$ (22)

We differentiate with respect to the independent variable and cancel out $\dot{u} = du/d\varphi$. We thus obtain

$$\ddot{u} + u = \frac{\alpha}{h^2} + 3\alpha u^2.$$ (23)

For comparison we treat the same problem by Newton's theory. We start with the energy equation (W = sum of kinetic and potential energy):

$$\frac{1}{2}\left\{\left(\frac{\partial r}{\partial t}\right)^2 + r^2\left(\frac{\partial \varphi}{\partial t}\right)^2\right\} - \frac{GM}{r} = \frac{W}{m}.$$ (24)

According to the law of equal areas we have, with the area constant denoted by hc:

$$r^2\frac{\partial \varphi}{\partial t} = hc.$$ (24a)

On the left side of (24) we factor out $(d\varphi/dt)^2 = h^2c^2/r^4$, put once more $u = 1/r$, and obtain after division by h^2c^2

$$\frac{1}{2}(\dot{u}^2 + u^2) - \frac{GM}{h^2 c^2}u = \frac{W}{mh^2 c^2}.$$

Also here we differentiate again with respect to φ, whereupon $\dot{u}$ cancels out, and find in view of (10a)

$$\ddot{u} + u = \frac{GM}{h^2 c^2} = \frac{\alpha}{h^2}. \tag{25}$$

The relativistic equation (23) differs from (25) only in the correction term $3\,\alpha\,u^2$. This has no appreciable influence on the size or shape of the orbit, but affects merely the position of the perihelion. To recognize this we place the direction $\varphi = 0$ at the perihelion, which may be defined by $u = u_{max}$ and hence $\dot{u} = 0$. Then u becomes an even function of φ. We may then, beginning with the solution of (25), expand this in a Fourier cosine series:

$$u = A + B \cos \varphi + \cdots ; \tag{26}$$

it is then found that the higher terms, indicated by $\cdots$, vanish. In (23) we substitute[1]

$$u = A + B \cos (\gamma\varphi) + C \cos(2\gamma\varphi) \cdots \tag{26a}$$

and find for the determination of the constant γ here introduced from (23) the equation

$$A + (1 - \gamma^2)B \cos (\gamma\varphi) + (1 - 4\gamma^2)C \cos (2\gamma\varphi) + \cdots$$

$$= \frac{\alpha}{h^2} + 3\alpha(A + B \cos (\gamma\varphi))^2$$

$$= \frac{\alpha}{h^2} + 3\alpha A^2 + 6\alpha AB \cos (\gamma\varphi) + \frac{3}{2}\alpha B^2(1 + \cos (2\gamma\varphi)).$$

Here we have already dropped the higher terms of the series in the correction term. A comparison of the coefficients yields

$$A = \frac{\alpha}{h^2} + 3\alpha A^2 + \frac{3\alpha}{2} B^2$$

$$(1 - \gamma^2)B = 6\alpha AB \tag{27}$$

$$(1 - 4\gamma^2)C = \frac{3}{2} \alpha B^2.$$

[1] In a manner similar as for the fine structure of the hydrogen atom; the following calculation may be clearer than the customary astonomical one (Eddington). It should be noted that every deviation from Newton's or Coulomb's law effects a motion of the perihelion of the Kepler ellipse. The motion brought about by the variation of mass is, however, much smaller (by a factor $\frac{1}{6}$) than that arising from the gravitational correction.

Since B drops out of the middle equation, it serves to determine γ:

$$1 - \gamma^2 = 6\alpha A, \qquad \gamma \cong 1 - 3\alpha A, \qquad 1 - \gamma = 3\alpha A. \qquad (28)$$

A may be determined geometrically in terms of the perihelion and aphelion distances (a and ε denote the major axis and the numerical eccentricity of the ellipse):

$$u_{\max} = \frac{1}{r_{\min}} = \frac{1}{a(1 - \varepsilon)} = A + B + \cdots \qquad \text{for} \qquad \gamma\varphi = 0,$$

$$u_{\min} = \frac{1}{r_{\max}} = \frac{1}{a(1 + \varepsilon)} = A - B + \cdots \qquad \text{for} \qquad \gamma\varphi = \pi,$$

so that

$$A = \frac{1}{a(1 - \varepsilon^2)}.$$

Hence we find from (28)

$$1 - \gamma = \frac{3\alpha}{a(1 - \varepsilon^2)}.$$

The precession $\delta\bar{\omega}$ of the perihelion in the course of one revolution is also determined geometrically, namely by the formula

$$\gamma(2\pi + \delta\bar{\omega}) = 2\pi, \qquad \delta\bar{\omega} = 2\pi \frac{1 - \gamma}{\gamma} \cong \frac{6\pi\alpha}{a(1 - \varepsilon^2)} \qquad (29)$$

if a term with α^2 is neglected. For Mercury the secular displacement of the perihelion is hence found to be $43''$, in agreement with observation.

With the aid of the preceding calculations the second test of the general theory of relativity, the *light deflection at the edge of the sun*, can also be readily treated. Light paths are geodetic lines, for which $ds = 0$. In the special theory of relativity they were the generatrices of the light cone $\Sigma dx_i^2 = 0$; now they are given by $\Sigma g_{ik} \, dx_i \, dx_k = 0$, i.e. in our case, according to Eq. (14), by $v = 0$. Hence we must set $h = \infty$ in Eq. (19). Eq. (23) then becomes

$$\ddot{u} + u = 3\alpha u^2.$$

In the integration it is permissible, as an approximation, to let γ approach 1. Then (27) leads to $\alpha A \to 0$, $C \to -\frac{1}{2}\alpha B^2$, $A \to \frac{3}{2}\alpha B^2$ (the last in view of $h = \infty$). Hence, by (26a)

$$u = \frac{3\alpha}{2} B^2 + B \cos \varphi - \frac{\alpha}{2} B^2 \cos (2\varphi). \qquad (30)$$

For $\varphi = 0$ the light is to be tangent to the edge of the sun ($r = R$). We must hence have

$$\frac{1}{R} = \frac{3\alpha}{2} B^2 + B - \frac{\alpha}{2} B^2 = B + \alpha B^2,$$

$$B = \frac{1}{R} \frac{1}{1 + \alpha B} \cong \frac{1}{R} \frac{1}{1 + \alpha/R} \cong \frac{1}{R},$$

since $\alpha \cong 1$ km is very small in comparison with R. With $x = r \cos \varphi$, $y = r \sin \varphi$ (30) then yields, after multiplication with rR:

$$R = \frac{3}{2} \frac{\alpha}{R} \sqrt{x^2 + y^2} + x - \frac{1}{2} \frac{\alpha}{R} \frac{x^2 - y^2}{\sqrt{x^2 + y^2}}$$

The light path comes to resemble a hyperbola, just as the path of the planet resembled an ellipse. With the assumption $|y| >> |x|$ we obtain

$$R = \pm \frac{3}{2} \frac{\alpha}{R} y + x \pm \frac{1}{2} \frac{\alpha}{R} y = x \pm \frac{2\alpha}{R} y \qquad (31)$$

The angle between the two asymptotes, which is equal to the deflection of the light from its original path, is $4\alpha/R = 1.75''$ and agrees well with the results of the solar eclipse expeditions. It is twice as large as the value obtained by an earlier more primitive calculation (Soldner as well as Einstein before 1915; see footnote on p. 311).

We note furthermore in this connection that in addition to the direction, the velocity of the light is changed by the gravitational field. In a radial direction, e.g. along the radius $\vartheta = 0$, taking account of $ds = 0$, it is

$$\frac{dr}{dt} = (1 - 2\alpha/r)c \text{ by (12)}. \qquad (31a)$$

Finally, the last of the enumerated tests of the theory, the *red shift of the spectral lines in the gravitational field*, can be understood without any calculation. Consider a point of the curved world and construct there the (gravity-free, Euclidean) tangential plane. Let the coordinate changes in the latter, $dX_1, \cdots, dX_4 = icdT$ coincide in direction with the coordinate changes $dr, \cdots, dx = icdt$ in the gravitational field. In view of the equality of the two line elements we then have for a particle at rest which is radiating light

$$- c^2 dT^2 = - c^2(1 - 2\alpha/r)dt^2. \qquad (32)$$

The measures of time dt and dT are hence different; the same applies to the frequencies ν and ν_0 (in absence of gravity), which are inversely proportional to these times. According to (32) we have

$$\nu = \sqrt{1 - 2\alpha/r}\,\nu_0 \cong (1 - \alpha/r)\nu_0$$

$$\nu - \nu_0 \cong -\frac{\alpha}{r}\,\nu_0. \tag{33}$$

The frequency is *reduced* by the gravitational field. In view of the meaning of α, given by Eq. (10a), the magnitude of the red shift is

$$\frac{1}{c^2}\frac{GM}{r} = \frac{|V|}{c^2}, \tag{34}$$

where V (Eq. (10)) is the gravitational potential. The spectrum of Sirius B and of other white dwarfs provides the experimental confirmation.

C. Unified Theory of Gravitation and Electrodynamics

Following Gauss and Riemann, Einstein *put metric first*, i.e., required the invariance of ds^2 and the tensor character of the $g_{\mu\nu}$. "Balance, rod, and clock" were the basic elements which he manipulated in the general, just as in the special theory of relativity. With them he was able to *geometrize gravitation*.

However, Maxwell's electrodynamics of vacuum also constitutes a complex of phenomena overshadowing material processes. The amazingly simple form which it assumes in the special theory of relativity and which may be transferred without appreciable changes to the realm of the general theory of relativity, covering arbitrary frames of reference, suggests similarly geometrization. However, the metric proves too restricted for this. Einstein attempted to broaden it by demanding, instead of the invariance of ds^2, merely that of $ds^2 = 0$ (i.e. that of the line elements of the light cone).[1] We are then concerned only with the ratios of the $g_{\mu\nu}$, rather than with the $g_{\mu\nu}$ themselves.

Hermann Weyl had recognized even a short time before this that it was simpler and more natural to drop the metric departure and to begin directly with Einstein's $\Gamma_{\mu\nu}^{\sigma}$. The resulting system is known as *affine world geometry*. It provides a rule for "parallelism at a distance", i.e. a prescription for proceeding along a world line without departing from the initial direction. *Both gravitation and electrodynamics* fit into this system quite naturally. The $\Gamma_{\mu\nu}^{\sigma}$ were here assumed symmetric in the μ and ν. The result of this theory is recorded in his classic book "Raum·Zeit·Materie", Springer, Berlin, 1918.[2]

However the system can be generalized even further: $\Gamma_{\mu\nu}^{\sigma}$ and $\Gamma_{\nu\mu}^{\sigma}$ can

---

[1] "On a reasonable extension of the basis of the general theory of relativity", Preuss. Akad. 1921, p. 261, as well as the following notes on the unified field theory: loc. cit. 1925, p. 414; 1928, p. 3; and 1929, p. 3.

[2] English edition: H. Weyl, "Space-Time-Matter," Methuen, London, 1922.

be chosen to be different. There results an *asymmetric affine* theory, which gives rise to a new antisymmetric tensor. It has been sketched by Erwin Schrödinger[1] and is being developed by him in friendly competition with Einstein.

The incentive was provided by the following: *Nuclear physics* has joined atomic physics as a younger sister science. Whereas atomic physics from the corpuscular standpoint rests on the electrodynamic interactions between electrons and protons, the forces of nuclear physics must be ascribed to the *mesons*, which, in the meantime, have come to be demanded by theory and have been discovered experimentally. (The name meson derives from the fact that these elementary particles have a mass intermediate between those of the electron and the proton). Nuclear physics hence is *meson theory*. It demands an antisymmetric tensor differing from that of electrodynamics. Such a tensor is furnished by the asymmetric affine world geometry, which thus would create a triple bond between gravitation, electrodynamics, and nuclear theory. Its detailed structure has not yet been determined, however. When at last it has been fully elaborated Maxwell's theory, too, will be revealed in its full beauty and symmetry.

---

[1] See his note in Nature, May 13, 1944, and, following it, several papers in the Proceedings of the Irish Academy for the years 1944–46, the last, with the title "The general affine field laws", in Vol. 51, p. 41.

SYMBOLS EMPLOYED THROUGHOUT THE TEXT AND THEIR DIMENSIONS

Note: As a matter of course all equations in this volume are written in a dimensionally consistent manner and hence not tied to any particular choice of units (e.g. M = meter, Q = coulomb). They are, to use a preferred current expression, "equations of quantities". The "numerical equations", which are correct only for a specific choice of units, are fortunately falling more and more into disuse even in engineering.

e, q	charge	Q, practical unit: 1 coulomb
ρ	charge density	QM^{-3}
ω	surface charge density	QM^{-2}
p, P	magnetic pole strength	QMS^{-1} (in §8 P serves as fifth independent unit)
ρ_m	magnetic density	$QM^{-2}S^{-1}$
ω_m	magnetic surface density	$QM^{-1}S^{-1}$
$\mathbf{E}$	electric field strength	newton/Q $= MKS^{-2}Q^{-1}$, 1 newton $= 10^5$ dynes
$\mathbf{D}$	electric excitation (displacement)	QM^{-2}, div $\mathbf{D} = \rho$, $D_n - D'_n = \omega$, Eq. (3.11)
$\mathbf{J}$	conduction current density	$QM^{-2}S^{-1} = AM^{-2}$, A = ampere
I	total conduction current	$QS^{-1} = A$, $I = \int J_n \, d\sigma$
$\dot{\mathbf{D}}$	displacement current density	$QM^{-2}S^{-1} = AM^{-2}$
$\mathbf{C}$	total current density $= \mathbf{J} + \dot{\mathbf{D}}$	$QM^{-2}S^{-1} = AM^{-2}$
V	electric potential difference $= \int \mathbf{E} \cdot d\mathbf{s}$	joule $Q^{-1} =$ volt $= V$;
$\mathbf{B}$	magnetic field strength (induction)	newton/P $= KS^{-1}Q^{-1} = VSM^{-2}$, div $\mathbf{B} = 0$; 1 gauss $= 10^{-4}$ VSM^{-2}
$\mathbf{H}$	magnetic excitation (ampere-turns per meter)	$PM^{-2} = QM^{-1}S^{-1} = AM^{-1}$, div $\mathbf{H}$ $= \rho_m$, $H_n - H'_n = \omega_m$, 1 oersted $= 10^3/(4\pi) \, AM^{-1}$
U	magnetic potential difference $= \int_1^2 \mathbf{H} \cdot d\mathbf{s}$	$QS^{-1} = A$
Φ	magnetic flux $= \int B_n \, d\sigma$	$M^2KS^{-1}Q^{-1} = VS$

$\Psi \begin{cases} \Psi_e & \text{electric potential} \\ \quad \mathbf{E} = -\text{grad } \Psi_e \\ \Psi_m & \text{magnetic potential} \\ \quad \mathbf{H} = -\text{grad } \Psi_m \end{cases}$

$\text{joule}/Q = V$

$QS^{-1} = A$

$\mathbf{A}$ vector potential, $\mathbf{B} = \text{curl } \mathbf{A}$ $MKS^{-1}Q^{-1} = VSM^{-1}$

$\mathbf{S}$ radiation vector $=$ Poynting   joule $M^{-2}S^{-1}$ = watt M^{-2}
vector $=$ energy flux density

W energy density,   joule M^{-3}

$$W = W_e + W_m, \qquad W_e = \tfrac{1}{2}\mathbf{D}\cdot\mathbf{E}, \qquad W_m = \tfrac{1}{2}\mathbf{H}\cdot\mathbf{B}$$

W_J Joule heat per unit volume   joule $M^{-3}S^{-1}$
$\quad = \mathbf{E}\cdot\mathbf{J}$

W energy in given volume   joule

ε dielectric constant $Q^2/(\text{joule } M) = SM^{-1}\Omega^{-1} =$
                                $\text{farad}\cdot M^{-1}$

μ permeability $MKQ^{-2} = \Omega SM^{-1} = \text{henry}\cdot M^{-1}$

ε_0, μ_0 vacuum constants $(\varepsilon_0\mu_0)^{-\frac{1}{2}} = \text{velocity of light, } MS^{-1}$
                               $(\mu_0/\varepsilon_0)^{\frac{1}{2}} = \text{wave resistance, } \Omega$

σ conductivity $M^{-3}K^{-1}SQ^2 = M^{-1}\Omega^{-1}$

$\varepsilon' = \varepsilon + i\sigma/\omega = \text{complex dielectric}$ $M^{-3}K^{-1}S^2Q^2 = SM^{-1}\Omega^{-1}$
constant

R resistance of a wire $\text{volt}/\text{ampere} = M^2KS^{-1}Q^{-2} = \Omega$

L selfinductance $\Omega S = \text{henry}$

$\mathbf{R}$ impedance $= R + i\omega L$ Ω

K capacitance $Q^2/\text{joule} = \Omega^{-1}S = \text{farad}$

$\mathbf{P}$ dielectric polarization $QM^{-2}, \mathbf{D} = \varepsilon_0\mathbf{E} + \mathbf{P}$, see p. 74

$\mathbf{M}$ magnetization $QM^{-1}S^{-1}, \mathbf{B} = \mu_0(\mathbf{H} + \mathbf{M})$, see p. 91

η electric susceptibility   pure number, $\mathbf{P} = \eta\varepsilon_0\mathbf{E}, \varepsilon = \varepsilon_0(1 + \eta)$

κ magnetic susceptibility   pure number, $\mathbf{M} = \kappa\mathbf{H}, \mu = \mu_0(1 + \kappa)$

ω angular frequency $S^{-1}, \omega = 2\pi/\tau, \tau = \text{period of vibration}$

k wave number in vacuum $M^{-1}, k = 2\pi/\lambda, \lambda = \text{wave-length}$

k_L wave number in conductor $M^{-1}, k_L^2 = \varepsilon\mu\omega^2 + i\mu\sigma\omega = \varepsilon'\mu\omega^2$

h wave number of surface waves M^{-1}
on cylindrical guide

κ in §20 to 25 $= \sqrt{\mu\sigma\omega/2}$ for al- $M^{-1}, 1/\kappa = d = \text{layer thickness in}$
ternating currents   skin effect

Additional Symbols in Parts III and IV

x_4 imaginary time coordinate $x_4 = ict$

ds world line element $ds^2 = \sum\limits_{j=1}^{4} dx_j^2$

$d\tau$ element of intrinsic time = $ds/(ic)$ $\qquad d\tau = \sqrt{1 - \beta^2}\, dt,\; \beta = v/c$

γ imaginary angle of rotation in the Lorentz transformation $\qquad \tan \gamma = i\beta$

$\mathbf{R}$ four-dimensional radius vector $\qquad \mathbf{R} = x_1,\, x_2,\, x_3,\, x_4$

$\mathbf{V}, \mathbf{W}$ four-vectors of velocity and acceleration $\qquad \mathbf{V} = d\mathbf{R}/d\tau,\; \mathbf{W} = d\mathbf{V}/d\tau$

$\mathbf{\Omega}$ four-potential = $\mathbf{A}, i\Psi/c$ $\qquad \mathrm{VSM}^{-1}$

$\mathbf{\Gamma}$ four-current = $\mathbf{J}, ic\rho$ $\qquad \mathrm{QM}^{-2}\mathrm{S}^{-1}$, in vacuum = $\rho(\mathbf{v}, ic)$

F six-vector of the field $\qquad \mathrm{VM}^{-1},\; F = c\mathbf{B},\, -\,i\mathbf{E} = c\,\mathrm{Curl}\,\mathbf{\Omega}$

f six-vector of the excitation $\qquad \mathrm{AM}^{-1}, f = \mathbf{H}, -ic\mathbf{D} = \dfrac{1}{\mu_0}\,\mathrm{Curl}\,\mathbf{\Omega}$

F^*, f^* dual six-vectors $\qquad F^* = -i\mathbf{E},\, c\mathbf{B},\, f^* = -ic\mathbf{D},\, \mathbf{H}$

Λ Lagrange density $\qquad \Lambda = \dfrac{1}{2c} f \cdot F = \dfrac{1}{2}\mathbf{H}\cdot\mathbf{B} - \dfrac{1}{2}\mathbf{D}\cdot\mathbf{E}$

M second invariant of the field $\qquad M = \dfrac{i}{2} F \cdot F^* = c\,\mathbf{B}\cdot\mathbf{E}$

$\mathbf{k}$ force density = $\dfrac{1}{c}\,\mathbf{\Gamma}\cdot F$ $\qquad k_{1,2,3} = \rho(\mathbf{E} + \mathbf{v} \times \mathbf{B}),\, k_4 = \dfrac{i\rho}{c}\,\mathbf{v}\cdot\mathbf{E}$

$\mathbf{K}$ Lorentz force, three-dimensional $\qquad \mathbf{K} = e(\mathbf{E} + \mathbf{v} \times \mathbf{B})$

$\mathbf{F}$ four-force, $\mathbf{F}\cdot\mathbf{V} = 0$ $\qquad F_{1,2,3} = \dfrac{\mathbf{K}}{\sqrt{1 - \beta^2}},\, F_4 = \dfrac{ie\,\mathbf{v}\cdot\mathbf{E}}{c\sqrt{1 - \beta^2}}$

T stress-energy tensor $\qquad T_{nm} = -\dfrac{1}{c}\sum\limits_{r=1}^{4} F_{nr} f_{mr} + \delta_{nm}\,\Lambda$

$\mathbf{G}$ momentum of point mass $\qquad \mathbf{G} = m(\mathbf{v}, ic),\, m = m_0/\sqrt{1 - \beta^2}$

E_0 rest energy $\qquad E_0 = m_0 c^2$

K kinetic potential $\qquad \mathrm{K} = m_0 c^2\,(1 - \sqrt{1 - \beta^2})$

$-\mathbf{\Gamma}\cdot\mathbf{\Omega}$ Schwarzschild invariant $\qquad \mathbf{\Gamma}\cdot\mathbf{\Omega} = \rho(\mathbf{v}\cdot\mathbf{A} - \Psi)$

$\mathbf{E}^*\big\}$ Abbreviations in the equations $\qquad \mathbf{E}^* = \mathbf{E} + \mathbf{v} \times \mathbf{B}$
$\mathbf{H}^*\big\}$ for moving media $\qquad \mathbf{H}^* = \mathbf{H} - \mathbf{v} \times \mathbf{D}$

$\dot{\mathbf{A}}$ Lorentz symbol $\qquad \dfrac{\partial \mathbf{A}}{\partial t} + \mathbf{v}\,\mathrm{div}\,\mathbf{A} - \mathrm{curl}\,(\mathbf{v} \times \mathbf{A})$

$\mathbf{J}_l$ conduction current density $\qquad \mathbf{J}_l = \mathbf{J} - \rho\mathbf{v}$

$\mathbf{R}$ in §36: reaction force of radiation $\qquad$ depending on the frame of reference also denoted by $\mathbf{R'}$ or $\mathbf{R}^*$

$\mathbf{\Gamma}\cdot\mathbf{\Omega} = \sum\limits_{n=1}^{4}\Gamma_n\Omega_n$ $\qquad$ scalar product of two four-vectors yields a scalar

$\mathbf{\Gamma}\cdot F_n = \sum\limits_{m=1}^{4}\Gamma_m F_{nm}$ $\qquad$ scalar product of a four- and a six-vector yields a four-vector

$f\cdot F = \dfrac{1}{2}\sum\limits_{n}\sum\limits_{m} f_{nm}F_{nm}$ $\qquad$ scalar product of two six-vectors yields a scalar

$\qquad = \sum\limits_{n<m}\sum\limits_{m=1}^{4} f_{nm}F_{nm}$

$(\mathbf{R} \times \mathbf{V})_{nm} = R_n V_m - R_m V_n$ vector product of two four-vectors yields a six-vector

$\text{Div } \mathbf{\Omega} = \sum_{n=1}^{4} \dfrac{\partial \Omega_n}{\partial x_n}$ divergence of a four-vector yields a scalar

$\text{Curl}_{nm}\, \mathbf{\Omega} = \dfrac{\partial \Omega_n}{\partial x_n} - \dfrac{\partial \Omega_n}{\partial x_m}$ curl of a four-vector yields a six-vector

$\mathbf{Div}_m\, T = \sum_{n=1}^{4} \dfrac{\partial T_{mn}}{\partial x_n}$ vector divergence of a (general or also antisymmetric) tensor yields a four-vector

$\mathbf{Div}_m^* F = \mathbf{Div}_m F^* = \sum_{n=1}^{4} \dfrac{\partial F_{mn}^*}{\partial x_n}$ dual vector divergence of a six-vector

$\mathbf{Div}^* F = 0$ results in $F = \text{Curl } \mathbf{\Omega}$

$\text{Div } \mathbf{Div}\, F = 0$ applies for every six-vector

$\mathbf{Div}\,\text{Curl } \mathbf{\Omega} = \text{Grad Div } \mathbf{\Omega} - \square\, \mathbf{\Omega} \qquad \square\, \Omega_m = \sum_{n=1}^{4} \dfrac{\partial^2 \Omega_m}{\partial x_n^2}$

Numerical Values, Results of Measurements, and Definitions

c = velocity of light in vacuum = $3.00 \cdot 10^8 \text{ MS}^{-1}$ (measurement)

μ_0 = permeability of vacuum = $4\pi \cdot 10^{-7} \Omega \text{SM}^{-1}$ (definition)

ε_0 = dielectric constant of vacuum = $10^7/(4\pi c^2)$ $\text{M}^{-1}\text{S}\Omega^{-1}$

    = $10^{-9}/(36.00\pi)\text{M}^{-1}\text{S}\Omega^{-1}$ (consequence)

$(\mu_0/\varepsilon_0)^{\frac{1}{2}}$ = wave resistance of vacuum = $120.0\pi\Omega$ (consequence)

e = electronic charge = $1.60 \cdot 10^{-19}$ Q (measurement)

e/m_0 = specific charge of the electron = $1.76 \cdot 10^{11}$ Q/K (measurement)

m_0 = rest mass of the electron = $0.90 \cdot 10^{-30}$ K (consequence)

eV = electron volt = $1.60 \cdot 10^{-19}$ joule (consequence)

$m_0 c^2$ = rest energy of the electron = $\frac{1}{2}$ million eV = $0.81 \cdot 10^{-13}$ joule (consequence)

PROBLEMS FOR PART I

I.1. The Boundary Conditions of Maxwell's Theory. Derive Eqs. (3.7a) to (3.12) for **E**, **B**, **H**, and **D** by the differential method. The transition from medium 1 to 2 must then be assumed to be continuous ("boundary layer" instead of "boundary surface"). Use a rectangular coordinate system x, y, z and let z be perpendicular to the boundary surface which, in the limit of infinite smallness, may be regarded as plane. In the boundary layer $(h \rightarrow 0)$ the derivatives with respect to z occurring in the differential equations (4.8) must be continuous in order that these equations may be meaningful.

I.2. The Magnetic Excitation Inside and Outside of an Infinitely Long Wire. Proof of Eqs. (4.10) to (4.13) from the differential equations.

I.3. The Magnetic Excitation within an Infinitely Long Solenoid. Proof of Eq. (4.14) from Maxwell's equations.

FIG. 44.

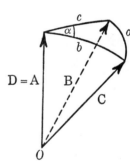

I.4. The Cosine Law of Spherical Trigonometry as Special Case of a General Vector Formula. Prove the vector formula:

$$(\mathbf{A} \times \mathbf{B}) \cdot (\mathbf{C} \times \mathbf{D}) = (\mathbf{A} \cdot \mathbf{C})(\mathbf{B} \cdot \mathbf{D}) - (\mathbf{A} \cdot \mathbf{D})(\mathbf{B} \cdot \mathbf{C})$$

and deduce from it, for the special case $\mathbf{D} = \mathbf{A}$ and with reference to Fig. 44, the cosine law

$$\cos a = \cos b \cos c + \sin b \sin c \cos \alpha.$$

PROBLEMS FOR PART II

II.1. The Charging Potential of a Conducting Ellipsoid of Revolution. Let a be the major axis, b the minor axis, and $c = \sqrt{a^2 - b^2}$ the linear eccentricity. For fixed c and variable a

$$\frac{x^2 + y^2}{a^2 - c^2} + \frac{z^2}{a^2} = 1$$

ιepresents the family of confocal ellipsoids with the separation of focal points $2c$. Show that on each of them the expression for Ψ given in Eq. (9.4) is constant (independent of x, y, z).

II.2. The Infinitely Long Rubbed Glass Rod and its Comparison with the Conducting Paraboloid of Revolution. Calculate the potential of an infinitely long uniformly charged straight line terminated at one end, and show that its equipotential surfaces are the same as those for the conducting paraboloid of revolution which is obtained from (9.4) by transition to the limit $c \rightarrow \infty$, $a \rightarrow \infty$.

II.3. Comparison of the Dielectric and the Conducting Sphere. For a dielectric sphere $r = a$ placed in an originally uniform electric field there always exists a concentric conducting sphere $r = b < a$, whose exterior field agrees, for $r > a$, with the exterior field of the dielectric sphere. Fig. 45 shows, for $r > a$, the field of the dielectric sphere, for $a > r > b$, not the (uniform) field within this sphere, but the analytical continuation of the exterior field, which is identical with the field of the conducting sphere of radius b. Prove that

$$b = a \sqrt[3]{\frac{\varepsilon - 1}{\varepsilon + 2}}.$$

The figure shows how the singularity of the equilibrium point of the conductor (see footnote concerning Fig. 9a) develops continuously from the regular behavior of the force lines for the nonconductor.

II.4. Edge Correction for the Plate Condenser According to Kirchhoff. Convince yourself that the relation

$$z = \frac{a}{2\pi} f(\zeta), \qquad f(\zeta) = 1 + \frac{2\pi i \zeta}{V} - \exp\left(\frac{2\pi i \zeta}{V}\right) \qquad \begin{cases} z = x + iy \\ \zeta = \Psi + i\Phi \end{cases}$$

represents the fringe field of the unilaterally terminated condenser in Fig. 46. $\Psi = $ const are the equipotential lines in the x, y-plane, $\Phi = $ const, the lines of force. Show that the two families of curves correspond qualitatively to the dotted lines in the figure, and that the line of force $\Phi = 0$ (drawn as full line in the figure) is an arc of a cycloid which joins the two edge points $x = 0$, $y = 0$ and $x = 0$, $y = a$.

II.5. The Capacitance of a Leyden Flask (Cylindrical Condenser). Let the dimensions be: height $h = 20$ cm, inner radius $r_1 = 5$ cm, wall thickness $d = 1$ mm. Let the dielectric constant of the glass be $6 \varepsilon_0$. Boundary corrections are to be neglected. The capacity is to be expressed in microfarads.

II.6. On the Definition of the Capacitance of Two Conductors with Equal and Opposite Charges. If in (10.15) we put $E_1 = -E_2 = E$ and $v = \Psi_1 - \Psi_2$ we obtain

$$2W = VE = (H_{11} + H_{22} - 2 H_{12})E^2$$

$$= K_{11}\Psi_1^2 + K_{22}\Psi_2^2 + 2K_{12}\Psi_1\Psi_2. \tag{1}$$

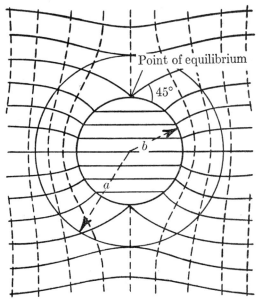

FIG. 45. The field of the *dielectric* sphere of radius $r = a$, which is produced by a uniform field on the outside, yields, when continued analytically into the interior, at the same time the field in the exterior of a *conducting* sphere of radius $r = b < a$ which is produced by the same uniform field.

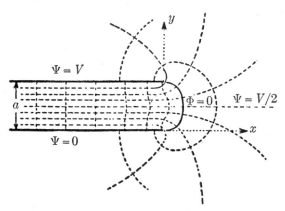

FIG. 46. Shape of the equipotentials $\Psi = \text{const}$ and lines of force $\Phi = \text{const}$ at the edge of a plate condenser.

Show by comparison with (10.11) that the following relation exists between the elementary definition of capacitance K and the coefficients H_{ij} and K_{ij} :

$$K = \frac{1}{H_{11} + H_{22} - 2H_{12}} = \frac{\Delta'}{K_{11} + K_{22} + 2K_{12}}, \qquad \Delta' = \begin{vmatrix} K_{11} & K_{12} \\ K_{21} & K_{22} \end{vmatrix} \qquad (2)$$

II.7. Characteristic Oscillations and Characteristic Frequencies of a Completely Conducting Cavity Bounded by a Rectangular Parallelepiped. Using Eqs. (24.9) and (24.10), represent the completely continuous field within a rectangular parallelepiped with the sides a, b, c, with the condition $E_{\text{tang}} = 0$ on the three pairs of bounding surfaces

$$x = \begin{cases} 0 \\ a \end{cases}, \qquad y = \begin{cases} 0 \\ b \end{cases}, \qquad z = \begin{cases} 0 \\ c \end{cases}.$$

II.8. Characteristic Oscillations and Characteristic Frequencies of the Interior of a Perfectly Conducting Circular Cylinder of Finite Length. Using Eqs. (24.6) and (24.7), represent the continuous field within a circular cylinder of radius a and length l, with the condition $E_{\text{tang}} = 0$ both on the mantel surface $r = a$ and on the two end surfaces $x = \begin{cases} 0 \\ l \end{cases}$.

II.9. Characteristic Oscillations within a Cavity Bounded by a Metal Sphere. As in §19, start with a Hertzian vector π which is directed along a diameter of the sphere $\left(\vartheta = \begin{cases} 0 \\ \pi \end{cases} \right)$, is periodic in t, and otherwise depends only on r. In contrast with §19 π must now be continuous also at the center of the sphere. The state then corresponds not to a spherical wave *emitted* from this point, but to a superposition of a *spherical wave radiated outward* and a *spherical wave* (reflected by the spherical surface) *radiated inward*. Determine the characteristic wave numbers k and the corresponding characteristic frequencies $\omega = kc$ from the boundary condition $E_{\text{tang}} = 0$ at the surface of the sphere $r = a$.

II.10. Determination of the Propagation Constants of Wire Waves from Kelvin's Telegraph Equation and from Rayleigh's Alternating Current Resistance a. for a Lecher two-wire line, b. for return conduction through the ground (let conduction in the forward direction be perfect).

PROBLEMS FOR PARTS III AND IV

III.1. The Lorentz Transformation for a Relative Motion Deviating from the x-Axis.

Let α be the angle between the relative motion $\mathbf{v}$ and the x-axis of the "system at rest". Let the xy-plane of the latter coincide with the plane through x and $\mathbf{v}$. We consider an "intermediate system" x_1, y_1, z_1, t_1 whos x_1 axis is to coincide with the $\mathbf{v}$-direction and whose x_1y_1-plane coincides with the $x\mathbf{v}$-plane. The transformation

$$x, y, z, t \to x_1, y_1, z_1, t_1, \qquad z_1 = z, \qquad t_1 = t \qquad (1)$$

is then an ordinary rotation through α in the xy-plane. Let a moving system x_1', y_1', z_1', t_1' be so placed that its x_1'- and y_1'-axes agree with the x_1- and y_1-axes of the intermediate system for $t = 0, t' = 0$. The transformation

$$x_1, y_1, z_1, t_1 \to x'_1, y'_1, z'_1, t'_1, \qquad y'_1 = y_1, \qquad z'_1 = z_1 \tag{2}$$

is then a special Lorentz transformation and is hence represented by Eq. (27.10).

If, finally, x'_1, y'_1, z'_1, t'_1 is rotated again through the angle $-\alpha$ in the $x'_1 y'_1$-plane, corresponding to the transformation

$$x'_1, y'_1, z'_1, t'_1 \to x', y', z', t', \qquad z'_1 = z', t'_1 = t' \tag{3}$$

the system of coefficients of the resulting total transition

$$x, y, z, t \to x', y', z', t' \tag{4}$$

is simplified. Convince yourself of the obvious fact that this transformation is orthogonal in four dimensions and of the not obvious fact that it may be represented by a three-dimensional vector formula.

III.2. On the Addition Theorem for Two Differently Directed Velocities. Prove Einstein's formula (27.19a) by the method of the Lorentz transformation.

III.3. The Field of an Electron in Uniform Motion. Transform the representation (30.6) by means of considerations of elementary geometry applied to Fig. 42 into the representation (28.14), (28.14a).

III.4. On the Relativistic Energy Theorem for the Electron. Derive the expression (32.7) for the kinetic energy and the energy theorem (32.6) from the equation of motion (32.5) of the electron by the usual method (scalar multiplication with the velocity).

III.5. The Electron in the Uniform Electrostatic Field. An electron enters a (vacuum) condenser with a transparent upper plate with the velocity v at an angle α. Let the plate separation be d, the potential difference of the upper with respect to the lower plate, V volts.

What curve does the electron describe in a non-relativistic treatment? How closely does it approach the lower plate?

For what velocity does it reach the lower plate?

(Example: $v = 5 \cdot 10^6$ meter/sec; $d = 10^{-2}$ meter; $V = 110$ volts.)

What potential field must an electron which is initially at rest traverse to attain the velocity $v = 5 \cdot 10^6$ M/S?

How do the conditions change for a relativistic treatment?

III.6. The Electron in a Uniform Magnetostatic Field. If the initial velocity of the electron is perpendicular to the lines of force a circular path is described. Determine its radius. If a component parallel to the lines of force is present, the path becomes a helix with circular projection. This applies in the relativistic just as in the non-relativistic case.

III.7. The Electron in a Uniform Electric Field and a Uniform Magnetic Field which is Parallel thereto. In Kaufmann's arrangement for the measurement of e/m the β-rays emitted by a radium sample pass first through a narrow aperture D and then cross a uniform electric field $\pm E$ and magnetic field B parallel thereto ($\pm E$ signifies reversal of polarity of the condenser). Assume that both fields begin at the aperture D and reach to the photographic plate, mounted perpendicular to the beam direction at a distance a from D. What curve is recorded on the plate if the β-rays are emitted with all possible velocities v? Neglect the change in the total velocity as compared with the large v, and represent the coordinates of the points of incidence as function of the parameter v.

III.8. The Electron in a Uniform Electric Field and a Uniform Magnetic field Perpendicular thereto. The path is a trochoid. Under the influence of

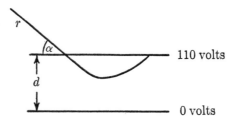

FIG. 47. The electron describes a ballistic parabola in the uniform condenser field.

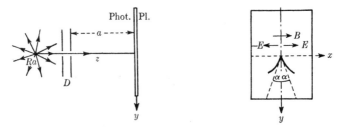

FIG. 48. Kaufmann's arrangement for the measurement of e/m. At the left: Lateral view. On the right: The pattern observed on the plate for β-rays with a continuous velocity spectrum emitted by the radium sample.

the electric field the circular motion produced by the magnetic field is converted into the motion of a point on a rolling circular disk. For what initial conditions is a *simple cycloid* obtained?

Motions of this type occur in the, "magnetron" electron tube.

III.9. The Characteristic of the Thermionic Diode According to Langmuir and Schottky. In practice a *cylindrical* configuration is generally employed: The cathode is a wire along the axis of cylinder whose mantel surface coincides with the anode. The *plane* configuration is mathematically simpler: Here the cathode at $x = 0$ and the anode at $x = l$ are plane circular disks separated at the edge by an insulating cylindrical tube. Let $V(x)$ be the potential at the point x between the cathode $V(0) = 0$ and the anode $V(l) = V$. Let the number of the electrons leaving the cathode per second be so large that it is permissible to assume a continuous space charge $-\rho$ of the electrons. Between cathode and anode the Poisson equation $\Delta V(x) = d^2 V(x)/dx^2 = \rho/\varepsilon_0$ is valid. The current density $\mathbf{J} = \rho\mathbf{v}$ transported through the tube is independent of x. v is determined from $mv^2/2 = eV(x)$. Integrate the Poisson equation by assuming a power law and deduce herefrom the socalled characteristic of the tube (I as function of the applied voltage V).

Convince yourself that the same method of assuming a power law is applicable also to the cylindrical configuration.

III.10. The Acceleration of an Electron in the Betatron. In the betatron[1] electrons are injected in the plane of symmetry between the axially symmetric pole pieces of an *alternating-current* electromagnet. The magnetic field forces them into a circular orbit. The pulsing of the magnetic field is accompanied by a vortex-like electric field which accelerates the electrons in their orbit. There is a radius $r = r_0$ of the path which remains unaltered with the pulsation of the magnetic field and with increasing electron velocity. After countless revolutions the electrons reach a velocity approaching that of light; they then resemble the beta-rays of radioactive materials, whence the name "betatron".

Let the axially symmetric magnetic field distribution $B(r, t)$ (its axial component) between the pole pieces, which decreases monotonically outwards, and the tangential initial velocity v_A of the electrons be given. We require

1. the attainable momentum mv of the electrons, their velocity, mass, and energy in eV (electron volts),

2. the radius r_0 of the equilibrium orbit,

3. the frequency of revolution at the end of the acceleration period and the total number of revolutions, and

4. the reaction force of the radiation at this point.

IV.1. The Field of Unipolar Induction. Let a bar be inserted in a uniform magnetic field **B** perpendicular to the lines of force and be displaced with uniform velocity along its axis. Compute

  a. the electric field in the interior and its potential,

  b. the voltage between its two sides,

  c. the external field, specifically for a circular cross section, and

  d. the surface charge.

  e. What change results if we pass from uniform translation of the bar

---

[1] This has also been called a rheotron, beam transformer, or electron centrifuge. The original idea of the device was given in the Aachen thesis of R. Wideröe in the year 1928.

along its axis to uniform rotation of a body of revolution about its axis of symmetry?

ANSWERS AND COMMENTS

I.1. The derivatives

$$\frac{\partial E_y}{\partial z} \quad \text{and} \quad \frac{\partial H_y}{\partial z}$$

occur in the x-components of Eqs. (4.8). These must remain finite in the transition to the limit $h \to 0$ so that the left sides of the equations in question do not become infinite, since this would make B_x and D_x infinite. In view of the y-components of Eqs. (4.8) this applies also to the derivatives

$$\frac{\partial E_x}{\partial z} \quad \text{and} \quad \frac{\partial H_x}{\partial z}.$$

The continuity of the *tangential components* E_x, E_y, H_x, H_y (see (3.9) and (3.8a)) follows herefrom.

In div $\mathbf{B} = 0$ (Eq. (4.4a)) there occurs the derivative $\partial B_z/\partial z$, which must also be continuous in the boundary layer: hence the condition for the continuity of the *normal component* B_z, Eq. (3.7a). The z-component of the Maxwell equation $\dot{\mathbf{B}} = -$ curl $\mathbf{E}$ does not suffice for this conclusion. It is true that on the right there occur only the tangential components E_x and E_y and differentiations with respect to x and y, so that the right side is continuous. However, the continuity of the left side, which may be deduced herefrom, would be consistent with a time-independent discontinuity of B_z. Hence the auxiliary condition div $\mathbf{B} = 0$ becomes necessary.

The same considerations, applied to Eq. (4.4b) for nonconductors, div $\mathbf{D}$ $= \rho$, lead in the limit $h \to 0$ to $\rho \to \infty$. This conclusion is by no means to be rejected, but indicates that at the boundary of two nonconductors there may exist a surface charge ω, which corresponds to an infinitely great charge density and is equal to the jump in the normal component of $\mathbf{D}$ on the two sides. By the z-component of the Maxwell equation $\dot{\mathbf{D}} =$ curl $\mathbf{H}$, this jump must be time-independe t. In the general case of a conductor and a nonconductor surface charge may also occur, in accord with Eq. (4.4c) or the z-component of the Maxwell equation $\dot{\mathbf{D}} + \mathbf{J} =$ curl $\mathbf{H}$; however, this surface charge need not be constant, but may decay as indicated by the current density J_x in the conductor.

I.2. We concern ourselves only with the magnetic field. Of the electric field, which we shall investigate in greater detail in §17, we need only know that it gives rise to a uniform current field J_z within the wire, $0 < r < a$, and to an also uniform current field J_{-z}, corresponding to an equal but oppositely directed total current I, in the return conductor:

$$I = \pi a^2 J_z = -\pi(c^2 - b^2)J_{-z}. \tag{1}$$

We employ polar coordinates r, φ, z with $r = 0$ as the axis of the wire. In view of the symmetry of the problem

$$\frac{\partial}{\partial \varphi} = \frac{\partial}{\partial z} = 0$$

for all three components H_r, H_φ, H_z. The φ- and z-components of the Maxwell equation curl $\mathbf{H} = \mathbf{J}$ then, with reference to the table in Problem I.3 of Vol. II, reduce to

$$\frac{dH_z}{dr} = 0 \quad (2) \quad \text{and} \quad \frac{1}{r}\frac{d}{dr}(rH_\varphi) = \begin{cases} J_z & ,0 < r < a \ , \\ 0 & ,a < r < b \ , \\ J_{-z} & ,b < r < c \ , \\ 0 & ,c < r < \infty \ , \end{cases} \quad (3)$$

whereas the r-component takes on the form $0 = 0$.

From (2) we obtain

$$H_z = \text{const} = 0,$$

the latter since H_z certainly must vanish for $r = \infty$. The fact that H_r must also vanish follows from the condition div $\mathbf{H} = 0$. The magnetic lines of excitation $\mathbf{H} = H_\varphi$ are hence coaxial circles about $r = 0$. We write $H_\varphi = H$ and tabulate the integration of (3) below:

Differential Equation		Solution	Determination of Constants
$0 < r < a$	$\dfrac{d}{dr}(rH) = J_z r$	$H = J_z \dfrac{r}{2} + \dfrac{A}{r}$	$A = 0$ since $H(0)$ is finite
$a < r < b$	$\dfrac{d}{dr}(rH) = 0$	$H = \dfrac{B}{r}$	$B = J_z a^2/2 = I/(2\pi)$ by continuity of H at $r = a$ and by Eq. (1)
$b < r < c$	$\dfrac{d}{dr}(rH) = J_{-z}r$	$H = J_{-z}\dfrac{r}{2} + \dfrac{C}{r}$	$C = \dfrac{I}{2\pi} - J_{-z}\dfrac{b^2}{2}$ $= \dfrac{I}{2\pi}\dfrac{c^2}{c^2 - b^2}$ by continuity of H at $r = b$ and by Eq. (1)
$c < r < \infty$	$\dfrac{d}{dr}(rH) = 0$	$H = \dfrac{D}{r}$	$D = 0$ by continuity of H at $r = c$ and by Eq. (1)

This solution agrees with Eqs. (4.10) to (4.13).

I.3. Coordinates r, φ, z and field symmetry $\partial/\partial\varphi = \partial/\partial z = 0$ as in I.2; a and b inner and outer radius of the solenoid. Now

$$J_r = J_z = 0, \quad \mathbf{J} = \begin{cases} J_\varphi & \text{for} \quad a < r < b, \\ 0 & \text{for} \quad r < a \ \text{and} \ r > b. \end{cases}$$

The differential equation curl $_r\mathbf{H} = 0$ is fulfilled throughout by symmetry, whereas the differential equation curl $_z\mathbf{H} = 0$ demands $d(rH_\varphi)/dr = 0$, $H_\varphi = A/r$, $A = 0$ because of the continuity of $\mathbf{H}$ at $r = 0$. Similarly $H_r = 0$ since div $\mathbf{H} = 0$. The table shows how H_z is to be evaluated:

Differential Equation		Solution	Determination of Constants
$0 < r < a$	$\dfrac{\partial H_z}{\partial r} = 0$	$H_z = $ const	const $= H = $ interior field
$a < r < b$	$\dfrac{\partial H_z}{\partial r} = -J_\varphi$	$H_z = -\displaystyle\int_a^r J_\varphi \, dr + A$	$A = H$ because of continuous joining with interior field
$b < r < \infty$	$\dfrac{\partial H_z}{\partial r} = 0$	$H_z = B$	$B = H - \displaystyle\int_a^b J_\varphi \, dr$ because of continuity at $r = b$

Since $H_z = 0$ for $r = \infty$ we must have $B = 0$ and therefore $H = \displaystyle\int_a^b J_\varphi \, dr$, which is equivalent with $H = N_1 I$ in Eq. (4.14): $\displaystyle\int_a^b J_\varphi \, dr$ is the total current which passes through the cross section of the solenoid of width $b - a$ and length 1 in the z-direction. This current is $N_1 I$, where N_1 is the number of turns per unit length.

I.4. The proof of the vector formula is obtained directly if the abbreviation $\mathbf{P} = \mathbf{A} \times \mathbf{B}$ is introduced, and the cyclic permutation rule $\mathbf{P} \cdot (\mathbf{C} \times \mathbf{D}) = \mathbf{C} \cdot (\mathbf{D} \times \mathbf{P})$, and the formula (6.2a), are employed.

To prove the cosine law set $\mathbf{D} = \mathbf{A}$. With $\mathbf{A}, \mathbf{B}, \mathbf{C}$ as radii of the unit sphere we have

$$\mathbf{A} \cdot \mathbf{B} = \cos c, \cdots , |\mathbf{A} \times \mathbf{B}| = \sin C, \cdots$$

The angle between the directions of $\mathbf{A} \times \mathbf{B}$ and $\mathbf{A} \times \mathbf{C}$ is equal to the angle α in the spherical triangle mapped out by $\mathbf{A}, \mathbf{B},$ and $\mathbf{C}$.

II.1. From the equation of the ellipsoid given in the Problem we calculate

$$x^2 + y^2 = (a^2 - c^2)\left(1 - \frac{z^2}{a^2}\right),$$

$$x^2 + y^2 + (z + c)^2 = a^2 + \frac{c^2}{a^2}z^2 + 2cz = \left(a + \frac{cz}{a}\right)^2.$$

In order to obtain the numerator and denominator occurring in the logarithm in (9.4) the root must be extracted so that it is positive for all $|z| \leq a$. This leads to

$$z + c + \sqrt{x^2 + y^2 + (z + c)^2} = z + c + a + \frac{cz}{a} = (a + c)\left(1 + \frac{z}{a}\right),$$

$$z - c + \sqrt{x^2 + y^2 + (z - c)^2} = z - c + a - \frac{cz}{a} = (a - c)\left(1 + \frac{z}{a}\right).$$

The quotient of the two is equal to $(a + c)/(a - c)$, i.e. a constant for each one of the confocal ellipsoids. Thus Eq. (9.4) is proved.

It is true that this constant changes if we proceed from an ellipsoid with principal axes $a_1 > a$, $b_1 > b$, c instead of from that with axes a, b, c. This does not, however, affect the identity of the fields in the exterior of the former ellipsoid, since no physical significance attaches to the numbering of the equipotential surfaces.

II.2. Let the rubbed glass rod have the constant charge λ per unit length and be infinitely thin. According to (7.5b) its potential is given by

$$4\pi\varepsilon\Psi = \lambda \int_0^\infty \frac{d\zeta}{\sqrt{x^2 + y^2 + (z - \zeta)^2}}$$

$$= \lambda \log \left(z - \zeta + \sqrt{x^2 + y^2 + (z - \zeta)^2}\right) \Big|_\infty^0$$

$$= \lambda \log \left\{\frac{z + \sqrt{x^2 + y^2 + z^2}}{x^2 + y^2}\right\} + \text{const.}$$

The constant here becomes infinite, namely in a sense equal to $-\lambda \log 0$.

Eq. (9.4), with $z' = z + c$, $E = 2c\lambda$ leads to the same expression (with z' replacing z) in the limit $c \to \infty$. The equipotential surfaces of the paraboloidal conductor agree with those of the glass rod of course only outside of the former, since within it $\Psi = \text{const.}$

II.3. According to Eqs. (9.13) and (9.14) the potential of the dielectric sphere, for $r > a$ (ε = relative dielectric constant of the sphere with reference to its surroundings), is:

$$\Psi_1 = -F\left(r - \frac{\varepsilon - 1}{\varepsilon + 2}\frac{a^3}{r^2}\right)\cos\Theta.$$

This is at the same time the analytical continuation of the potential into the interior of the sphere. For $r = b < a$ it yields

$$\Psi_1 = -Fb\left(1 - \frac{\varepsilon - 1}{\varepsilon + 2}\frac{a^3}{b^3}\right)\cos\Theta.$$

If here we set

$$b = a\sqrt[3]{\frac{\varepsilon - 1}{\varepsilon + 2}},$$

Ψ_1 becomes independent of Θ, namely, as we must demand for the grounded conducting sphere, $\Psi_1 = 0$.

II.4. This problem is mathematically related to the conformal mapping problems in Vol. II, §29, 30, 31, and makes use of the identity of two-dimensional potential theory and the theory of functions of a complex variable $z = x + iy$ elucidated in Vol. II, §19. The combination of the potential Ψ and the stream function Φ to form the complex variable ζ of the present problem was discussed at that point. We give z as function of ζ, rather than ζ as function of z, because z is a single-valued function of ζ.

The proof of the mapping function $f(\zeta)$ rests on the following:

If we set $\Psi = \begin{Bmatrix} 0 \\ V \end{Bmatrix}$, we find $f(\zeta) = \begin{Bmatrix} 0 \\ 2\pi i \end{Bmatrix} + 1 - \varphi - e^{-\varphi}, \varphi = \dfrac{2\pi\Phi}{V}$;

Hence $y = \begin{Bmatrix} 0 \\ a \end{Bmatrix}, x = \dfrac{a}{2\pi} (1 - \varphi - e^{-\varphi}) \leqq 0$ for $-\infty < \Phi < +\infty$.

The $\begin{Bmatrix} \text{lower} \\ \text{upper} \end{Bmatrix}$ condenser plate $y = \begin{Bmatrix} 0 \\ a \end{Bmatrix}, x \leqq 0$ is thus at the potential

$$\Psi = \begin{Bmatrix} 0 \\ V \end{Bmatrix}.$$

Also the bisecting plane of the condenser is an equipotential surface. If we set $\Psi = V/2$, we find $f(\zeta) = \pi i + 1 - \varphi + e^{-\varphi}$; hence $y = a/2, x = \dfrac{a}{2\pi} (1 - \varphi + e^{-\varphi})$. However, now $x \gtrless 0$ as φ varies between $-\infty$ and $+\infty$. We are therefore dealing not with a semi-infinite straight line or plane, but with a bilaterally infinite line or plane.

The boundary points $x = 0, y = 0$ and $x = 0, y = a$, corresponding to $\Psi = 0, \Phi = 0$ and $\Psi = V, \Phi = 0$, respectively, are branching points of the conformal mapping. The line of force $\Phi = 0$, which joins the two boundary points, is given in parametric form (with $\psi = 2\pi\Psi/V$) by

$$x = \frac{a}{2\pi} (1 - \cos \psi), \qquad y = \frac{a}{2\pi} (\psi - \sin \psi).$$

This is the equation of the simple cycloid (see e.g. the quite similar representation in Vol. I, Eq. (17.2)).

The lines of force in the interior of the condenser and at a large distance from the boundary points belong to the parameter values

$$\varphi \gg 1, \qquad 0 < \psi < 2\pi.$$

Since here $\exp (i\psi - \varphi)$ vanishes with increasing φ, we obtain simply

$$x + iy = \frac{a}{2\pi} (1 + i\psi - \varphi), \quad \text{i.e.} \quad x = \frac{a}{2\pi} (1 - \varphi), \quad y = \frac{a}{2\pi} \psi.$$

Since $\Phi = $ const on these lines of force, we have also $\varphi = $ const and hence $x = $ const. On the other hand, Ψ varies on them between 0 and V, and hence

ψ between 0 and 2π and y between 0 and a. The lines of force hence approximate closer and closer to straight lines perpendicular to the condenser plates, as is to be expected.

II.5. The differential equation of the potential in the cylindrical co-ordinates r, φ, z is, when independent of φ and z,

$$\frac{1}{r}\frac{d}{dr}r\frac{d\Psi}{dr} = 0.$$

It yields

$$\frac{d\Psi}{dr} = -E_r = \frac{A}{r}, \qquad D_r = \frac{-\varepsilon A}{r}.$$

Hence the surface density on the inner and outer electrode is

$$\omega_1 = \frac{-\varepsilon A}{r_1}, \qquad \omega_2 = \frac{\varepsilon A}{r_2}$$

and the charge per unit length of the z-coordinate

$$e_1 = 2\pi r_1\omega_1 = -2\pi\varepsilon A, \qquad e_2 = 2\pi r_2\omega_2 = +2\pi\varepsilon A = -e_1.$$

A second integration leads to

$$\Psi = A\log r + B, \qquad \Psi_1 - \Psi_2 = V = -A\log\frac{r_2}{r_1} = \frac{e_1}{2\pi\varepsilon}\log\frac{r_2}{r_1},$$

so that

$$K_1 = \frac{e_1}{V} = 2\pi\varepsilon\bigg/\log\frac{r_2}{r_1} = \text{capacity per unit length.}$$

For $d \ll r_1$ we find

$$\log\frac{r_2}{r_1} = \log\left(1 + \frac{d}{r_1}\right) \cong \frac{d}{r_1}$$

and, neglecting end corrections,

$$K = K_1h = \frac{2\pi r_1 h\varepsilon}{d} = \frac{\text{surface}}{\text{separation}} \cdot \text{dielectric constant.}$$

In MKSQ-units:

$$r_1 h = 100 \text{ cm}^2 = 10^{-2}\text{ M}^2, \qquad d = 1\text{ mm} = 10^{-3}\text{ M},$$

$$\varepsilon = 6\varepsilon_0 = \frac{1}{6\pi}\cdot 10^{-9}\frac{Q^2}{\text{joule M}}, \qquad K = \frac{1}{3}\cdot 10^{-8}\text{ farad} = \frac{1}{3}\cdot 10^{-2}\text{ microfarad.}$$

II.6. The first of the relations given in Eq. (2) of the Problem is obvious in view of Eq. (1). The second is obtained as follows:

By (10.14):

$$E = K_{11}\Psi_1 + K_{12}\Psi_2, \tag{1}$$

$$-E = K_{21}\Psi_1 + K_{22}\Psi_2.$$

Hence

$$0 = (K_{11} + K_{21})\Psi_1 + (K_{12} + K_{22})\Psi_2. \tag{2}$$

In addition, the following linear relation exists between Ψ_1 and Ψ_2:

$$V = \Psi_1 - \Psi_2. \tag{3}$$

From (2) and (3) we compute

$$\Psi_1 = \frac{K_{12} + K_{22}}{K_{11} + 2K_{12} + K_{22}} V, \qquad \Psi_2 = -\frac{K_{11} + K_{12}}{K_{11} + 2K_{12} + K_{22}} V.$$

Substitution in (1) yields

$$E = \frac{K_{11}K_{22} - K_{12}^{2}}{K_{11} + 2K_{12} + K_{22}} V.$$

The factor of V is the capacity in the elementary sense. Hence also the second relation (2) of the Problem has been proved.

II.7. If the phase factor which must be thought of as added to (24.9), namely exp (ihx), is replaced by $\genfrac{}{}{0pt}{}{\cos}{\sin} \pi l \frac{x}{a}$, where l is an integer (standing instead of travelling wave) and if $\genfrac{}{}{0pt}{}{\cos}{\sin}$ is chosen with due regard of the boundary conditions prescribed for the individual **E**-components, a particular characteristic electromagnetic oscillation of the interior of the parallelepiped is obtained for which $H_x = 0$; similarly, proceeding from (24.10), one for which $E_x = 0$ is obtained.

However these are not yet the general characteristic oscillations of the parallelepiped, as is evident from the specific values $H_x = 0$ and $E_x = 0$, respectively. The general system, which has complete symmetry with respect to the three axes, is

$$E_x = A \cos\left(\pi l \frac{x}{a}\right) \sin\left(\pi n \frac{y}{b}\right) \sin\left(\pi m \frac{z}{c}\right).$$

$$E_y = B \sin\left(\pi l \frac{x}{a}\right) \cos\left(\pi n \frac{y}{b}\right) \sin\left(\pi m \frac{z}{c}\right), \tag{1}$$

$$E_z = C \sin\left(\pi l \frac{x}{a}\right) \sin\left(\pi n \frac{y}{b}\right) \cos\left(\pi m \frac{z}{c}\right),$$

$$\sqrt{\frac{\mu_0}{\varepsilon_0}}\, H_x = A' \sin\left(\pi l\,\frac{x}{a}\right)\cos\left(\pi n\,\frac{y}{b}\right)\cos\left(\pi m\,\frac{z}{c}\right),$$

$$\sqrt{\frac{\mu_0}{\varepsilon_0}}\, H_y = B' \cos\left(\pi l\,\frac{x}{a}\right)\sin\left(\pi n\,\frac{y}{b}\right)\cos\left(\pi m\,\frac{z}{c}\right), \qquad (2)$$

$$\sqrt{\frac{\mu_0}{\varepsilon_0}}\, H_z = C' \cos\left(\pi l\,\frac{x}{a}\right)\cos\left(\pi n\,\frac{y}{b}\right)\sin\left(\pi m\,\frac{z}{c}\right)$$

Since div $\mathbf{E} = 0$ the A, B, C must satisfy the condition

$$A\,\frac{l}{a} + B\,\frac{n}{b} + C\frac{m}{c} = 0, \qquad (3)$$

whereas the A', B', C' are determined from the A, B, C by the equations

$$ikA' = \frac{n\pi}{b}\,C - \frac{m\pi}{c}\,B,$$

$$ikB' = \frac{m\pi}{c}\,A - \frac{l\pi}{a}\,C, \qquad (4)$$

$$ikC' = \frac{l\pi}{a}\,B - \frac{n\pi}{b}\,A.$$

This system will play an important role in the problem of black-body radiation in Vol. V, just as the general system of elastic characteristic vibrations in §44 of Vol. II was of importance for the problem of specific heats.

II.8. We find from (24.6) with $h = \pi m/l$ (m = integer, l = length of the cylinder), if, again, exp (ihx) is replaced by $\genfrac{}{}{0pt}{}{\cos}{\sin}\, hx$ and care is taken, by the proper choice of the cosine or sine, to fulfill the boundary conditions $E_{\text{tang}} = 0$,

$$E_x = \frac{\sqrt{k^2 - h^2}}{h}\, J_n(\rho)\,\cos\,(n\varphi)\,\cos\,(hx), \qquad \sqrt{\frac{\mu_0}{\varepsilon_0}}\, H_x = 0$$

$$E_r = \qquad -J'_n(\rho)\,\cos\,(n\varphi)\,\sin\,(hx),$$

$$\sqrt{\frac{\mu_0}{\varepsilon_0}}\, H_r = \frac{k}{ih}\,\frac{n}{\rho}\, J_n(\rho)\,\sin\,(n\varphi)\,\cos\,(hx), \qquad (1)$$

$$E_\varphi = \qquad \frac{n}{\rho}\, J_n(\rho)\,\sin\,(n\varphi)\,\sin\,(hx),$$

$$\sqrt{\frac{\mu_0}{\varepsilon_0}}\, H_\varphi = \frac{k}{ih}\, J'_n(\rho)\,\cos\,(n\varphi)\,\cos\,(hx).$$

Here, as in (24.6), $\rho = \sqrt{k^2 - h^2}\, r$ and $\sqrt{k^2 - h^2}\, a = w_\nu$ is one of the infinitely many roots of $J_n(w) = 0$. The characteristic wave number k and the characteristic frequency ω are accordingly given by

$$k^2 = h^2 + \frac{w_\nu^2}{a^2}, \qquad \omega = kc \,(c = \text{velocity of light}). \tag{2}$$

The system of characteristic functions represented by (1) is triply infinite and is ordered by the numbers n, ν, and the integer m contained in h. Convince yourself that (1) satisfies not only the boundary conditions, but also the relations between $\mathbf{E}$ and $\mathbf{H}$ demanded by the Maxwell equations.

Similarly we obtain from (24.7), with the same meaning of h,

$$\sqrt{\frac{\mu_0}{\varepsilon_0}}\, H_x = \frac{\sqrt{k^2 - h^2}}{h}\, J_n(\rho)\, \cos\,(n\varphi)\, \sin\,(hx), \qquad E_x = 0,$$

$$\sqrt{\frac{\mu_0}{\varepsilon_0}}\, H_r = J'_n(\rho)\, \cos\,(n\varphi)\, \cos\,(hx),$$

$$E_r = -\frac{k}{ih}\frac{n}{\rho}\, J_n(\rho)\, \sin\,(n\varphi)\, \sin\,(hx), \tag{3}$$

$$\sqrt{\frac{\mu_0}{\varepsilon_0}}\, H_\varphi = -\frac{n}{\rho}\, J_n(\rho)\, \sin\,(n\varphi)\, \cos\,(hx),$$

$$E_\varphi = -\frac{k}{ih}\, J'_n(\rho)\, \cos\,(n\varphi)\, \sin\,(hx).$$

The characteristic wave number and the characteristic frequency are now given by

$$k^2 = h^2 + \frac{w_\nu'^2}{a^2}, \qquad \omega = kc \tag{4}$$

where w_ν' is one of the infinitely many roots of $J'_n(w') = 0$. The series of characteristic oscillations (3) is again triply infinite.

Do (1) and (3) supply the complete system of characteristic vibrations of the interior of the cylinder?

II.9. Except for a multiplying constant the appropriate solution, continuous at $r = 0$, of the differential equation (19.16) is

$$\Pi = \frac{\sin\,(kr)}{r}\, e^{-i\omega t}.$$

The separation into two parts

$$\Pi = \frac{1}{2ir}\,(e^{ik(r-ct)} - e^{-ik(r+ct)})$$

indicates the superposition, mentioned in the Problem, of a spherical wave radiated outwards and one radiated inwards. Eqs. (19.17) and the following equations yield, without the time factor

$$\varepsilon_0 E_r = \cos \vartheta \left(\frac{d^2}{dr^2} + k^2 \right) \frac{\sin (kr)}{r}, \qquad \varepsilon_0 E_\vartheta = - \sin \vartheta \left(\frac{1}{r} \frac{d}{dr} + k^2 \right) \frac{\sin (kr)}{r},$$

$$E_\varphi = 0, \qquad H_r = H_\vartheta = 0, \qquad H_\varphi = \frac{i\omega}{r} \sin \vartheta \left(\frac{d}{dr} \sin (kr) - \frac{\sin (kr)}{r} \right).$$

The boundary condition $E_\vartheta = 0$ demands for $r = a$

$$\cos(ka) - \frac{\sin (ka)}{ka} [1 - (ka)^2] = 0.$$

Therefore the transcendental equation

$$\tan x = \frac{x}{1 - x^2}, \qquad x = ka.$$

Its graphical solution yields a first root x_1 which is somewhat smaller than π and an infinite series of additional roots which asymptotically approach the value $x_\nu = \nu\pi$.

In addition to this singly infinite system of characteristic functions, for which the electric lines of force lie in the meridional plane $\varphi = $ const and the magnetic lines are perpendicular thereto, there are ∞^2 less symmetric characteristic functions, with a Legendre function dependence in ϑ and φ.

II.10. The telegraph equation (18.19), with $G = 0$ and the assumption $I = I_0 \exp i(hx - \omega t)$, yields

$$h^2 = \omega K(\omega L + i\mathbf{R}). \tag{1}$$

For large ω, more exactly, for $\omega L \gg |\mathbf{R}|$, this leads to

$$h - k = \frac{i}{2} \frac{\mathbf{R}}{\mathbf{z}}; \tag{2}$$

Here

$$k = \omega\sqrt{KL} \tag{2a}$$

is the wave number of the perfectly conducting line and

$$\mathbf{z} = \sqrt{\frac{L}{K}} \tag{2b}$$

is its wave impedance. $\mathbf{R}$ is the impedance operator from Eq. (20.19), composed of the real resistance R and the inner inductive reactance ωL_i (numerically equal to the resistance in the presence of the skin effect) to form a complex quantity:

$$\mathbf{R} = R - i\omega L_i = (1 - i)R. \tag{2c}$$

It may be noted, incidentally, that $\mathbf{R}$ may be replaced directly by R in (1) if we interpret L as the sum $L_a + L_i$ of the outer and inner selfinductance and not, as was done in the telegraph equation, as the outer selfinductance L_a alone. We assume that ω is large enough that a fully developed skin effect occurs. According to (20.12) the depth of penetration is then given by

$$d = 1/\sqrt{\mu\sigma\omega/2} \qquad (3)$$

We shall furthermore assume that the wave amplitude may vary slowly along the circumference of the wire so that the validity of the solution originally obtained for the plane problem in §20 B is not impaired. The alternating current resistance of a surface strip of length 1 and width 1, measured in the direction of the circumference, is then by (20.15a)

$$R_1 = \frac{1}{\sigma d}. \qquad (4)$$

The current through this metal strip (lying directly underneath) is equal to the line integral of $\mathbf{H}$ about this strip, which in our case reduces to the value of $\mathbf{H}$ at the surface. Hence we obtain for the Joule heat developed in the strip, utilizing (4),

$$R_1 \mathbf{H}^2 = \frac{1}{\sigma d} \mathbf{H}^2$$

and for the Joule heat developed in unit length of the conductor as a whole (ds = line element of the circumference, $\oint$ = integration over the circumference):

$$\frac{1}{\sigma d} \oint \mathbf{H}^2 \, ds = RI^2 \quad \text{with} \quad I = \oint \mathbf{H} \cdot d\mathbf{s}. \qquad (5)$$

R and I are resistance and total current of this unit length. From (5) we compute

$$R = \frac{1}{\sigma d} \oint \mathbf{H}^2 \, ds \Big/ \left(\oint \mathbf{H} \cdot d\mathbf{s} \right)^2. \qquad (6)$$

$\mathbf{H}$ is to be obtained from the *quasistationary* field in the dielectric. This yields also the external selfinductance L_a of unit length of the conductor and its capacity K, as well as its wave impedance $\mathbf{z}$ in Eq. (2b).

a. *The Lecher Two-wire Line.* Wire radius a, separation of wire axes $2b$, separation of the two source lines $2\zeta_0$ (see Fig. 36); u, v bipolar coordinates for representing the field in the dielectric, u = const magnetic, v = const electric lines of force;

$$ds = g \, dv, \qquad g = \frac{\zeta_0}{\cosh u - \cos v}, \qquad \zeta_0 = \sqrt{b^2 - {}^2}. \qquad (7)$$

On the circumference of the wire $u = u_0$, $\cosh u_6 = b/a$.

Since the bipolar coordinate v signifies the magnetic potential directly, the desired magnetic field component at the periphery of the wire is

$$\mathbf{H} = \frac{dv}{ds} = \frac{1}{g} = \frac{1}{\zeta_0} (\cosh u_0 - \cos v). \tag{8}$$

From this the integrals occurring in (6) may be calculated:

$$\oint \mathbf{H} \cdot d\mathbf{s} = \int_0^{2\pi} \frac{1}{g} g\, dv = 2\pi,$$

$$\oint \mathbf{H}^2 ds = \int_0^{2\pi} \frac{1}{g^2} g\, dv = \frac{1}{\zeta_0} \int_0^{2\pi} (\cosh u_0 - \cos v)\, dv = \frac{2\pi}{\zeta_0} \cosh u_0 = \frac{2\pi}{\zeta_0} \frac{b}{a}$$

and hence (6) leads to (the factor 2 to be added to (6) arises from the two-wire line):

$$R = \frac{2}{\sigma d} \frac{2\pi}{\zeta_0} \frac{b}{a} \Big/ (2\pi)^2 = \frac{1}{\pi\sigma\zeta_0 d} \frac{b}{a} \Omega M^{-1}. \tag{9}$$

Since furthermore the bipolar coordinate u determines the electric potential, the difference in potential between the two wires is

$$2u_0 = 2 \operatorname{arccosh} \frac{b}{a} = 2 \log \frac{b + \zeta_0}{a}.$$

From this we find for the capacity and selfinductance per unit length

$$\frac{\varepsilon_0}{K} = \frac{L}{\mu_0} = \frac{2u_0}{2\pi} = \frac{1}{\pi} \log \frac{b + \zeta_0}{a} \tag{10}$$

and for the wave impedance

$$\mathbf{z} = \sqrt{\frac{\mu_0}{\varepsilon_0}} \frac{1}{\pi} \log \frac{b + \zeta_0}{a} \Omega. \tag{11}$$

Taking account of (2c), (9) and (11) yields for the propagation constant in Eq. (2)

$$h - k = \frac{1 + i}{2} \sqrt{\frac{\varepsilon}{\mu_0} \frac{1}{\sigma\zeta_0 d} \frac{b}{a}} \Big/ \log \frac{b + \zeta_0}{a}. \tag{12}$$

This expression agrees with the value calculated in (25.20). To realize this it is merely necessary to substitute the value of d from Eq. (3) in (12), and to express the conductivity σ by the complex dielectric constant $\varepsilon' = \varepsilon + i\sigma/\omega \cong i\sigma/\omega$, and to note the meaning of the abbreviation $p(1/p = (b + \zeta_0)/a)$ in (25.20).

b. *Return through ground.* The earth's surface now takes the place of the plane of symmetry $u = 0$ of the bipolar coordinates. In formula (6) for R the factor 2 is now to be omitted, since the forward conduction through the wire is to be assumed to be resistance-free, so that R refers only to the return conduction through ground. On the other hand K is to be doubled, L and $\mathbf{z}$ to be halved. b now signifies the height of the wire above ground, ζ_0 the height of the source line above ground, which does not differ appreciably from b. We then obtain from (9), (11), and (12)

$$R = \frac{1}{2\pi\sigma bd}, \qquad \mathbf{z} = \sqrt{\frac{\mu_0}{\varepsilon_0} \frac{1}{2\pi}} \log \frac{2b}{a}, \tag{13}$$

$$h - k = \frac{1+i}{2} \sqrt{\frac{\varepsilon_0}{\mu_0} \frac{1}{\sigma bd}} \Big/ \log \frac{2b}{a}. \tag{14}$$

Numerical example: frequency in the range of radiotelephony $10^6 \ S^{-1}$, σ (earth) $= 10^{-2} - 10^{-4} \Omega^{-1} M^{-1}, b = 10 \ M, a = 1 \ mm$. We compute from (3)

$$d = 50.4 \ M \ \text{to} \ 5.04 \ M$$

and obtain from (13) and (14)

$$R = 3.2 \ \text{to} \ 0.32 \ \Omega/M$$

$$|h - k| = 38 \ \text{to} \ 3.8 \cdot 10^{-4} \ M^{-1}.$$

At low frequencies the penetration is so great compared with the usual height of the wire b that (in contrast with that in §25) our present method of calculation fails.

For the single wire without return conductor the external field can no longer be calculated in a quasistationary manner, so that the above approximate method no longer constitutes a simplification as compared with §22. Our numerical example indicates, by the way, how greatly the field of the single wire is disturbed even by a non-metallic return conductor; see in this connection the note at the beginning of §22 regarding the failure of Hertz's original experiments with wire waves and the influence of the laboratory walls.

III.1. The transformation (4) of the Problem is to be built up out of the transformations (1), (2), and (3) in the following manner:

$$L' = DLD^{-1} \tag{a}$$

(L' and L = Lorentz transformations, D = rotation, D^{-1} = inverse rotation). Here

D: $\qquad\qquad x_1 = x \cos \alpha + y \sin \alpha, \qquad z_1 = z,$

$\qquad\qquad\qquad y_1 = -x \sin \alpha + y \cos \alpha, \qquad t_1 = t;$

L: $\qquad\qquad x_1' = \dfrac{x_1 - t_1 \beta c}{\sqrt{1 - \beta^2}}, \qquad y_1' = y_1,$

$\qquad\qquad\qquad t_1' = \dfrac{t_1 - x_1 \beta/c}{\sqrt{1 - \beta^2}}, \qquad z_1' = z_1;$

D^{-1}:
$$x' = x_1' \cos \alpha - y_1' \sin \alpha, \qquad z' = z_1',$$
$$y' = x_1' \sin \alpha + y_1' \cos \alpha, \qquad t' = t_1'.$$

By their successive combination we find, with the abbreviation
$$\eta = (1 - \beta^2)^{-\frac{1}{2}}$$
the following system of coefficients for L':

	x	y	z	ict
x'	$1 + (\eta - 1) \cos^2 \alpha$	$(\eta - 1) \cos \alpha \sin \alpha$	0	$i\beta\eta \cos\alpha$
y'	$(\eta - 1) \cos \alpha \sin \alpha$	$1 + (\eta - 1) \sin^2 \alpha$	0	$i\beta\eta \sin\alpha$
z'	0	0	1	0
ict'	$-i\beta\eta \cos \alpha$	$-i\beta\eta \sin \alpha$	0	η

It is four-dimensionally orthogonal (sum of the squares equal to 1, sum of the products equal to 0, both in the horizontal rows and the vertical columns) and hence can be read just as well from top to bottom as from left to right.

The velocity $\mathbf{v}$ of transformation (2) forming an angle α with the x-axis has, in the x, y, z-system, the components $v \cos \alpha$, $v \sin \alpha$, 0. With $\mathbf{r} = x$, y, z; $\mathbf{r}' = x'$, y', z' we therefore obtain

$$\frac{\mathbf{v}}{v} \cdot \mathbf{r} = x \cos \alpha + y \sin \alpha, \qquad \frac{\mathbf{v}}{v} \cdot \mathbf{r}' = x' \cos \alpha + y' \sin \alpha.$$

With these abbreviations our system, read from left to right, yields

$$\mathbf{r}' = \mathbf{r} + \frac{\mathbf{v}}{v} \left\{ (\eta - 1) \frac{\mathbf{v}}{v} \cdot \mathbf{r} - \beta\eta ct \right\} \tag{b}$$

$$t' = \eta \left\{ t - \frac{\beta}{c} \frac{\mathbf{v}}{v} \cdot \mathbf{r} \right\}$$

and, read from top to bottom,

$$\mathbf{r} = \mathbf{r}' + \frac{\mathbf{v}}{v} \left\{ (\eta - 1) \frac{\mathbf{v}}{v} \cdot \mathbf{r}' + \beta\eta ct' \right\} \tag{c}$$

$$t = \eta \left\{ t' + \frac{\beta}{c} \frac{\mathbf{v}}{v} \cdot \mathbf{r}' \right\}.$$

Whereas (b) follows from symbol (a), (c) corresponds to the inversion of (a):
$$L = D^{-1}L'D. \tag{d}$$

By taking the derivative of (b) with respect to t' or of (c) with respect to t, respectively, we obtain for the three-dimensional velocity vectors $\mathbf{q}' = d\mathbf{r}'/dt'$ and $\mathbf{q} = d\mathbf{r}/dt$

$$q' = \frac{q - v\{\eta - (\eta - 1)v\cdot q/v^2\}}{\eta(1 - v\cdot q/c^2)}, \tag{e}$$

$$q = \frac{q' + v\{\eta + (\eta - 1)v\cdot q'/v^2\}}{\eta(1 + v\cdot q'/c^2)}. \tag{f}$$

III.2. The Problem can be formulated in the following manner: Let a system x_1, t_1 move with respect to a system x, t with the velocity $\beta_1 c$ along the x-axis. In system x_1, t_1 let a point P move with the velocity $\beta_2 c$ at an angle α with respect to the x_1-axis (and at the same time, the x-axis). What is the resultant velocity of point P as observed from system x, t?

Apart from additive constants the motion of P in the x_1, t_1-system is described by

$$x_1 = \beta_2 c t_1 \cos \alpha, \qquad y_1 = \beta_2 c t_1 \sin \alpha. \tag{1}$$

At the same time the Lorentz transformation

$$x_1 = \frac{x - \beta_1 ct}{\sqrt{1 - \beta_1^2}}, \qquad y_1 = y, \qquad t_1 = \frac{t - \beta_1 x/c}{\sqrt{1 - \beta_1^2}} \tag{2}$$

applies for every point of the x_1, t_1-system.

Substitution of x_1 from (1) and t_1 from (2) leads to

$$x - \beta_1 ct = \beta_2 c \cos \alpha \, (t - \beta_1 x/c)$$

or, after collecting terms with x and t,

$$x(1 + \beta_1\beta_2 \cos \alpha) = (\beta_1 + \beta_2 \cos \alpha)ct.$$

Hence

$$\frac{dx}{dt} = \frac{\beta_1 + \beta_2 \cos \alpha}{1 + \beta_1 \beta_2 \cos \alpha} c; \tag{3}$$

in addition, (2) and (1) lead to

$$y = \frac{\beta_2 c \sin \alpha}{\sqrt{1 - \beta_1^2}} (t - \beta_1 x/c),$$

$$\frac{dy}{dt} = \frac{\beta_2 c \sin \alpha}{\sqrt{1 - \beta_1^2}} \left(1 - \frac{\beta_1}{c}\frac{dx}{dt}\right),$$

and, in view of (3),

$$\frac{dy}{dt} = \frac{\beta_2 c \sin \alpha}{\sqrt{1 - \beta_1^2}} \left(1 - \frac{\beta_1^2 + \beta_1\beta_2 \cos \alpha}{1 + \beta_1\beta_2 \cos \alpha}\right) = \frac{\beta_2 c \sin \alpha \sqrt{1 - \beta_1^2}}{1 + \beta_1 \beta_2 \cos \alpha}. \tag{4}$$

The resultant in the x, t-system is

$$q = \sqrt{\left(\frac{dx}{dt}\right)^2 + \left(\frac{dy}{dt}\right)^2}.$$

If we put $q/c = \beta$, we obtain

$$\beta^2 = \frac{(\beta_1 + \beta_2 \cos \alpha)^2 + \beta_2^2(1 - \beta_1^2) \sin^2 \alpha}{(1 + \beta_1 \beta_2 \cos \alpha)^2}$$

$$= \frac{\beta_1^2 + 2\beta_1\beta_2 \cos \alpha + \beta_2^2 - \beta_1^2 \beta_2^2 \sin^2 \alpha}{(1 + \beta_1 \beta_2 \cos \alpha)^2} \quad \text{Q.E.D.}$$

(5)

III.3. In (30.6) $\mathbf{r}$ denotes the vector $L \to P$, and r, hence, one side of the triangle LOP in Fig. 42. Let O be the position of the electron at the time of the field observation at P. Let the velocity $\mathbf{v}$ of the electron be directed, unlike §28, along the positive x-axis, leading to a specialization of the formulas in §30. The length LO then becomes equal to $v\tau$ with $\tau = r/c$, so that

$$LO = v\tau = \beta r.$$

The length OP is the separation between electron and point of reference at the time of the observation and will be designated by r',

$$r' = \sqrt{x'^2 + y'^2 + z'^2},$$

where x', y', z' are the coordinates of P relative to O. ϑ and ϑ' are the angles at L and O shown in the figure and we have

$$r' \cos \vartheta' = x', \qquad r \cos \vartheta = x' + \beta r. \tag{1}$$

Furthermore, by the Pythagorean theorem,

$$r^2 = r'^2 + (\beta r)^2 + 2r'\beta r \cos \vartheta', \tag{2}$$

so that

$$r^2(1 - \beta^2) - 2\beta r x' = r'^2.$$

The solution of this quadratic equation for r yields

$$r = \frac{\beta x'}{1 - \beta^2} + \frac{1}{\sqrt{1 - \beta^2}} \sqrt{x'^2 + y'^2 + z'^2 + \frac{\beta^2}{1 - \beta^2} x'^2}. \tag{3}$$

As in (28.13a) we denote the square root on the right by s:

$$s = \sqrt{x'^2 \left(1 + \frac{\beta^2}{1 - \beta^2}\right) + y'^2 + z'^2} = \sqrt{\frac{x'^2}{1 - \beta^2} + y'^2 + z'^2}$$

and obtain by (3)

$$r = \frac{s}{\sqrt{1 - \beta^2}} + \frac{\beta}{1 - \beta^2} x'. \tag{4}$$

Now, $v_r = v \cos \vartheta$ and, in view of (1),

$$r \frac{v_r}{c} = \beta r \cos \vartheta = \beta x' + \beta^2 r;$$

Accordingly, by (4),

$$r\left(1 - \frac{v_r}{c}\right) = r(1 - \beta^2) - \beta x' = \sqrt{1 - \beta^2}\, s. \qquad (5)$$

Thus the quotient occurring in (30.6) becomes

$$\frac{1 - v^2/c^2}{r^3(1 - v_r/c)^3} = \frac{1 - \beta^2}{(1 - \beta^2)^{3/2}} \frac{1}{s^3}. \qquad (6)$$

At the same time the vectorial factors multiplied herewith in view of the assumed direction of $\mathbf{v}$, are resolved into their components:

$$\mathbf{r} \times \mathbf{v} = 0, z'v, - y'v. \qquad (7)$$

$$r\frac{v_x}{c} - r_x = r\beta - r\cos\vartheta = r\beta - x' - r\beta = -x',$$

$$\qquad (7a)$$

$$r\frac{v_y}{c} - r_y = -y', \qquad r\frac{v_z}{c} - r_z = -z'.$$

If all of this is substituted in Eq. (30.6) taking due account of the changed sign of v, Eqs. (28.14) and (28.14a) are obtained.

The equation $s = $ const defines the family of mutually similar "Heaviside ellipsoids", flattened in the direction of motion (see §28C); the electric lines of force in x', y', z' space are the orthogonal trajectories of the family.

III.4. Scalar multiplication with $\mathbf{v}$ of the left side of (32.5) results in

$$\mathbf{v} \cdot \frac{d}{dt} \frac{m_0 \mathbf{v}}{\sqrt{1 - \beta^2}} = m_0 \frac{\mathbf{v} \cdot \dot{\mathbf{v}}}{\sqrt{1 - \beta^2}} + m_0 v^2 \frac{d}{dt} \frac{1}{\sqrt{1 - \beta^2}}.$$

The second term on the right is equal to

$$m_0 v^2 \frac{\boldsymbol{\beta} \cdot \dot{\boldsymbol{\beta}}}{(1 - \beta^2)^{3/2}} = m_0 \beta^2 \frac{\mathbf{v} \cdot \dot{\mathbf{v}}}{(1 - \beta^2)^{3/2}}.$$

Together with the first term this yields

$$m_0 \frac{\mathbf{v} \cdot \dot{\mathbf{v}}}{\sqrt{1 - \beta^2}} \left(1 + \frac{\beta^2}{1 - \beta^2}\right) = m_0 \frac{\mathbf{v} \cdot \dot{\mathbf{v}}}{(1 - \beta^2)^{3/2}} = \frac{d}{dt} \frac{m_0 c^2}{\sqrt{1 - \beta^2}}. \qquad (1)$$

On the right side of (32.5) scalar multiplication with $\mathbf{v}$ yields

$$\mathbf{v} \cdot \mathbf{K} = \mathbf{v} \cdot \mathbf{E} + \mathbf{v} \cdot (\mathbf{v} \times \mathbf{B}) = \mathbf{v} \cdot \mathbf{E} + \mathbf{B} \cdot (\mathbf{v} \times \mathbf{v}) = \mathbf{v} \cdot \mathbf{E}. \qquad (2)$$

Hence, setting (1) and (2) equal to each other, we obtain

$$\frac{d}{dt} \frac{m_0 c^2}{\sqrt{1 - \beta^2}} = \mathbf{v} \cdot \mathbf{E}, \qquad (3)$$

as in Eq. (32.6).

III.5. The path is of course a ballistic parabola with the acceleration

$$g = \frac{e}{m}\frac{V}{d}.$$

The distance of its vertex from the upper and lower plate, respectively. is

$$h = \frac{(v \sin \alpha)^2}{2g} = \frac{m}{e}\frac{d}{V}\frac{v^2 \sin^2 \alpha}{2}, \qquad d - h = d\left(1 - \frac{m}{e}\frac{v^2 \sin^2 \alpha}{2V}\right).$$

The lower plate is reached with

$$v \sin \alpha = \sqrt{\frac{e}{m}\,2V}.$$

For a velocity $v = 5 \cdot 10^6$ M/S this is not satisfied even for $\alpha = \pi/2$ With the value e/m_0 from Eq. (33.8) we have then instead

$$\frac{d - h}{d} = 1 - \frac{25 \cdot 10^{12}}{2 \cdot 110 \cdot 1.76 \cdot 10^{11}} = 0.35.$$

The voltage required to produce v is found to be, from $eV = mv^2/2$,

$$V = \frac{1}{2}\frac{25 \cdot 10^{12}}{1.76 \cdot 10^{11}} = 70 \text{ volts.}$$

The "electron volt" ev is a unit of energy much used in atomic physics, particularly in the form "million electron volts", Mev. Since $e = 1.60 \cdot 10^{-19}$ Q we have in our units

$$1 \text{ Mev} = 10^6 \cdot 1.60 \cdot 10^{-19} \text{ joule} = 1.60 \cdot 10^{-13} \text{ joule.}$$

One half of this is almost exactly equal to the rest energy of the electron, i.e.

$$m_0 c^2 = 0.80 \cdot 10^{-13} \text{ joule}, \quad m_0 = 0.90 \cdot 10^{-30} K.$$

If, in the condenser field, the velocity suffers changes which are comparable with c, the constancy of the x-momentum (x parallel to the plates) results in the fact that v_x cannot be constant, and hence x cannot be proportional to t. Correspondingly, the equation of motion for the y-direction shows that y is not proportional to t^2. Hence the path is not a parabola, but a transcendental curve (catenary). Similarly in §32d the Kepler orbit, which in the limiting case of an infinitely distant center of attraction becomes the ballistic parabola, was not an ellipse, but a transcendental curve (ellipse with precessing perihelion).

To compute extremely high velocities from the number z of the corresponding Mev we may use the energy equation

$$\frac{1}{\sqrt{1 - \beta^2}} - 1 = z\frac{\text{Mev}}{m_0 c^2}.$$

For an energy of 200 Mev, such as occurs in cosmic radiation, we have

$$\frac{1}{\sqrt{1-\beta^2}} = 1 + 400 \cong 4 \cdot 10^2, \qquad \beta = 1 - \frac{1}{32} \cdot 10^{-4}.$$

III.6. we consider immediately the case of high velocities (the familiar case $v \ll c$ is contained therein).

Let the direction of $\mathbf{B}$ be the z-direction. In the plane perpendicular thereto let s be the projection of the direction of motion, n the direction perpendicular to s, s, n, and z forming a right-handed system. We then have always

$$v_n = 0, \; (\mathbf{v} \times \mathbf{B})_z = 0, \qquad (\mathbf{v} \times \mathbf{B})_s = v_n B = 0, \qquad (\mathbf{v} \times \mathbf{B})_n = -v_s B.$$

Hence the momenta in the s- and z-directions are constant:

$$\frac{v_s}{\sqrt{1-\beta^2}} = C, \qquad \frac{v_z}{\sqrt{1-\beta^2}} = C'.$$

Since $v_n = 0$, squaring and adding leads to $\dfrac{\beta^2}{1-\beta^2} = \text{const}$, so that also β, v_s, and v_z are constant.

The equation of motion for the n-direction is (the charge of the electron is negative!)

$$\frac{d}{dt} \frac{\dot{v}_n}{\sqrt{1-\beta^2}} = \frac{\dot{v}_n}{\sqrt{1-\beta^2}} = \frac{-e}{m_0}(\mathbf{v} \times \mathbf{B})_n = \frac{e}{m_0} v_s B;$$

Hence

$$\dot{v}_n = e \frac{\sqrt{1-\beta^2}}{m_0} v_s B = \frac{e}{m} v_s B.$$

$\dot{v}_n$ is the centrifugal acceleration, and as such equal to v_s^2/ρ, where ρ is the radius of curvature of the path projected on the s, n-plane. Therefore

$$\frac{1}{\rho} = \frac{e}{m}\frac{B}{v_s}.$$

The same formula applies also for the non-relativistic calculation, where however $m = m_0 = \text{const}$. The curvature $1/\rho$ vanishes nonrelativistically only for $v_s = \infty$, whereas relativistically it becomes zero for $\beta = 1$, i.e. $v_s^2 + v_z^2 = c^2$. The product ρB (commonly written ρH) is the experimental measure of the "stiffness" of the cathode ray.

III.7. If x is the common direction of the electric and the magnetic field and z is the direction of the β-ray leaving D, the equations of motion of the β-particle are, with Lorentz's expression $\mathbf{F} = -e(\mathbf{E} + \mathbf{v} \times \mathbf{B})$ for the force (negative sign because of the negative charge of the electron)

$$\frac{d}{dt}\frac{v_x}{\sqrt{1-\beta^2}} = \mp\frac{eE}{m_0},$$

$$\frac{d}{dt}\frac{v_y}{\sqrt{1-\beta^2}} = -\frac{e(\mathbf{v}\times\mathbf{B})_y}{m_0} = -\frac{ev_z B}{m_0}, \qquad \frac{d}{dt}\frac{v_z}{\sqrt{1-\beta^2}} = \frac{ev_y B}{m_0}.$$

Since v_x and v_y can be neglected in comparison with $v_z \cong v$, we have $\beta^2 \cong v^2/c^2$, and from the third equation of motion to the same approximation $v \cong$ const. Hence the first two equations of motion can be integrated directly and we obtain, if the time t is measured from the moment of passage through D so that the instant of incidence on the photographic plate may be set equal to $t = a/v$,

$$x = \mp\frac{eE}{m_0}\sqrt{1-\frac{v^2}{c^2}\frac{a^2}{2v^2}},$$

$$y = -\frac{eB}{m_0}\sqrt{1-\frac{v^2}{c^2}\frac{a^2}{2v}}.$$

This is the parametric representation of the two branches of the curve which result when the polarity of the electric field is reversed.

If v^2 is neglected in comparison with c^2 elimination of the parameter v leads to the two branches of a parabola

$$y^2 = \mp Cx, \qquad C = \frac{e}{m_0}\frac{B^2}{E}\frac{a^2}{2}.$$

They touch at the point $x = y = 0$ with a vertical tangent. This point corresponds to the value of the parameter $v = \infty$.

If the relativity factor $\sqrt{1-v^2/c^2}$ is retained, elimination of v leads to the curve of the fourth order

$$y^4 + D^2 y^2 = C^2 x^2, \qquad D = \frac{eB}{m_0 c}\frac{a^2}{2} \ (C \text{ as above}),$$

which takes the place of both branches of the above parabola. At the point $x = y = 0$, which now corresponds to the parameter $v = c$, it has a cusp; the two tangents at this point have the two distinct directions

$$\frac{dy}{dx} = \pm\frac{C}{D} = \pm\frac{Bc}{E},$$

and form accordingly a finite angle 2α (see Fig. 48, on the right) with each other. This is clearly evident from Kaufmann's photographic records. However it was not possible to arrive at a definite decision between Lorentz's and Abraham's variation of mass (see the beginning of §33), as intended by Kaufmann, although this should be possible in principle from the complete shape of the curve; the fields employed were not uniform and their distribution had to be established by laborious probe measurements.

III.8. The two mutually perpendicular fields $E_y = E$ and $B_z = B$ give rise to the Lorentz force

$$F_x = -e \frac{dy}{dt} B, \qquad F_y = -e \left(E - \frac{dx}{dt} B \right), \qquad F_z = 0.$$

For not too great velocities ($m = m_0$) the equations of motion are

$$m \frac{d^2x}{dt^2} + e \frac{dy}{dt} B = 0,$$

$$m \frac{d^2y}{dt^2} - e \frac{dx}{dt} B = -eE.$$

If the second equation is multiplied by i, and if we set $\zeta = x + iy$, we obtain by addition

$$\ddot{\zeta} - i\alpha = -i \frac{e}{m} E, \qquad \alpha = \frac{e}{m} B.$$

The general integral is

$$\zeta = A e^{i\alpha t} + \frac{E}{B} t + C.$$

The time $t = 0$ can be so chosen that for it $dy/dt = 0$, so that $\dot{\zeta}_0 = \dot{x}_0$ becomes real. We then have $\dot{x}_0 = i\alpha A + E/B$, $\zeta_0 = A + C$ and hence

$$\zeta - \zeta_0 = \frac{i}{\alpha} \left(\dot{x}_0 - \frac{E}{B} \right) (1 - e^{i\alpha t}) + \frac{E}{B} t.$$

Separating real and imaginary parts we obtain with $\varphi = \alpha t$, $a = E/(\alpha B)$, $b = (\dot{x}_0 - E/B)/\alpha$:

$$x - x_0 = a\varphi + b \sin \varphi, \qquad y - y_0 = b(1 - \cos \varphi).$$

This is the equation of the general cycloid or trochoid (overlapping or stretched, depending on $a \lessgtr b$). For $\dot{x}_0 = 0$ (i.e. $b = -a$) we obtain a representation of the ordinary cycloid, such as occurred, with the same notation, in Vol. I, Eq. (17.1) in connection with the cycloidal pendulum (where however we had put $x_0 = y_0 = 0$).

III.9. Since we are dealing with a stationary state, **J** is constant in both time and space as

$$\text{div } \mathbf{J} = \frac{\partial \mathbf{J}}{\partial x} = 0. \qquad (1)$$

At the same time v and ρ are constant in time but not constant in space since

$$v = \sqrt{2\frac{e}{m}V(x)}, \qquad \rho = J/v. \tag{2}$$

Poisson's equation becomes

$$\sqrt{V(x)}\,\frac{d^2V(x)}{dx^2} = C, \qquad C = J/(\varepsilon_0\sqrt{2e/m}). \tag{3}$$

It may be integrated by putting

$$V(x) = Ax^\alpha. \tag{4}$$

(3) then leads to

$$A^{3/2}\alpha(\alpha - 1)x^{\alpha/2+\alpha-2} = C,$$

i.e.

$$\frac{\alpha}{2} + \alpha - 2 = 0, \qquad \alpha = \frac{4}{3}; \qquad A^{3/2}\frac{4}{9} = C, \qquad A = \left(\frac{9}{4}C\right)^{2/3}. \tag{5}$$

For $x = l$ we obtain from formula (4) and the meaning of C in (3)

$$V = Al^\alpha = \left(\frac{9}{4}\frac{l^2}{\varepsilon_0\sqrt{2e/m}}\right)^{2/3}J^{2/3}, \tag{6}$$

$$J = \frac{4}{9}\frac{\varepsilon_0\sqrt{2e/m}}{l^2}V^{3/2}. \tag{7}$$

The total current $I = \pi a^2 J$ (a = radius of cathode and anode) becomes

$$I = \frac{4\pi}{9}\varepsilon_0\sqrt{2e/m}\,\frac{a^2}{l^2}V^{3/2}. \tag{8}$$

This is the desired equation of the characteristic. The reader may convince himself that owing to our factor ε_0 it is correct dimensionally, namely has the dimension Q/S also on the right side. We note expressly that, according to (3) and (4), V does not increase linearly with x, and that $\partial V/\partial x$ is equal to zero at the cathode. Here, according to (1), $\rho = \infty$ and $v = 0$. The last corresponds to the fact that, in the statement of the problem, we have neglected the (small) velocity of emission of the thermionic electrons as compared with the velocity impressed on them by the field.

For the cylindrical arrangement (radius of the hot-filament cathode: $r = 0$, radius of the cylinder-mantel anode: $r = a$, length of the cylinder mantel: l) the preceding equations change as follows:

$$\text{div } \mathbf{J} = \frac{1}{r}\frac{\partial(r\mathbf{J})}{\partial r} = 0, \qquad rJ = \text{const} = \frac{I}{2\pi l} \qquad (1')$$

$$\rho = J/v = I \Big/ \left(2\pi l r \sqrt{2\frac{e}{m}V(r)}\right) \qquad (2')$$

$$\sqrt{V(r)}\,\frac{d}{dr}\left(r\,\frac{dV(r)}{dr}\right) = C, \qquad C = I/(2\pi\varepsilon_0 l\sqrt{2e/m}). \qquad (3')$$

$$V(r) = Ar^\alpha \qquad (4')$$

$$\alpha = \frac{2}{3}, \qquad A = \left(\frac{9}{4}C\right)^{2/3} \qquad (5')$$

$$\text{for } r = a, \qquad V = Aa^\alpha = \left(\frac{9}{8\pi}\frac{a}{\varepsilon_0 l\sqrt{2e/m}}\right)^{2/3} I^{2/3} \qquad (6')$$

$$I = \frac{8\pi}{9}\,\varepsilon_0\sqrt{2e/m}\,\frac{l}{a}\,V^{3/2}. \qquad (8')$$

At the cathode dV/dr now becomes infinitely large according to $(4')$ because of $r = 0$, in contrast with dV/dr for the plane configuration. Nevertheless the total charge on the filament approaches zero with vanishing r; this is the reason for the absence of the logarithmic singularity of the potential occurring otherwise for a charged wire, whereas by $(4')$ V vanishes for $r = 0$. For this reason $(8')$ applies not only for $r = 0$, but also for wires of small finite thickness with sufficient accuracy.

III.10. The magnetic flux through the electron path of radius r

$$\Phi = 2\pi \int_0^r B(r, t) r\, dr$$

yields

$$\frac{\partial\Phi}{\partial r} = 2\pi r B(r, t), \qquad \frac{\partial\Phi}{\partial t} = 2\pi \int_0^r \dot{B}(r, t) r\, dr. \qquad (1)$$

According to the law of induction we have

$$2\pi \int_0^{r_0} \dot{B}(r, t) r\, dr = -2\pi r_0 E(r_0, t). \qquad (2)$$

By multiplication with the absolute value e of the charge of the electron we obtain herefrom as accelerating force in the orbit $r = r_0$:

$$-eE(r_0, t) = \frac{e}{r_0}\int_0^{r_0} \dot{B}(r, t) r\, dr = \frac{e}{2\pi r_0}\frac{\partial\Phi}{\partial t}. \qquad (3)$$

The equation for the change in momentum of the electron can then be integrated with respect to t and yields

$$mv - (mv)_A = \frac{e}{2\pi r_0} (\Phi - \Phi_A),$$ (4)

where Φ_A denotes the magnetic flux for the initial state $v = v_A$, $m = m_A$. Accordingly, for given initial momentum, mv is to be regarded as known. We hence calculate for the answer to question (1):

$$\beta = \frac{v}{c} = \frac{1}{\sqrt{1 + m_0^2 c^2/(m^2 v^2)}}, \qquad m = m_0 \sqrt{1 + \left(\frac{mv}{m_0 c}\right)^2},$$ (5)

$$eV = (m - m_0)c^2.$$

Numerical example: For a path diameter $2r_0 = 10^{-1}$M a flux amplitude $\Phi_{max} = 10^{-2}$ VS is readily attainable in practice. Let the initial flux be very small, $\Phi_A \cong 0$. We can also assume $v_A \cong 0$ since the initial velocity is insignificant in comparison with the great final velocity. We then find from (4)

$$\frac{(mv)_{max}}{m_0 c} = \frac{e}{2\pi r_0} \frac{\Phi_{max}}{m_0 c} = \frac{1.6 \cdot 10^{-18}}{\pi} \frac{10^{-2}}{0.9 \cdot 10^{-30} \cdot 3 \cdot 10^8} = 18.8,$$

$$\beta_{max} = \left(1 + \frac{1}{18.8^2}\right)^{-1/2} = 1 - \frac{1}{710}, \quad m_{max} = 18.8 \, m_0,$$ (6)

$$m_{max} - m_0 = 17.8 \, m_0, \qquad eV_{max} = 17.8 \, m_0 c^2.$$

Since, from the discussion of Problem III.5, $m_0 c^2$ is equal to $\frac{1}{2} \cdot 10^6$ electron volts, we have

$$eV_{max} \cong 9 \cdot 10^6 \, ev.$$ (7)

On question 2.: The orbit $r = r_0$ was assumed to be known till now; it will now be computed. On every circular orbit there must be equilibrium between the centrifugal force and the force of Biot-Savart:

$$\frac{mv^2}{r} = evB(r).$$ (8)

This signifies according to Eq. (1)

$$mv = \frac{e}{2\pi} \frac{\partial \Phi}{\partial r}.$$ (9)

Substitution from Eq. (4) with $\Phi_A = 0$, $v_A = 0$ yields

$$\frac{\Phi}{r} = \frac{\partial \Phi}{\partial r}.$$ (10)

Plot ordinate B, as a function of the abscissa r, as a monotonically decreasing curve which eventually may have to be determined experimentally. Multiplication with $2\pi r$ then yields, by Eq. (1), the curve for $\partial \Phi/\partial r$, and integration with respect to r that for Φ. Its ordinates are to be divided by

r and the resulting curve must be pursued to its intersection with the curve for $\partial \Phi / \partial r$. The abscissa of the point of intersection is the desired value $r = r_0$.

In order that this orbit may be *stable* the field distribution must satisfy certain conditions, which, for example, have been clearly set forth by Gans.[1]

On question 3.: From β_{max} we obtain for the frequency of revolution

$$\nu_{max} = \frac{\beta_{max} c}{2 \pi r_0} \simeq 10^9 \, \mathrm{S}^{-1}. \tag{11}$$

In order to be able to compute the number of revolutions with ease, we assume that the flux Φ does not increase sinusoidally, but linearly from the initial state $\Phi_A = 0$ to the final state Φ_{max}. For 500-cycle alternating current in the windings of the electromagnet the time of rise (= a quarter period) is then $1/2000$ S. We hence obtain

$$\frac{\partial \Phi}{\partial t} = 2000 \, \Phi_{max} \, \mathrm{S}^{-1} = 20 \, \mathrm{V}. \tag{12}$$

Since by (3) the accelerating force is then also constant in time, this is at the same time the gain in energy in one revolution, measured in electron volts. Since by (7) the maximum kinetic energy measured in this manner was $9 \cdot 10^6$, the number of revolutions becomes

$$\frac{9 \cdot 10^6}{20} = 450000. \tag{13}$$

On question 4.: The state of the maximum number of revolutions is reached for $\Phi = \Phi_{max}$, i.e. $\partial \Phi / \partial t = 0$. According to Eq. (3) we then have $E(r_0) = 0$, i e. $\dot{v}_{tang} = 0$. From the equation of the circle

$$\mathbf{r} = r_0 e^{i \omega t}$$

we find for the magnitude and direction of the derivatives of $\mathbf{r}$:

$$\mathbf{v} = i \omega r_0 e^{i \omega t}, \qquad \dot{\mathbf{v}} = -\omega^2 r_0 e^{i \omega t}, \qquad \ddot{\mathbf{v}} = -i \omega^3 r_0 e^{i \omega t}. \tag{14}$$

We conclude herefrom: $\ddot{\mathbf{v}}$ is opposite in direction to $\mathbf{v}$; $\ddot{\mathbf{v}} = -\omega^2 \mathbf{v}$, furthermore, $\mathbf{v} \cdot \dot{\mathbf{v}} = 0$, $\mathbf{v} \cdot \ddot{\mathbf{v}} = -\omega^2 v^2$.

For the reaction force of radiation we find hence by (36.26) (the symbols $\mathbf{v}'$, $\mathbf{v}''$ given there have the same meaning as our $\dot{\mathbf{v}}$, $\ddot{\mathbf{v}}$):

$$
\begin{aligned}
\mathbf{R}^* &= \frac{e^2}{6 \pi \varepsilon_0 c^3} \frac{v}{1 - \beta^2} \left(-\omega^2 - \frac{\omega^2 v^2}{c^2 (1 - \beta^2)} \right) \\
&= -\frac{e^2}{6 \pi \varepsilon_0 c^3} \frac{v \omega^2}{1 - \beta^2} \left(1 + \frac{\beta^2}{1 - \beta^2} \right) = -\frac{e^2}{6 \pi \varepsilon_0 c^3} \frac{v \omega^2}{(1 - \beta^2)^2} \tag{15}
\end{aligned}
$$

$$|\mathbf{R}^*| = \frac{e^2 v^3}{6 \pi \varepsilon_0 r_0^2 c^3} (1 - \beta^2)^{-2}.$$

[1] R. Gans, Zeits. f. Naturforschung, Vol. 1, p. 485, 1946.

In our numerical example we find with $v \cong c$, $\beta = 1 - 1/710$:

$$| R^* | = \frac{e^2 (710)^2}{24 \pi \varepsilon_0 r_0^2}.$$

With $2r_0 = 10^{-1} \text{M}$ and $36 \pi \varepsilon_0 = 10^{-9} \dfrac{Q^2}{\text{joule M}}$ (Eq. (7.18)):

$$| R^* | = \frac{6(710)^2 e^2 V}{10^{-11} MQ} = 6 \cdot 7.1^2 \cdot 1.6 \cdot 10^{-4} \frac{eV}{M} \cong 0.048 \frac{eV}{M}. \qquad (16)$$

This may be compared with the force of the electric circulating field, which by (3) and (12) is

$$\frac{20}{2\pi r_0} \frac{eV}{M} = \frac{200}{\pi} \frac{eV}{M}.$$

The reaction force increases with the fourth power of the particle energy. As the reaction force becomes comparable with the accelerating force of the circulating field the balance between increase in particle mass and centripetal force is upset. Thus the reaction force sets a limit to the maximum energy which the betatron can impart to an electron.

The principal purpose of the betatron is the production of x-rays of very great hardness. Since their limiting energy $h\nu$ is given by the maximum energy of the betatron electrons it depends, in accord with Eq. (5), on the momentum mv which can be attained. By Eq. (4) this is determined by the ratio Φ/r_0. For a proportional increase of all of the dimensions of the magnet ($B_{\max}$ and hence also B_{av} are fixed by the saturation of the magnet) Φ increases quadratically, Φ/r_0 hence linearly. We found for our path diameter $2r_0 = 10^{-1} \text{M}$ in (7)

$$eV_{\max} = 9 \cdot 10^6 \text{ eV} = 9 \cdot 1.602 \cdot 10^{-6} \text{ erg} = 1.45 \cdot 10^{-5} \text{ erg}.$$

Hence, for x-rays

$$1.45 \cdot 10^{-5} \text{ erg} = h\nu = \frac{hc}{\lambda} = 2 \cdot 10^{-16} \frac{\text{erg} \cdot \text{cm}}{\lambda}, \qquad \lambda = 1.4 \cdot 10^{-11} \text{cm} = 1.4 \text{ X}.$$

The X-unit $= 10^{-11}$ cm here introduced is the unit of length customary in x-ray spectroscopy; thus the K-radiations of the heaviest elements have wave-lengths of about 100 X-units. We thus find ourselves with our betatron of relatively modest dimensions in a domain far beyond the short-wave-length limit of ordinary x-ray spectra and even beyond that of the natural γ-rays which is reached, for ThC, at $\lambda = 4.7$ X. By increasing the betatron dimensions the limit can be lowered still further and the energy of 9 Mev, found above, be increased.

IV.1. We proceed from the fact that for an observer moving with it, the field within the rod is both free of current and free of charge: $\mathbf{J}' = 0$

and $\rho' = 0$. It then follows from Eqs. (34.6) with the definition of the conduction current in (34.9a), except for correction terms of the order β^2

$$\mathbf{J}_l = \mathbf{J} - \rho\mathbf{v} = 0, \qquad \rho = 0. \tag{1}$$

The rod hence has neither conduction current nor volume charge for an observer at rest in the laboratory as well. However, it possesses surface charge ω and a Rowland current (which, in fact, is demanded by $\mathbf{J}_l = 0$) $\mathbf{J} = \omega\mathbf{v}$.

Viewed from the laboratory, the field thus consists of the superposition of a stationary electric and magnetic field, the latter added to the original uniform field and derived from the Rowland current $J = \omega v$. We therefore have

$$\text{curl } \mathbf{E} = 0, \qquad \mathbf{E} = -\text{grad } \Psi, \qquad \text{curl } \mathbf{H} = \mathbf{J}. \tag{2}$$

Show that this agrees with the general Eqs. (34.13), if in them the meaning of $\mathbf{E}^*$, $\mathbf{H}^*$ is substituted from (34.8) and that of $\dot{\mathbf{B}}$, $\dot{\mathbf{D}}$ from (34.12). For $\partial/\partial t = 0$ and the auxiliary conditions from (34.11a) they then become

$$-\text{curl}(\mathbf{v} \times \mathbf{B}) = -\text{curl}(\mathbf{E} + \mathbf{v} \times \mathbf{B}), \text{ so that curl } \mathbf{E} = 0,$$

$$\rho\mathbf{v} - \text{curl}(\mathbf{v} \times \mathbf{D}) + \mathbf{J} - \rho\mathbf{v} = \text{curl}(\mathbf{H} - \mathbf{v} \times \mathbf{D}), \text{ so that curl } \mathbf{H} = \mathbf{J}.$$

By (34.7) we infer from $\mathbf{J}_l = 0$, $\mathbf{E}^* = 0$, i.e. to sufficient accuracy ($\mathbf{B} = \mathbf{B}_0 = $ original field):

$$\mathbf{E} = -\mathbf{v} \times \mathbf{B}_0, \qquad E = -vB_0, \qquad \Psi = vB_0 x + C. \tag{3}$$

We have here assumed that $\mathbf{B}_0$ has the z-direction, $\mathbf{v}$, the y-direction, and that x, y, z constitute a right-handed system. C is a constant of integration which is independent of x and, in view of the symmetry of the problem, also of y and z. Question a. is thus answered.

On question b.: Consider two points x_1, x_2 on the periphery of the rod, e.g. $x_1 = $ point of entrance, x_2 point of exit of the x-axis. The difference of potential is then by (3):

$$V = \Psi_1 - \Psi_2 = vB_0(x_1 - x_2). \tag{4}$$

As a difference of potential this is independent of the path (connecting wire of infinitely high resistance) by which we imagine points 1 and 2 to be joined; this path may be imagined either in the exterior or the interior of the body.

c. We consider the external field. Here the boundary conditions (34.15) take effect. They demand continuity of the tangential component of $\mathbf{E}$ (not of $\mathbf{E}^*$) and are equivalent to continuity of the potential at the surface of the rod, whereas nothing is stated regarding the normal derivative. Since Ψ is known in the interior by (3), the surface values $\overline{\Psi}$ of Ψ are also known. We hence must solve a *boundary-value problem* for the exterior, with the normalization condition $\Psi = 0$ at infinity. A solution can be attained for any shape of the rod. It becomes elementary for the circular cross section, to which, by conformal mapping, every other cross section may be reduced, transferring the boundary values prescribed for the latter. It hence suffices to deal with the circular cross section.

If r, φ are ordinary polar coordinates, $r = 0$ is the center of the circle, $r = a$ its periphery, and φ is measured from the x-axis, (3) yields for the interior and the periphery of the circle

$$E = E_x = -vB_0, \qquad E_r = E_x \cos \varphi = -vB \cos \varphi,$$

$$E_\varphi = -E_x \sin \varphi = vB_0 \sin \varphi. \tag{5}$$

The potential in the exterior can generally be expressed as Fourier series

$$\Psi = \sum \left(\frac{a}{r}\right)^n (A_n \cos (n\varphi) + B_n \sin (n\varphi));$$

In view of the boundary condition only the term with A_1 differs from zero. Hence

$$\Psi = A_1 \frac{a}{r} \cos \varphi, \qquad E_\varphi = -\frac{1}{r} \frac{\partial \Psi}{\partial \varphi} = A_1 \frac{a}{r^2} \sin \varphi, \tag{6}$$

and, in view of the continuity of E_φ for $r = a$

$$A_1 = avB_0, \qquad \Psi = \frac{a^2}{r} vB_0 \cos \varphi, \qquad \frac{\partial \Psi}{\partial r} = -\left(\frac{a}{r}\right)^2 vB_0 \cos \varphi. \tag{7}$$

d. To determine the surface charge we must pass from **E** to **D**. We may utilize for this purpose in the interior of the rod Eq. (34.5), whose right side vanishes in view of $\mathbf{E}^* = 0$. Thus within the rod we do not have $\mathbf{D} = \varepsilon\mathbf{E}$, but

$$\mathbf{D} = -\frac{1}{c^2} \mathbf{v} \times \mathbf{H} = -\varepsilon_0 \mu_0 \mathbf{v} \times \mathbf{H} \cong -\varepsilon_0 \mathbf{v} \times \mathbf{B}_0 = \varepsilon_0 \mathbf{E}. \tag{8}$$

Since this ε_0 is derived from the general relation $\varepsilon_0\mu_0 = 1/c^2$, ε_0 represents the dielectric constant of vacuum and is not, in general, identical with the dielectric constant of the surroundings. It is characteristic and satisfying that in an exact application of Minkowski's theory in (8) there appears the well-defined dielectric constant of vacuum rather than the somewhat problematical and scarcely measurable dielectric constant of the metal. We conclude from (8) to begin with for the whole interior of the rod (since $\mathbf{E} = \text{const}$):

$$\text{div } \mathbf{D} = \rho = 0,$$

which agrees with the initial equation (1). The interior of the rod is free from space charge, even as observed from the laboratory.

At the surface of the rod, as judged from its interior, we have by (5) and (8),

$$D_n = \varepsilon_0 E_n = -\varepsilon_0 E_r = \varepsilon_0 v B_0 \cos \varphi$$

if n denotes the normal directed toward the interior.

For the sake of simplicity we set the dielectric constant of the exterior (air) also equal to ε_0. Then, by (7), we have for $r = a$, as seen from the outside (n denoting the normal directed outwards),

$$D_n = -\varepsilon_0 \frac{\partial \Psi}{\partial r} = \varepsilon_0 v B_0 \cos \varphi.$$

The sum of these two D_n yields the surface divergence of $\mathbf{D}$ at the surface, i.e. the surface charge

$$\omega = 2 \varepsilon_0 v B_0 \cos \varphi. \tag{9}$$

It varies from place to place and has its maximum values for $\varphi = 0$ and $\varphi = \pi$, i.e. $\pm 2\varepsilon_0 v B_0$.

The electric lines of force, which in the interior are straight lines and perpendicular to the axis of the rod, are bent in the exterior from the points of positive surface charge to those of negative surface charge along the shortest possible paths, particularly in the neighborhood of the two points $\varphi = \pi/2$ and $\varphi = 3\pi/2$. Only for $\varphi = 0$ and π are the lines of force perpendicular to the surface and flow off to infinity.

e. If the straight rod is bent into a circular ring, and this is rotated about the axis of symmetry perpendicular to its midplane, every section of the ring is subject to approximately the same conditions as the corresponding section of the straight rod, provided only that the radius of curvature of the ring is large compared to the radius of its cross section. The same applies for a circular disk ring, provided that the radius of its inner bounding cylinder is not too small. Since however the velocity is small in the excluded section of the disk, and the phenomenon of unipolar induction becomes insignificant at small velocities, this restriction may be overlooked and our results be extended to the whole disk and eventually also to an arbitrary body of revolution. We can then apply our Eqs. (3) and (8) also to the field in its interior:

$$\mathbf{E} = -\mathbf{v} \times \mathbf{B}_0, \qquad \mathbf{D} = \varepsilon_0 \mathbf{E} \tag{10}$$

and deduce therefrom the corresponding values of the voltage V and the interior potential Ψ, whereas the potential on the outside must be obtained by the solution of a complex three-dimensional boundary-value problem.

However, the following interesting difficulty arises: If the general (or, rather, too specialized) rule $\rho = \text{div } \mathbf{D}$ is employed to compute the space charge within the rotor, we obtain by (10), since now $\mathbf{v}$ and hence also $\mathbf{D}$ vary in space,

$$\rho = -\varepsilon_0 \text{ div } (\mathbf{v} \times \mathbf{B}_0) = -\varepsilon_0 \mathbf{B}_0 \cdot \text{curl } \mathbf{v} = -2\varepsilon_0 \dot\varphi B_0 ,$$

($\dot\varphi$ = angular velocity of the rotation). This is not zero, as was the case for the translation and as we might have expected from the standpoint of the observer rotating with the body. This contradiction is, however, no objection to *Minkowski's theory of moving media*, which (see footnote 3 at the beginning of §34) is based on the Lorentz transformation of uniform translation, but merely an indication that it is not directly applicable to problems involving rotation.

Author Index

A

Abraham, M., 155, 273, 292, 300, 353
Ampère, A. M., 4, 119
Anderson, C. D., 306
Arago, D. F., 288

B

Barkla, C. G., 155
Barnett, J. S., 98
Becker, R., 100
Bessel, F. W., 1, 313
Bethe, H., 265
Blackett, P. M. S., 306
Boltzmann, L., 3, 18
Bitter, F., 100
Bopp, F., 301
Born, M., 303, 306
Braun, F., 143
Brillouin, L., 198
Broglie, L. de, 5, 198
Bucherer, A. H., 273

C

Cady, W. G., 78
Cartan, E., 310
Christoffel, E. B., 310
Cohn, E., 49, 280
Curie, P., 78, 90

D

Davy, H., 3
Debye, P., 73, 190, 193
Dirac, P. A. M., 148, 301, 305
Döring, W., 100
Dolezalek, F., 177

E

Eddington, A. S., 261, 314, 315, 318
Eichenwald, A., 284, 285
Einstein, A., 98, 212, 227, 228, 229, 233, 234, 236, 264, 265, 280, 301, 307, 309, 310, 311, 320, 321, 322
Eötvös, R., 313
Euler, H., 306
Ewing, J. A., 97

F

Faraday, M., 3, 105, 256, 259, 288
Frank, Ph., 224
Fues, E., 47

G

Gans, R., 358
Gauss, K. F., 2, 42, 52, 308, 309, 321
Gentile, G., Jr., 199
Giorgi, G., 45
Goethe, J. W. von, 311
Goudsmit, S., 98

H

Haas, W. J. de, 98
Hahn, O., 265
Heaviside, O., 2, 43, 77, 241
Heisenberg, W., 98, 229, 306
Heitler, W., 301
Helmholtz, H. von, 5, 101, 236, 266, 285
Henry, J., 105
Herglotz, G., 249
Hertz, H., 2, 5, 36, 49, 52, 112, 149, 151, 152, 154, 177, 185, 280, 285
Hilbert, D., 1
Hittorf, W., 2
Hondros, D., 185, 190, 193, 209
Hu, N., 301

I

Infeld, L., 303
Ives, H. E., 228

J

Jacobi, C. G. J., 1
Janet, M., 310
Jaumann, G., 199
Jaumann, J., 85, 199
Joliot-Curie, F., 306
Joliot-Curie, I., 306
Joos, G., 53

K

Kalantaroff, P. L., 45
Kaluza, T., 212

Subject Index